T0178033

BRITISH AND FOREIGN

BUILDING STONES

BRITISH AND FOREIGN
BUILDING STONES

A DESCRIPTIVE CATALOGUE OF THE
SPECIMENS IN THE
SEDGWICK MUSEUM, CAMBRIDGE

BY

JOHN WATSON

Cambridge :
at the University Press
1911

CAMBRIDGE
UNIVERSITY PRESS

University Printing House, Cambridge CB2 8BS, United Kingdom

Cambridge University Press is part of the University of Cambridge.

It furthers the University's mission by disseminating knowledge in the pursuit of education, learning and research at the highest international levels of excellence.

www.cambridge.org
Information on this title: www.cambridge.org/9781107505780

First published 1911
First paperback edition 2015

A catalogue record for this publication is available from the British Library

ISBN 978-1-107-50578-0 Paperback

PREFATORY NOTE

PREFATORY NOTE

IN putting this first edition into the hands of the student, I desire to emphasise two facts. First, that the collection is very far from being exhaustive, although it is hoped that in due time it may be fairly representative of the building stones of the world. Secondly, that the descriptive notes, which precede the catalogue, do not in any degree aspire to the character of a geological exposition : they are only an attempt to weave together a few facts, which may be useful to the student.

I am greatly indebted to Professor Hughes and his colleagues, indeed to all who are officially connected with the Sedgwick Museum, for their unfailing kindness, and the friendly assistance they have at all times afforded me. Special thanks are due to Mr A. Harker in undertaking to be responsible for the petrological designation which appears on the labels of the Igneous rocks.

I wish to express my acknowledgments to Professor Grenville A. J. Cole of Dublin for the kind help he has given me whilst collecting specimens to illustrate the building stones of Ireland; and to Dr G. P. Merrill of the United States National Museum at Washington, for his advice and ready assistance in my endeavour to form a representative collection of specimens from the United States.

Lastly, I desire to record my deep sense of gratitude to Professor T. G. Bonney, who has kindly revised my manuscripts. I need hardly add that his revision has attached a scientific value to the contents of this publication which otherwise it would not have possessed.

Through the courtesy of the Woodwardian Professor, sufficient space has been allocated in the museum for this collection*, and ample room still remains for additions to it. It is therefore hoped that whenever students and others find that they can contribute fresh examples, they will avail themselves of the opportunity. By following this suggestion, the donors will have the satisfaction of knowing that not only are they enhancing the value of the collection for educational purposes, but also that their names will be enrolled among the many benefactors of the ancient and renowned University of Cambridge.

J. W.

BRACONDALE,
    CAMBRIDGE, 1911

---

* The Sedgwick Museum is open to the public, and the collection is exhibited in the Economic Section on the ground floor.

CONTENTS

                                          PAGE
Introduction   1

DESCRIPTIVE NOTES

BRITISH, COLONIAL AND FOREIGN BUILDING STONES

Igneous Rocks (Plutonic)   16
    „   „   (Volcanic)   77
Metamorphic Rocks   94
Sedimentary Rocks
    Cambrian and Silurian   .   .   .   98
    Devonian and Old Red Sandstone   .   .   110
    Carboniferous   117
    Permian   142
    Trias   152
    Jurassic   165
    Cretaceous   193
    Tertiary and Recent   216
        Eocene   217
        Miocene   227
        Pliocene   234
        Pleistocene and Recent   .   .   .   236

CATALOGUE

BRITISH BUILDING STONES

		PAGE
Igneous Rocks (Plutonic)	. . . .	247
„ „ (Volcanic)	. . . .	259
Metamorphic Rocks	. . . . .	260
Sedimentary Rocks		
Cambrian and Silurian	. . .	261
Devonian and Old Red Sandstone	.	264
Carboniferous	. . . .	269
Permian	. . . . . .	290
Trias	. . . .	295
Jurassic	. . . . . .	304
Cretaceous	. . . . .	315
Pleistocene and Recent	. . .	319

COLONIAL AND FOREIGN
BUILDING STONES

		PAGE
Igneous Rocks (Plutonic)	. . . .	323
„ „ (Volcanic)	. . . .	356
Metamorphic Rocks	. . . . .	367
Sedimentary Rocks		
Cambrian and Silurian	. . .	369
Devonian and Old Red Sandstone	.	373
Carboniferous	. . . .	375
Permian	. . . . .	380
Trias	. . . . . .	383
Jurassic	. . . . .	391
Cretaceous	. . . . .	399
Tertiary and Recent		
Eocene	. . . . .	411
Miocene	. . . .	418
Pliocene	. . . .	424
Pleistocene and Recent	. .	426
INDEX	. . . .	431

INTRODUCTION

Specimens of Building Stones brought together to form a collection are usually arranged according to their lithological character irrespective of their age.

Possibly this method may have been found to be the most convenient for architects and builders, otherwise it would not have been so generally adopted. Bearing in mind however that the present collection is primarily for the use of students of geology at Cambridge University, it has been thought that the specimens will be more helpful to them, if they are arranged in stratigraphical order, hence this method has been adopted, except in the case of the igneous rocks.

For convenience of reference, the examples representing the building stones of the British Isles have been arranged separate from those obtained from the Colonies and from foreign countries.

The specimens are chiefly in the form of 4½-inch cubes, the sides of which are dressed in the usual style adopted for the purposes for which the stone is generally used in the region from which the specimen comes. One face of the cube is polished, when it admits of this treatment, and another side is usually left rough, in order that the grain and texture of the rock may be clearly seen.

A label is affixed to the front of each specimen, bearing the following particulars:—

1. The name by which the stone is best known in commerce.

2. The stratigraphical position of the stone, or, in the case of an igneous rock, its petrological designation[1].

3. The name and the locality of the quarry whence the specimen was obtained.

4. The name and address of the donor of the example[2].

These details also appear in this catalogue, as well as particulars of the colour, texture, and other physical characteristics of the stone, and its chemical composition, so far as these have been obtainable.

The descriptive notes which precede the catalogue itself give the names of important buildings in which the stone has been employed, with the date of their erection, when available, together with brief remarks on special points of interest. The description of the general appearance of the stone itself is necessarily brief, owing to the small space available; the details are therefore somewhat curtailed, but it is hoped that they are fairly accurate.

Whilst care has been exercised in gathering these particulars, freedom from inaccuracy, in all cases, cannot be guaranteed. Quarry owners, architects and others, are not always agreed as to the name by which a particular rock is recognised in the building stone industry, nor is the name adopted by them always in accordance with geological nomenclature; for example, stones are frequently distinguished as granites in commerce, which have no affinity whatever with those rocks, the *Petit Granit* of Belgium belonging to the Carboniferous Limestone being a striking instance. The name on the label is that which appeared to be the least likely to give rise to misunderstanding. On the other hand the geological formation ought seldom to be open to doubt, but so rapid are the strides made

---

[1] As mentioned in the preface, Mr A. Harker, F.R.S., has kindly determined the petrological names of the igneous rocks.

[2] Where the donor's name does not appear, the specimen has been presented by the compiler of these notes.

in the advancement of science and the consequent changes in geological classification and nomenclature, that a name which is correct to-day, may not be so to-morrow.

It is hoped however that the name and the locality of the respective quarries are fairly correct and comprehensive. It is also expected that no errors occur in the names and addresses of the numerous donors, to whom deep gratitude is due for their generous contributions to this collection.

The chemical composition is given in those cases where it is known. Chemical tests and analyses are pronounced by some practical men as unreliable guides to determine the quality of a building stone[1]. Perhaps it is true that knowledge of the chemical composition is not in most cases as useful as that of its physical properties; nevertheless it is of great importance. For example the atmosphere in and around large cities and centres of industry is usually charged with gases impregnated with various acids. These acids when caught by the falling rain are deposited in solution on the buildings, and often by chemical action cause a decay of the stone. It is therefore important to ascertain the chemical composition of a rock, and thus arrive at its susceptibility to decay under certain atmospheric influences.

It is also most important to ascertain the chemical composition of a stone that is destined to be used for maritime engineering work. "The acidity and salinity of sea water may, and often does, bring about molecular changes in minerals containing soluble salts; certain compounds of lime are decomposed and softened by sea water, and they often give rise to the formation of other compounds which tend to destroy the cohesion of the material of which they are the ingredients, by producing cracks and fissures[2]."

---

[1] *Practical Masonry*, W. R. Purchase, p. 160.
[2] *Harbour Engineering*, B. Cunningham, p. 85.

Considering the importance attaching to the physical properties of a building stone, it is regrettable that a complete statement of the weight, specific gravity, crushing strain, and porosity, of the rock represented by the specimens, has in so few cases been procurable.

When stone is required for vaulting, or other similar work, the selection of one light in weight, or of low density, is very important. On the other hand in the building of quay or dock walls materials of high density are essential, because, when immersed in water, a stone loses a very considerable part of its effective weight. " Furthermore if, compared with its weight, the stone possesses a very large bulk, it presents a correspondingly large surface to wave action, thus increasing the scope or field of disturbing force. The two factors of weight and volume, must therefore be taken into joint consideration, they shew that the smaller the surface area of a stone, and the greater its unit weight, the less likelihood there is of disturbance. In other words, the higher the specific gravity, the greater the stability[1]."

To arrive at the weight of a cubic foot of stone, it is only necessary to multiply its specific gravity by 62·23 lbs.—the weight of a cubic foot of ordinary water. But it will be observed, on referring to the catalogue, that in many instances the weight per cubic foot is tabulated, whilst the specific gravity is not given. It is therefore obvious that the weight has not been arrived at by first ascertaining the specific gravity; but the method has been adopted, which is often resorted to by quarrymen and others, of weighing a cubic foot of stone. If care is taken to see that the measurements are accurate, and that the interstitial water, which is always present to a greater or less degree in every stone, is eliminated before weighing, the result is generally sufficiently correct for ordinary practical purposes.

[1] *Harbour Engineering*, B. Cunningham, p. 84.

The test of compression, or crushing strain, is treated by some architects and workers of stone with less consideration than it deserves. It is true that the resisting power of most building stones greatly exceeds the strain that they are required to support, nevertheless the test is useful as an aid for determining the weathering property of a stone. The durability of a rock depends either on the manner of contact of the mineral particles of which it is composed or on the strength of the particles themselves. Moreover, in the latter case, the durability will be determined by the strength of the weakest material which is abundantly distributed throughout the mass. Thus, if the interstitial matter between the grains of a sandstone is weak or lacks cohesion, the stone will be disintegrated when subjected to abrasion, or exposed to the influences of the weather, however strong the grains themselves may be. A rock may consist mainly of quartz grains, and yet be so soft as to crumble when compressed between the fingers, while another rock, having the same general composition, but in which the grains are firmly cemented together, may be one of the most durable of stones. The test of compression is undoubtedly a convenient method by which to determine the physical strength of a stone, although in estimating its weathering property the chemical composition, as well as other factors, must not be neglected.

The test of porosity, or ratio of absorption, is a valuable one, and ought to be considered together with other physical features. Walls built of a stone having a high ratio of absorption are liable to be damp, and this is especially the case if the stone is fine in texture as well as porous, for water penetrates further, and is retained longer in small cavities, than in larger ones. Such a stone will probably, on microscopic examination, be found to contain minute fissures along which water will be absorbed to a considerable depth by capillary attraction. Moreover, Mr Buckley points out that the danger from frost depends, not so much upon the

amount of absorption, as upon the size of the pores.
Rocks with large pore spaces stand frost better than
those with small ones, because they do not retain the
water which they absorb[1].

Trustworthy information as to the kinds of stone
used in the construction of any particular building,
has been collected with difficulty, especially when it
relates to ancient ecclesiastical structures.   The wear-
ing and decaying influences of time, have, in many
cases, obliterated the evidence necessary to identify
the various materials of which old abbeys and cathedrals are built.   Treatment, tending to render identification impossible, has unfortunately in some cases
been carried out.   An instance to substantiate this
statement may be cited.   When the restoration of Exeter
Cathedral was begun by Sir Gilbert Scott in 1871, it
was found that the stone work of the interior of the
greater part of the building had been covered by former
generations with coats of yellow wash.   When this
was removed, it was found that whilst all the groining
ribs were composed of the beautiful white Beer Stone,
a specimen of which is in this collection[2], the surface
of the vault had been filled in with a variety of other
Devonshire stones[3].

The despoiling hand of time, and the obliterating
propensities of man, render it absolutely necessary, in
most cases, to depend on documentary evidence for
identification.

Historians however rarely record anything that
gives the reader the slightest clue to the locality from
which the material used in building came; indeed the
nature of the material itself is seldom referred to.
The difficulty of identification is probably the reason
for this omission.

[1] "Building and Ornamental Stones of Wisconsin," E. R.
Buckley, *Bul. Wisconsin Survey*, No. IV. p. 396.
[2] See page 200.
[3] *Cathedrals of England and Wales*, T. F. Bumpus, Vol. II.
p. 235.

When it has been possible to associate a building with the stone represented by the specimen, the date of erection of the building is in most cases recorded.

Special reference is made to buildings in and around Cambridge constructed with the stone which the specimen represents; for it is manifest that it is of more value to examine the large amount of stone visible in a building, than merely to look at a specimen on the shelves of a museum.

While it is hoped that the information which it has been practicable to tabulate under this head will be instructive, it is well to remember that a certain class of stone having proved itself to be durable in a particular building, is not a guarantee that the same rock will give equally good results if employed for structural work in another district. Geographical and meteorological variations have to be considered. Sudden change of temperature is a powerful agent in the destruction of rocks, for stones as a rule possess low conductivity for heat and slight elasticity. Frequently rocks reveal a tendency to decay when placed in a position which is much exposed to the sun, whereas the building remains perfect if it is shaded or has a northern aspect. There are many instances at present existing which substantiate this statement, and two striking examples may be referred to. Somerset House in London, built of Portland Stone from the upper division of the Jurassic system in Dorsetshire, is much more decayed on the south side than on the north. York Minster, which is built of Magnesian Limestone from Nottinghamshire and South Yorkshire, is also more decayed on its southern than on its northern aspect[1].

Again, the presence of corrosive substances in the air, or even of much moisture, will cause a rock to disintegrate more rapidly than it would in a purer or drier atmosphere, and a very changeable climate may also affect a stone. Further, buildings in the country are often more durable than those in populous and smoky

[1] *Masons' Practical Guide*, R. Robson, p. 6.

towns, although constructed with the same stone, owing to a growth of lichens which frequently covers the entire surface in situations favourable to vegetation, and which seems to afford a protection against the ordinary causes of decomposition of the stone[1].

These brief notes, it is hoped, will have made it patent to the student of Economic Geology, that the details of the chemical composition and physical characteristics of a stone, and the record of its use in buildings, which are given in the descriptive notes and in the catalogue, are not without value.

Whilst emphasizing the importance of recording the data just referred to, it must be borne in mind that they will require frequent revision. It is not necessary to remind the geological student of the numberless causes which serve to produce physical and chemical alterations and variations in the same series of rocks. Nothing exists in nature that is not liable to change its condition and manner of being. Reference to the specimens themselves is sufficient to prove that the character of a rock may differ considerably within a very small area, there being frequently several varieties in one stratum, so that alterations in the quality of the stone are liable to occur as the quarrying proceeds. It is therefore important that the characters which have at one time been ascertained and recorded, should be occasionally checked in the laboratory and testing room, and fresh investigations made, to ensure accuracy[2].

No attempt has been made when compiling this descriptive catalogue to offer suggestions, or tender advice, as to the superiority claimed for any particular stone represented in the collection. On the contrary, great care has been exercised that no remarks made

---

[1] Guilt's *Encyclopaedia of Architecture*, p. 450.

[2] A suggestion has been made, which seems to carry weight, that a University such as Cambridge, where scientific knowledge is always available, would be a fit place for a National Bureau, where building stone could be examined, tested, and reported upon, from time to time, and authenticated certificates issued.

can be construed as a recommendation to adopt any special kind of stone, the desire being that the collection should not in any sense be looked upon as an advertising medium.

Whilst adhering strictly to this line of procedure, it may be useful to mention one or two general principles which ought to be observed in utilising a rock for structural purposes, since they may be of help to those students who desire to gain the requisite knowledge for selecting a building stone, and using it to the best advantage.

It is admitted by most workers of stone, that nearly all stratified rocks used for construction should be placed in the building with the bedding of the stone as nearly horizontal as possible, and that the exposed surface should be formed by a cross section of the natural divisional planes. The stone thus placed is best able to resist atmospheric disintegration and occupies in the artificial structure a similar position to that which it originally occupied in nature[1], although of course the bedding planes are not necessarily horizontal in the quarry, the strata being often inclined or even upright, owing to their having been tilted by earth movements. This principle is acknowledged to be of extreme importance, it is therefore incumbent upon those responsible for the carrying out of structural work in stone that they should be able if possible to discriminate with accuracy the plane of the bedding of any building stone. In many rocks the direction of the bedding is easily distinguishable, but in others it is exceedingly difficult to ascertain, because of their evenness of grain. In this latter category are included the rocks which are known as "freestones," a name generally given to those which can be freely worked in any direction and sawn with a toothed saw. The chances of discrimination are minimised, if, as is often the case, the stone immediately after being quarried is partially trimmed down to a shape

[1] *Building Stones*, E. Hull, p. 306.

approximately rectangular, a manipulation which is popularly known among quarrymen as "scappling," and often carried out previous to the stone having been selected or purchased for any particular work. It is true that an experienced mason knows that in working a stone, it invariably tears with the bed, and that by the application of the "drag," a somewhat primitive implement, he can generally distinguish the line of bedding, but as such an artificer is not always available, it is hazardous to rely upon this method if an easier and a safer way can be found. Some workers in stone content themselves by observing the lie of the micaceous flakes in a rock, as a sufficient guide to determine the direction of the bedding. In many cases however the grains of mica are so minute that it is difficult to detect even their presence, while some rocks do not contain any micaceous ingredients.

As there are so very few exceptions to the rule that a stone should be placed with the plane of bedding in a horizontal position in the building[1], many architects and others wisely stipulate that *all* sedimentary stones, however free they appear to be from lamination, must be placed in the same position that they originally occupied when in their natural bed, and that *the plane of bedding must be distinctly marked on every stone before it leaves the quarry*. Some practical men insist that this rule is only applicable to sandstones, and not to the limestones that are usually employed for building purposes, as most of the latter do not possess the kind of lamination observable in sandstones. Nevertheless varieties exist, such as those that are commonly termed "shelly limestones," which have a coarsely laminated structure, and the same rule should be followed in placing such in a building.

[1] It may be well to note that there are a few exceptions where this rule can be ignored with impunity. Stone required for cornices and weatherings may be face-bedded, indeed it is preferable to have it so placed, but the proportion of stone necessary for this class of work in an ordinary building is very small.

Besides offering more resistance to the action of the weather, there is another reason why the bedding planes should be horizontal. A stone can support a weight better when its laminae are at right angles to the pressure, than when the stress is applied in the same straight line as the grain of the stone, thus rendering it liable to split[1]. So numerous are the instances throughout the British Isles (Cambridge not excepted) where this important principle has obviously been neglected, that one is constrained to emphasize the desirability of the student of economic geology giving it his special attention.

Building stone is frequently placed on the market bearing a distinguishing trade name which belonged to a rock no longer obtainable, the original quarry having long been worked out. The new rock, thus passing under the same name, may be equal in quality to its predecessor, but such a practice is to be deprecated and is apt to lead to confusion.

Most rocks suitable for building, especially among the limestones and sandstones, undergo a process of hardening by being exposed to the atmosphere after they are quarried. This hardening is explained by Newberry and others on the supposition that the water with which the stones are permeated, holds in solution, or at least in suspension, a small amount of siliceous, calcareous, ferruginous, or clayey matter. On exposure to the atmosphere this "quarry water," as it is technically called, is drawn by capillarity to the surface of the block and evaporated. Thus the dissolved or suspended material is deposited and serves as an additional cementing constituent to bind the grains more closely together[2]. In other words the outside of the stone is hardened by this process, sometimes to a

---

[1] *Building Construction*, C. F. Mitchell, p. 49.
[2] "Building and Ornamental Stones," G. P. Merrill, *Report of the U.S. National Museum*, 1886, p. 339.

considerable depth, and it thus resists the weathering influences which are always at work. It is obvious therefore that the process of seasoning, which means hardening, is important.

It is advantageous in most cases to have the stone dressed and carved before the process of seasoning commences. The quarry water just referred to as present in every stone, to a greater or less degree, includes the water of saturation which is given off with comparative readiness, and the water of imbibition which is retained with considerable tenacity. The water of imbibition is said to carry a much larger percentage of mineral cement in solution than does the water of saturation, and as the former is the last to leave the stone, the "crust" is not likely to be formed until the stone is thoroughly seasoned[1], and when once formed, any disturbance, by dressing or carving, naturally destroys the advantage sought to be gained by the process of seasoning. Therefore it is advantageous to dress and finish the stone soon after it is quarried, not only from an economical point of view, the manipulation being much easier when the stone is soft, but also from the standpoint of future durability. The exigences of trade frequently prevent this important feature of seasoning being carried out in its entirety, while sometimes it is wholly ignored, which leads to disastrous consequences. A wise builder does not furnish the interior of his house with fittings made from green timber, neither should he construct the exterior with "green" stone. It is recorded that Sir Christopher Wren rejected any stone for the erection of St Paul's Cathedral that had not been quarried and exposed to the atmosphere for three years.

In the selection of a stone for building, colour is an important factor from an architectural point of view. The colours to choose from are very numerous, as the specimens in the collection clearly demonstrate, and

[1] "Building Stones of Wisconsin," E. R. Buckley, *Bul. Wisconsin Survey*, No. IV. p. 30.

although the variety may be limited in some localities, facilities for transport are now so great and the cost so little, that a choice of colour can be secured for almost any neighbourhood. In large cities and towns the walls of all the buildings become so begrimed with smoke and dust, that it is almost impossible to distinguish the original colour of the stone, and it is therefore immaterial whether it was light or dark, but in the suburbs, where this happily does not occur, a judicious choice of colour is desirable, and a variety of light and dark buildings frequently adds to the beauty of the landscape; moreover, a careful blending of colours in the buildings themselves may be carried out with advantage. But although from a decorative point of view the beauty of stone depends on its effective appearance, to the architect and builder, as well as to the student of economic geology, there is a beauty in a good stone beyond that which appeals openly to the eye, a beauty more subtle, requiring a scientific, as well as an artistic appreciation—the impression it gives of strength, and suitability for its purpose[1].

These few remarks only touch the fringe of the numerous important factors connected with the general principles bearing upon the utilisation of rocks suitable for construction. There still remain very many details to be considered, far too numerous even to summarise here, and beyond the aim of these notes to discuss, but it is hoped that what has been said will induce the student to study more carefully than has been customary, the character of the rocks in relation to their special uses and the circumstances in which it is proposed to employ them.

How many practical workers in stone or students of architecture know, or have ever been shewn, how to determine the nature of the cementing material of a sandstone, with a view to learning something of its

[1] *The Beauty and Use of Irish Building Stones*, G. A. J. Cole, p. 3.

durability ? The simple chemistry and the structural details of building stones should be familiar to all who have an interest in their sale or purchase[1]. It is true that a thorough knowledge of the constituent minerals of a rock, their mode of contact, and the extent to which they are liable to be affected by atmospheric influences, can seldom be attained without a careful and exhaustive examination of the stone, aided by the microscope, and a complete chemical analysis; but the knowledge is none the less essential, and ought to be gained, either by personal investigation, or through the medium of an acknowledged expert. "A high authority has observed, that in modern Europe, and particularly in Great Britain, there is scarcely a public building of recent date, that will be in existence a thousand years hence[2]." This is a formidable indictment, but it is hoped that the architect of the future, with the help of scientific knowledge combined with practical experience, will be able to erect buildings which will be monuments of durability as well as of artistic beauty.

The student may detect, when examining the collection, that more than one of the groups of rocks composing the "earth's crust" have been omitted. The Basalts and other basic lavas, with few exceptions, do not appear in the collection of igneous rocks. Although these eruptive masses are occasionally utilised by the builder for rubble masonry in the immediate neighbourhood of the quarries[3], they are chiefly employed for making paving setts and road metal, for which their hard and tough nature render them specially suitable. Roofing slates and marbles are also excluded from the collection. The important positions that these three groups of rocks hold in the

[1] *The Beauty and Use of Irish Building Stones*, G. A. J. Cole, p. 4.

[2] Guilt's *Encyclopaedia of Architecture*, p. 449.

[3] Occasionally, as in Auvergne, the Eifel, and Siebenberge, they are used in (or exported for) the masonry of important buildings.

building industry, is practically the reason why they do not find a place among the other specimens. The immense variety of these rocks and the widespread area over which they occur, make the subjects so large as to call for independent treatment. Specimens of marbles, arranged under the head of " Ornamental and Decorative Rocks," as well as " Stones suitable for paving and road making," and " Roofing slates and Flags," will be found in other parts of the museum, in positions specially allotted to these classes of material, and it is hoped that these collections will some day have catalogues of their own.

IGNEOUS ROCKS (PLUTONIC).

Some petrologists recognise an intermediate group of dyke rocks between the Plutonic and the Volcanic Igneous divisions, to which they give the name of "Hypabyssal." Others however are content to divide the igneous rocks into the two divisions just mentioned. The latter method has been adopted, as being more convenient in the classification of this collection, especially since many of the rocks classed as Hypabyssal are called "Granites" in commerce.

Placing the specimens of granite and other igneous rocks first is manifestly a departure from the stratigraphical classification adopted in this collection. That granite should be regarded as the primitive rock of the earth's crust, forming the floor of all stratified deposits, is a theory that has long been abandoned. It is now known that granite is intrusive into stratified rocks belonging to different geological periods, and that the granitic intrusions are therefore of different ages.

Notwithstanding this it has been considered more convenient to group all the specimens of igneous rocks together in this collection, rather than to place them among the specimens of the various formations with which they are now supposed to be contemporaneous.

BRITISH ISLES. The most important granitic intrusions in England which yield stone for structural purposes are in Devonshire and Cornwall. Great plutonic masses extend in the form of a broken chain from Dartmoor to the Scilly Islands. The granite from this district has been used for building construction and for monumental purposes from time immemorial. The ancient cromlechs, monoliths, and innumerable Celtic crosses, dotted over the bleak and elevated moors of both counties, and the Hut Dwellings, of which remains are frequently seen, were all constructed of granite. In these examples of ancient architecture the building material was derived from blocks of granite left on the surface as a result of the denudation of the Tors, and described many years ago by Professor Sedgwick[1].

It was not until early in the 18th century that the rock was quarried[2]. The De Lank quarry, near Bodmin in Cornwall, was one of the first to be developed, and from it Smeaton procured the material for building the exterior of his famous Eddystone Lighthouse (1756). Devonshire and Cornish granite was then known as "Moor Stone," presumably on account of the chief supply coming from moorland tracts. In this connection, it may be useful to refer to Smeaton's narrative of the Eddystone Lighthouse published in 1793. He writes, "Having now examined all the places I could hear of where Moor Stone was worked, which lie tolerably convenient for water carriage to Plymouth, I became convinced of the necessity of making use of Portland, or some other free working stone for the interior work. For the granite though preferable to all others in point of duration, yet being at best of a stubborn nature, and the working thereof confined to a few hands, I perceived there would be no possibility of procuring so great a quantity in any reasonable time at a moderate expense[3]." These remarks of Smeaton also point to the fact that the facilities for procuring granite at that period were very limited.

At a later date Devonshire granite from Dartmoor,

[1] *Life and Letters of Sedgwick*, Clark and Hughes, Vol. I. p. 286.

[2] *Granites*, G. F. Harris, p. 33.

[3] Smeaton's *Eddystone Lighthouse*, 1793, Book II. Chap. II. p. 53.

was sent to London for the construction of Waterloo Bridge (1817), one of Rennie's masterpieces of engineering. The next recorded important work in which Devonshire granite was employed is the existing London Bridge, opened for traffic in 1831. This granite also came from Dartmoor, and was chiefly obtained from the Princetown quarries. It is interesting to note that in the year 1902, the owners of these quarries undertook the important task of widening London Bridge, and used the granite from the same quarries as the original stone had come from.

The Prison quarries at Princetown yield a light grey porphyritic stone which is employed exclusively for extensions and repairs to the well-known government buildings on Dartmoor. Swell-Tor quarry, another of the Princetown group, and situated about two miles west of the town, furnished a large portion of the granite used in the construction of the Thames Embankment, London (1864–7). The same stone was employed for building the new bridge across the Thames at Vauxhall in 1903, and for constructing the Bristol Water Works in 1902. This granite was also selected for the Royal Mausoleum at Frogmore.

The Tor quarries near Merrivale, a little further to the west, yield a light grey biotite-granite which was used for the foundations of the Houses of Parliament in 1840. The Gunnislake quarries near Tavistock, furnished the granite for constructing the new Admiralty Harbour at Dover in 1900; it was also used for building Milford Docks.

There are several quarries situated on the east borders of Dartmoor, the principal one being that of Haytor Rock, about four miles west of Bovey Tracy. These quarries have ceased working except for the supply of granite for local purposes, but a large quantity of this stone was used in the construction of London Bridge in 1831. The stone was conveyed by carts over the moor to the river Teign and thence by sea to London. The roads across the moor can still be traced; they were specially prepared for the conveyance of the granite by paving the cart tracks with large blocks of the stone.

The specimen of greenish grey muscovite-biotite-granite with long porphyritic crystals of orthoclase, from Blackenstone quarries near Moreton Hampstead, represents a description of stone which is largely used in Exeter and elsewhere for monumental work as well as for ordinary building.

The Nelson Monument in Trafalgar Square, London, is constructed of Devonshire granite.

The numerous specimens in the collection, from Devon and Cornwall, shew that most of the granite is light grey in colour, with distinct black and white crystals. Many of the varieties are porphyritic, the large felspars being popularly called "Horses' Teeth" by the quarrymen. This characteristic is displayed in a striking degree in the façade of Messrs Pryor's premises in the Petty Cury at Cambridge. This granite came from the Colcerrow quarries near St Austell, Cornwall. The same stone was used for constructing Plymouth Breakwater, and Pembroke Docks. A large quantity of Cornish granite has been used in the recent additions to Keyham Docks, being derived chiefly from the Carsnew quarries near Penryn, and from the Lamorna quarries near Penzance. The specimen from Lamorna indicates that this variety also contains unusually large crystals of felspar. Carsnew granite was employed for building Putney Bridge, London. The base of the Duke of Wellington's tomb in St Paul's Cathedral is made of Cornish granite from the Cheesewring quarries, near Liskeard, and the piers of Westminster Bridge are made of the same stone. Beachy Head Lighthouse is a good example of a building constructed with Cornish granite. The lighthouse was erected in 1828, and was built of granite from the same quarry as that used for Smeaton's lighthouse. This rock was also employed in constructing the Tower Bridge in London.

Curious to relate, the introduction of steam, as a propelling power for ships, gave a stimulus to the granite industry, which was specially beneficial to Cornwall. About 1840 the Government then in office recognised that steam would become the motive power of the future in the Royal Navy, and consequently the appliances which had sufficed for sailing ships would soon be superseded. A large steam basin and docks were forthwith designed, and it was decided that they should be constructed at Keyham, adjacent to the dock-yard at Devonport. The quantity of granite required for these works was very large, and many of the quarries existing in the neighbourhood of Penryn owe their origin to this development at Devonport.

Another important group of English granite quarries, yielding stone suitable for structural work, is that of Shap, situated on the wild and picturesque fells of Westmorland. Shap granite, as the specimens shew, contains large flesh-coloured orthoclase crystals, which impart a warm rich tinge to the rock, and it admits of a very fine polish. It is interesting to record that numerous boulders of Shap granite are found at great distances from the site of the quarries, for instance they are seen in the valleys of the north-eastern moorlands of Yorkshire, and even as far as the coast, having been transported to that district during the glacial period. There are two distinct shades in the granite from this group of quarries, which divide the stone into two classes, known in commerce as "Light Shap" and "Dark Shap." Examples of both will be found in the collection. In Cambridge this porphyritic stone can be seen in the pillars supporting the building occupied by Messrs Sayle in St Andrew's Street, nearly opposite to the University Offices.

A striking characteristic of Shap granite is the occurrence of distinct patches of dark grey stone, much finer in the texture than the surrounding rock. These inclusions, like those of the normal granite, frequently contain porphyritic felspars. They tend to diminish

the value of the mass for decorative purposes, although
in some cases the stone has been employed for important structural work notwithstanding the presence of
these inclusions. The pillars of the Midland Grand
Hotel, St Pancras Railway Terminus, London, are an
example[1]. The shafts and flanking arches of the west
portal of St Mary's Cathedral, Edinburgh (1879), are
composed of Shap granite.

Masses of grey, medium-grained biotite-granite
occur in Cumberland, which are quarried for building
purposes, as well as for paving setts. The specimen
from the Eskdale quarries is a good example of the
variety usually employed for structural work.

The igneous intrusions of Leicestershire, and of
North Wales, are extensively quarried, but owing to
the fine grain and tough nature of these rocks, they
are difficult to manipulate, especially when required
for decorative work; they are therefore seldom used
by the builder, but are chiefly employed for paving and
road making[2]. There are however one or two exceptions. The Mountsorrel quarries in Leicestershire yield
a handsome hornblende-biotite-granite which is largely
employed in building construction, and it seems to be
well suited for the purpose. It possesses a warm rose-
coloured tint and is extremely hard, as the crushing
strain indicated in the catalogue proves[3]. The dark
red fine-grained specimen from the Croft quarries is
another example of the granite of that county which
is occasionally used for building.

The specimen of granite-porphyry from the Llanbedrog quarries in Carnarvonshire, is typical of the
Welsh granitic rocks that are used for structural
purposes. The Llanbedrog rock is often employed
for forming dock walls.

---

[1] *Quart. Journal Geol. Society*, Vol. XLVII. p. 280, A. Harker
and J. E. Marr.

[2] Numerous specimens of these rocks will be found in the
cabinets devoted to specimens of rocks suitable for paving and
road making.

[3] See page 250.

ISLE OF MAN. This Island yields several varieties
of rocks which are utilized by the Manxman both for
ordinary building purposes and for the construction of
bridges. There are two important granite masses in
the Island: the Foxdale granite and the Dhoon granite[1].
The quarries at Foxdale near St John's are the source
from which the chief supply is drawn. The specimens
indicate that it is a light grey, medium-grained, useful
looking stone. St John's Church is an example of a
building erected with Foxdale granite. On the other
side of the Island, a little to the north-east of the
celebrated lead-mines of Laxey, the Dhoon granite is
worked, a fine-grained biotite rock harder and closer
than the Foxdale stone. The granite of the Dhoon is
quarried for road making as well as for building. A
dark blue-grey gabbro is quarried at Poortown, also a
close-grained aplite at Crosby near Douglas, both of
which yield a useful building stone. "In the Northern
Drift plain the absence of solid rock has led to the use
of glacial boulders both for road making and building.
For the latter purpose the chief rock is the abundant
Criffel granite the large blocks of which are blasted into
pieces and trimmed into shape; a good example of its
use can be seen in Bride Church[2]."

LUNDY ISLAND. The principal part of this Island
is composed of granite, only the south-east corner consisting of Devonian slate, in which the former is
intrusive.

Along the top of the precipitous cliffs on the western
side of the Island are a series of tors, locally known as
"Cheeses," formed of a number of horizontal layers of
granite, with their edges rounded, similar to those
frequently observed on Dartmoor in Devonshire. The
most important granite mass however is on the east of
the Island, where there are extensive quarries yielding

[1] *Text Book of Petrology*, F. H. Hatch, 1909, p. 284.
[2] *Geology of the Isle of Man*, G. W. Lamplugh, p. 433.

valuable building stone[1]. The specimens in the collection shew that this rock is chiefly of a light grey colour, and that it varies in fineness, one specimen being very coarse, and distinctly porphyritic. It resembles in many respects the granites of Devonshire and Cornwall.

Although Lundy Island granite is said to enjoy a good reputation among architects and engineers, the demand for it seems for· some reason to have almost ceased, the quarries being mostly idle. Large quantities of this stone were used for the construction of the Thames Embankment in London between 1864 and 1870; the marine wall, facing the river, is largely composed of immense blocks of this granite. It was also used by the Government in the construction of the fortifications on St Catherine's Island, off the coast of South Wales, near Tenby.

SCOTLAND. Granite for many years has been closely associated with Scotland's prosperity. The Aberdonian is justifiably proud of his " Granite City," and he is fortunate in his environment if credence can be given to John Ruskin's assertion that "granite forms delightful and healthy countries[2]." It is no exaggeration to say that Aberdeen owes its existence to granite. The city is built on it, and of it: granite is everywhere.

In 1764 Aberdeen granite was first used in London for paving the streets, and about 30 years later large granite blocks were sent for constructing the Docks at Portsmouth. This was practically the first use of this material for building purposes in England.

Having in view the important position Aberdeen occupies in the building world as a producer of granite, we may refer briefly to the method of working its

[1] "Geology of Lundy Island," T. M. Hall, *Devon Assoc. of Science*, 1871.

[2] *Modern Painters*, J. Ruskin, Vol. iv. Pt. v. p. 144.

quarries. A recent author[1] writes, "The Aberdeen-shire quarries have a characteristic oval outline in plan, with a deep pit-like section of which Rubislaw and Kemnay quarries may be taken as typical. These are the largest granite quarries in the kingdom, and have each a depth of over 300 feet, which is gradually being increased. The commercial granite is found in 'posts' or isolated masses cut off from each other by 'bars' of inferior and often worthless rock, and the joints are highly irregular, features which make the quarrying of Aber-deenshire granites a difficult matter, and the principles of quarrying strictly indefinable. Overlying the granite is a more or less thick covering of exceedingly hard boulder clay, and for some distance down, the surface rock is usually of inferior quality, owing chiefly to the decay of the felspar. The overburden is thus costly to remove, and the top rock unremunerative. Under these circumstances the main principle underlying the de-velopment of the Aberdeenshire quarries is to work downwards, after a sufficient area has been opened out in which to locate the relative position of good rock, 'bars,' and 'master' joints, and to afford ample room to carry on quarrying operations. As a rule the quality of rock also improves with the depth, and there is thus a temptation to deepen without a proportionate surface area. Where this has been done the quarry has assumed the form of a conical pit with a small floor, difficult and costly to work, and in some cases finally abandoned for these reasons."

The prevailing colour of the granitic rocks quarried in and around Aberdeen is grey, but the surrounding districts, as the specimens shew, produce granite of almost every variety and shade of colour.

The granite from the Dyce quarries, as the specimen shews, is a dark grey, even-grained stone. It is ex-tensively employed for all sorts of structural work, both at home and abroad. Examples of it can be seen in

[1] "Granite quarrying in Aberdeen," W. Simpson, *Quarry*, Sept. 1907.

many of the Banks of London where it is used for the internal structures. It was employed in the Bank of Australia in Melbourne.

The light silvery-grey specimen from the Kemnay quarries, near Inverurie, is an example of the granite that has been selected for forming the base of the Queen Victoria Memorial in London. It was employed for building the piers of the Forth Bridge in 1885.

The fine-grained blue-grey specimen from the Rubislaw quarries, is typical of the celebrated grey granites of Aberdeen, of which the city is almost entirely built. The group of quarries from which a large proportion of this well-known building material comes, is practically in Aberdeen. Rubislaw, which claims to be one of the oldest quarries in the county, is just outside the city. On the other side are the Dancing Cairns and Sclattie quarries, whilst on the opposite side of the valley is Persley. This last-mentioned quarry yields a granite slightly lighter in colour than the others, but, as the specimens shew, they are all similar in colour and texture. The Bell Rock Lighthouse, erected in 1806, was built of Rubislaw granite, and it was used with other granites in the construction of Waterloo Bridge, London, in 1817.

The rock from the Correnie quarries near Alford, possesses a warm red tint, and the specimen shews that the colour is even more brilliant than that of the Red Peterhead. It is largely in demand for decorative work, as in the Municipal Buildings of Glasgow, erected in 1887. This granite was also used for the railway bridge which crosses the Tay near Dundee. A bluish-grey biotite-granite is also quarried in the Alford district, which is used largely for all kinds of structural work. The specimen from Tillyfourie quarries is an example of this stone.

The example from Tyrebeggar quarries is typical of the bright pink fine-grained biotite-granite of Aberdeenshire. This stone is extensively employed for decorative work as well as for ordinary building. The specimen from Crathie quarries is another example of the pink variety quarried in the county.

The specimens from the neighbourhood of Peterhead illustrate the handsome granitic rocks quarried in that district, which are famous for their rich and varied shades of colour and degrees of texture.

They are usually divided into two classes, and are known commercially as Red and Blue Peterhead granite, The former is the more popular of the two varieties, and it has been used over a long period for ornamental construction as well as for monumental work, not only in the British Isles but all over the civilised world. The pillars in the Fishmongers' Hall and those of Carlton Club, London, are examples, also the columns of St George's Hall, Liverpool. There are two good examples in Cambridge of the use of Red Peterhead granite. The columns in the Entrance Hall of the Fitzwilliam Museum in Trumpington Street are of Red Peterhead[1], and the piers which subdivide the transept arches in the Chapel of St John's College have shafts of the same stone[2].

The Blue Peterhead granite is a medium-grained rock, bluish grey in colour, varying in shade in the different quarries. The specimen from the Rora quarries near Peterhead represents a dark variety. This stone, although not so popular as the Red Peterhead granite, is much esteemed for all kinds of decorative building as well as for ornamental purposes. It may be seen in Cambridge in the pillars supporting the roof of the building covering the ice-making

[1] The only documentary evidence bearing on this statement is a Grace proposed to the Senate of the University dated 22nd May, 1846, authorising Mr Cockerell (the architect) to contract for Red Granite Columns for the Hall of the Fitzwilliam Museum (*Architectural History of Cambridge*, Willis and Clark, Vol. III. p. 216), and on comparing the specimens of Peterhead Granite in the collection with the pillars, there is little doubt as to their origin.

[2] "Geology of the College Chapel of St John's College," T. G. Bonney, *Eagle*, March, 1907.

machinery belonging to the Messrs Pryor in Petty
Cury.

The specimen from Cairngall quarries represents
the lighter coloured Blue Peterhead granite. This
muscovite-biotite rock has been extensively employed
for all kinds of important decorative work. The massive
sarcophagus of the late Prince Consort at Frogmore,
which weighs thirty tons, consists of Blue Peterhead
from the Cairngall quarries.

There are many other examples of Aberdeenshire
granites which display a great variety of colour, far
too numerous to refer to individually, neither does
the collection represent all the shades that can be
obtained, although the specimens shewn are fairly
typical. An interesting example of the use of Scotch
granites can be seen in the church of Crathie near
Balmoral. Nineteen columns made from the various
Aberdeenshire granites are incorporated in the pulpit
of the church. This work was executed by the express
desire of the late Queen Victoria.

Granite possessing a delicate pink tint is quarried
near Banchory in Kincardineshire. It is in demand for
monumental work, as well as for ordinary building. The
specimen from the Hill o' Fare quarries is an example.

The granite quarried in Inverness-shire for building
purposes has a more brilliant shade and is finer in the
grain than that from the Ross of Mull. The specimen
from the Abriachan quarries is a typical example.

Argyllshire yields a valuable granite, but it is not
extensively quarried. This granite almost became
famous in the middle of the last century. The Royal
Commissioners, appointed in 1839 to consider the
question of the stone to be used for the New Houses
of Parliament, were offered by the Earl of Breadalbane,
as a free gift to the nation, as much granite from his
Oban estates as might be required for this purpose.
The cost of transport, combined with the extra expense
of using this hard description of stone for building
purposes, where decorative architecture was essential,

induced the Commissioners to refuse this munificent offer[1].

The biotite-granite quarried on the Ross of Mull, was employed in the construction of the Holborn Viaduct in London, also Westminster Bridge. The specimen shews it to be a warm red coarse-grained rock.

Granite is extensively quarried in Kirkcudbrightshire, and used for purposes of construction. There are three principal granitic intrusions in this county, which are collectively known as the Galloway granites[2]. That distinguished as the Criffel mass, furnishes most of the material for building. The colour of the rock is chiefly grey, as the specimens indicate. The example from the Dalbeattie quarries possesses a pinkish shade, the felspar crystals being tinged with that colour. Doubtless this is due to the presence of oxide of iron in a marked degree, as the chemical analysis discloses. It will be observed that the specimen from the Creetown quarries is free from the brown tint. These granites have been extensively employed for constructing docks in Liverpool and Swansea.

IRELAND is rich in granites, most of which are useful for structural purposes, but comparatively little of it is developed, and many of the quarries are idle. If the collection contained a specimen from every quarry in Ireland, capable of producing a granite suitable for building, its size would have been greatly increased. But since the collection is intended mainly to include examples of rocks which are at the present time being worked for building purposes, Ireland is but poorly represented.

The counties of Dublin, Wicklow, and Wexford, forming part of the great intrusive mass of Leinster, are the chief centres of the granite industry in Ireland. The specimen from the Dalkey quarries in Co. Dublin is typical of the granite from that county. A special

[1] *Parliamentary Reports*, 1839, Vol. xxx.
[2] *Text Book of Petrology*, F. H. Hatch, 1909, p. 286.

interest is attached to this specimen, as it represents one of the earliest rocks developed for building purposes in Ireland. It is on record that it was quarried as early as 1680, and the quarries have been regularly worked ever since, although not so actively now as formerly. This granite was used in the construction of Kingstown Harbour. The specimen from the Glencullen quarries is another example of Co. Dublin granite.

Granitic intrusions form the chief mountains of Co. Wicklow, extending as far south as Wexford. As the specimens shew, most of the granites in these districts are grey muscovite rocks, some of them possessing a delicate pink tint, which is greatly esteemed in Dublin for decorative work. An example is the Wellington Monument in Phœnix Park. The stone was supplied from the Ballyknockan quarries, and was extensively employed for building Trinity College in 1832, and the Museums in 1889.

The specimens from the quarries at Newtownbarry are typical of the granite worked in Co. Wexford for local building purposes.

Co. Down is largely composed of granitic rocks. Those forming the picturesque mountain ranges of Mourne, produce a granite which is sometimes employed for building purposes, but owing to the numerous cavities it contains, and the distinctively crystalline structure of the stone, it is not extensively used[1]. On the other hand the granites of Newry, as the specimens demonstrate, are close-grained rocks of even texture, and are much sought after for structural purposes in the north-east of Ireland, as well as in other parts of the country. It is on record that the late Queen Victoria selected grey granite of Newry, from among several others, for the base and pedestal of the Albert Memorial in Hyde Park, London[2]. The stone was supplied from the Glenville quarries. The specimen in the collection shews that this granite has a pink tint.

[1] *Building and Ornamental Stones*, E. Hull, p. 44.
[2] *Ibid.* p. 45.

The specimens from Co. Galway shew that there is
no lack of variety in the granites from this county,
and that they are in every respect well adapted for all
kinds of architectural work. Some of the examples
display features of great beauty, which compare favourably with any specimens of granite in the collection,
and it is to be regretted that so little has been done to
utilize this handsome and valuable building material.
The pink, fine-grained specimen from the Shantalla
quarries is particularly pleasing to the eye. This stone
has been employed for the superstructure of the Parnell
Monument, lately erected in Dublin. The base of the
monument is composed of granite from the Barna
quarries, which, as the specimen shews, is a medium-
grained, dark grey stone with a pink tint.

Next in importance, as regards area of distribution,
are the granitic intrusions of Co. Donegal, but here
again only a small proportion of the available material
is worked, notwithstanding that the physical properties
of the granite, with its handsome pink orthoclase
crystals, as displayed in the specimen, recommend it
for all kinds of ornamental work.

JERSEY and GUERNSEY have for long been celebrated for their granitic rocks.

A large proportion of the hornblende variety quarried
in Guernsey is transported to London and utilized for
road making and for paving purposes, but although, as
the specimens indicate, the grain of this stone is very
fine-grained and the texture tough, rendering it expensive to work, a considerable quantity is used for
building purposes.

As the numerous specimens shew, the granitic rocks
of Jersey are much coarser in grain than those of
Guernsey. The hornblendic rocks of Jersey are distinguishable by their pink tint, which colour is displayed
in many shades. The granite from the La Moie quarries
near St Helier was largely used in the construction of
Chatham Docks.

AUSTRIA. On the flanks of the Noric Alps, which

traverse Upper and Lower Austria, are plutonic intrusions that yield a granite suitable for structural
purposes. The quarries near Gmünd in Lower Austria
produce a grey coarse-grained rock with large white
porphyritic crystals of felspar. Those near Freistadt
and Mauthausen in Upper Austria, also yield a granite
suitable for building. The specimens indicate that the
granite from the quarries of Upper Austria is much
finer in grain than that of Lower Austria. Immense
quantities of stone are sent to Vienna from both
districts to be used for building and monumental work.
An example of the granite from the Gmünd quarries
can be seen in the Central Railway Station of Vienna.
The Mauthausen granite, which is believed to be the
granite first used for building in that city, was employed in the foundations of the Parliament Houses.

At Mrakotin in Moravia a useful grey granite is
quarried. This stone is also extensively employed for
building in Vienna. The specimen indicates that it is
coarser in grain than the granite of Upper Austria.

The specimen from quarries near Pilsen is an example of the hard blue-grey rock obtained in Hungary
and highly esteemed by architects and builders in that
country for structural purposes.

DENMARK. The specimens in the collection from
Denmark demonstrate that granitic rocks of various
shades of colour and grades of texture occur in that
country, all suitable for building, many being very
handsome in appearance and capable of taking a high
polish.

Denmark proper however does not yield any granite:
it comes from the mountainous surface and steep rocky
shores of the Island of Bornholm. The numerous
quarries may be divided into three groups: those near
Allinge at the northern extremity; the quarries in the
neighbourhood of Rönne, on the west side of the Island;
and those bordering on the east coast, in the district
of Nexö. Most of the quarries on the Island are

conveniently situated near the coast, rendering the shipment of the granite easy and inexpensive.

The specimen of hornblende - granite from the Mosselökke quarries, one mile north-west of Allinge, is typical of the pinkish grey stone present in the northern group. It is shipped to Copenhagen, and there extensively employed for all kinds of building purposes. It was used in the construction of the Town Hall between 1892 and 1900, also for the premises of the Royal Society of Denmark in 1899. A dark grey coarse-grained stone of the same petrological class as the granite just mentioned, is quarried three miles south - east of Allinge. The specimen shews that there is a slight pink shade of colour in this rock; large veins of pegmatite are said to traverse it. The State Life Insurance Offices, in Copenhagen, were built of this stone in 1906.

Among the igneous intrusions in the Rönne district is a very dark grey diorite which is quarried at Klippegaard, three miles east of Rönne, and used for building and monumental work. The specimen shews that this rock is practically black, and makes a handsome decorative stone. It has been used in the building of the Town Hall of Copenhagen (1895).

Besides the pink coarse-grained granite quarried at Helvedsbakkerne, three miles north of Nexö, a dark grey syenite occurs in that district which is streaked with blue veins. Both these rocks were used in the construction of the Copenhagen Town Hall between 1892 and 1900.

FRANCE. Granite is extensively employed for building purposes in France in the several districts where the rock is found.

In the south-east, the builders inhabiting the Departments of the Alps, employ the handsome protogine, which will be described more in detail when referring to the granites of Switzerland.

Farther north, in the Department of Vosges, is

a handsome grey porphyritic rock which is popular for building as well as monumental work. Alexander III Bridge in Paris is constructed with the granite from the Raon l'Étape quarries in this Department.

The adjoining Department of Haute Saône produces a granite peculiar in its colour as well as its texture, which the specimen indicates.

Specimens from quarries in the Department of Maine et Loire present various shades of colour and degrees of fineness, one possessing a rich crimson tint. These granites are largely employed both in that region and in Normandy for ecclesiastical work and for ordinary construction.

In the north-west of France, notably in Normandy and Brittany, granitic rocks are largely quarried for building purposes. The specimen from Laber in Brittany is a typical example of the attractive porphyritic rock of that district. This granite resembles, in a striking degree, the porphyritic granite of Shap in Westmorland. The large felspathic crystals, as will be seen by comparison of the two rocks, are very similar both in size and colour.

The blue-grey specimen from Montjoie quarries near Vire is typical of the granite quarried for structural purposes in the Departments of Manche and Calvados. This stone is used sometimes for ordinary building, but chiefly for monumental work. It is also in demand for the millstones of oil and chocolate mills.

The green and red mottled, coarse-grained specimen from quarries near Ajaccio, is an example of the granitic rocks of the Island of CORSICA, where they are worked for structural purposes. Although this granite takes a high polish, and is attractive in appearance, as the specimen indicates, it is seldom used by the Corsican as a decorative stone, being chiefly employed for foundations and ordinary building; many of the houses in Ajaccio are constructed of it.

GERMANY is largely dependent on Saxony for her supply of granite for building purposes.

Important quarries are worked near Schmöllen; these yield a hornblendic rock, of various shades of grey. The specimen from the Ratschken quarries is typical of the light coloured rock, and the cube from the Grund quarries represents the dark grey granite.

There are extensive quarries near Kamenz, about twenty miles south-east of Dresden, which furnish a useful building material. The specimens indicate that the rock from these quarries is of the same mineral composition as the Schmöllen granite but much coarser in the grain, in some cases developing the large felspathic crystals characteristic of the Cornish granite. This stone is extensively employed for building in Dresden, and the comparative proximity of these quarries to the Elbe furnishes a ready and cheap method of transport to the numerous populous districts on either side of the river, extending as far as Hamburg, from which port the granite is exported to Holland and other countries.

Among the numerous buildings where Saxon grey granite has been employed are: the University (1889), also the Market Hall (1890), and the General Post Office (1906) in Leipzig; the Market Hall (1892) and the Technical Schools (1901) in Dresden; the Imperial Post Office (1892) and the Potsdam Railway Station (1893) in Berlin; the Town Hall (1890) and the Jews' Synagogue (1905) in Hamburg.

On the borders of the Black Forest in Würtemberg there are granitic intrusions which yield a stone resembling in colour the granites of Saxony, but differing a little in mineral composition. The cube from quarries near Alpirsbach is a typical example. It is employed for general construction in Würtemberg, also in the towns on either side of the Rhine.

ITALY is well supplied with marbles, and the climatic conditions of the country are favourable to their use in building operations. This necessarily retards the development of the granite industry.

There are however several quarries in the northern provinces of Piedmont which yield a granite, useful both for ordinary building and decorative construction. The specimens of Baveno granite are good examples of the stone produced from the quarries near that town on the shore of the beautiful Lake Maggiore. The delicate pink tint of this granite has secured for it a popularity among architects, not only in Italy, but in all parts of the world. The polished columns adorning the exterior of the Bank of Africa in Durban are composed of Baveno granite. Beautiful examples of decorative building in Italian granite can be seen in the churches and cemeteries of many of the towns of Italy. The magnificent mausoleums erected by the Milanese and Genoese aristocracy are mainly constructed of Baveno and other Italian granites. On each side of the main porch of the interior of Milan Cathedral there is a huge monolithic column of polished Baveno granite[1]. The quarries lie in the south-west nook of the west bay of the lake, and extend along the slopes of the hills between Fariolo and Baveno. Most of the rough hewn blocks are transported by water to the opposite shore of the bay, where lies the pretty town of Suna. There numerous artificers are at work fashioning the stone into various designs, destined to adorn new examples of architecture in all parts of the civilised world.

The specimen from the Alzo quarries on the shore of Lake Orta is a good example of the very light, nearly white granite of Italy. With the exception of the Bethel granite from Vermont in the United States[2], this is the lightest coloured specimen in the collection.

The example from the Montorfano quarries, on the river Tosa, near to Gravellona, is typical of the grey granite of Italy. It will be noticed that a very slight pink tint is present, although not nearly so pronounced

[1] *Building and Ornamental Stones*, E. Hull, p. 48.
[2] See p. 72.

as that in the Baveno granite[1]. This stone has been much used for all kinds of decorative building. The large monolith columns in St Paul's Church, Rome, are composed of it, also the pillars supporting the main entrance to the Waverley Hotel, London (1894).

The dark brown hornblendic specimen from the Piedmont quarries represents the Black Granite of Italy, which is extensively used for building and ornamental work. Examples can be seen in the Waverley Hotel, London. The pedestal of the monument to the Black Prince in Leeds is composed of it.

In NORWAY there exist immense masses of eruptive rocks, most of which yield stone useful for building purposes. True granite occurs, and is largely quarried on the east side of the Christiania Fjord.

Extensive tracts are met with along the coast of Romsdal. The whole of the Lofodens and Vesteraalen, together with the outermost islets, holms and skerries along the coast of Nordland consist almost exclusively of granite[2].

Besides the numerous and beautiful specimens of Norwegian granites, there are various specimens of the Devonian augite-syenites known as "Larvikites" from quarries in and about Larvik near Christiania[3]. These so-called Norwegian labradorites have recently taken a prominent position among building stones for ornamental work, a place which they owe chiefly to the beautiful play of iridescent colouring exhibited on the polished surface of the felspar crystals, at first supposed to be labradorite from its resemblance to the chatoyant felspar common in Labrador. The mineral is now known to be a soda-orthoclase.

Examples of this ornamental building material can often be seen in the façades of restaurants in London

---

[1] In this district the pink and the grey varieties pass one into the other.

[2] *Encycl. Britannica*, 9th Ed. Vol. XVII. p. 578.

[3] *Petrology for Students*, A. Harker, p. 51.

and elsewhere. The exterior pillars of Messrs Macintosh and Son's business premises on Market Hill in Cambridge are composed of Norwegian Larvikite.

Since these rocks have been generally introduced into the English market distinctive trade names have been given to the different varieties. The blue-grey, coarse-grained stone is known as "Royal Blue Granite," the dark greenish grey rock as "Emerald Pearl Granite," and the brownish blue stone as "Bird's Eye Granite."

Recently large quantities of Norwegian granite have been imported into the British Isles. The geographical position of the quarries in Norway, many of them being practically on the sea-board, the absence of "overburden," and cheap labour, are advantages which enable the Norwegian to compete successfully against the less favourably placed British quarry master. The light grey specimen from Hov quarries, and the pinkish grey rock from Ulleberg, both in the Christiania district, are examples of the granite that has been largely employed in the construction of the Keyham Dock extensions at Devonport. The fine-grained grey specimen from Lilholt quarries near Frederikshald represents the granite that was used in building the Stock Exchange, Manchester. This stone is known in the trade as "Grey Royal Granite."

PORTUGAL. The granitic rocks of Portugal, that yield material for building purposes, occur on the slopes of the Serra de Marão, which on the left bank of the Tamega shelter the celebrated wine districts of Traz-os-Montes. The lofty range of Serra de Lonsão in the province of Beira is entirely composed of granite, and contributes its share of building material.

As the specimens shewn indicate, there is but little variation in the colour of these granites, the prevailing tint being an unprepossessing light brownish grey; moreover, the texture of the rock prevents it from yielding kindly to the polisher, consequently it is not much sought after by the decorative builder. The granite is however a useful material for ordinary construction and for engineering work where stability is essential. The quay walls forming the harbour of Laxoes, near Oporto, are built of granite from the St Gens quarries, which are situated between these two towns. Considerable quantities of this granite are also employed as training standards or poles in the vineyards situated near the quarries.

The yield of some quarries is almost exclusively given over to the manufacture of paving setts, which are largely exported to South America.

RUSSIA. The plutonic intrusions of Russia are widely distributed in Finland, Northern Russia, the Ural Mountains, and the Caucasus; they also form the back-bone of the extension of the Carpathians through Southern Russia. They consist for the most part of red and grey gneisses, and granulites, with subordinate masses of granite.

The chief supply of material suitable for structural purposes, is obtained from quarries in Finland. The numerous handsome specimens in the collection from this Province indicate the immense variety of colour possessed by these rocks.

There is a constant demand in St Petersburg for the biotite-granites, gneisses, and gabbros, which furnish an inexhaustible supply of attractive building material for all kinds of architectural work.

With the exception perhaps of Aberdeen, granite is. used for building purposes more freely in St Petersburg than in any other European city. Most of the Imperial Palaces are built entirely of it, as are also the bridges across the Neva.

The black, fine-grained specimen from the Svartä quarries near Ekanäs is worthy of special notice, being an example of the black Russian granite which rivals the well-known " Black Granite " of Sweden.

The light blue-grey specimen from the quarries of

Hangö is an example of the granite employed for constructing the Russalka Monument at Reval.

The warm red, fine-grained specimen from the same locality is typical of the stone used for building the New Cathedral at Warsaw, and the Cathedral at Narva.

The dark red specimen from quarries on a small island off Helsingfors is typical of the Finnish granite chiefly employed for monumental work in Russia.

The specimen from the Wyborg quarries in Finland is an example of the granite which forms the pedestal supporting the statue of Peter the Great on the English Embankment overlooking the Neva, which is believed to be one of the largest masses of dressed granite in the world. This immense piece of stone was not brought from the Wyborg quarries direct, but was cut out of an "erratic block," said to have been found embedded in a bog between St Petersburg and Cesterbeck[1].

The huge sarcophagus beneath the dome of the Invalides in Paris (the gift of Emperor Nicholas of Russia), in which are deposited the remains of the Emperor Napoleon Bonaparte, is composed of red granite from Finland.

The specimen of fine-grained red granite from a quarry on one of the small islands in Lake Ladoga is representative of the granite intrusions which form these islets. This stone is largely used for the construction of facades, and for monumental work in St Petersburg as well as in other parts of Russia.

There is also a dark grey diorite quarried on another of these islands. The specimen denotes that it is not so handsome as the biotite-granite just mentioned, but it is said to be a useful stone for general building, and a large quantity is shipped to St Petersburg for this purpose. The granite of these small islands is transported to St Petersburg by numerous steamers

---

[1] *Building and Ornamental Stones*, E. Hull, p. 51.

plying during the summer on the Neva between the
city and the lake.   In winter, quarrying ceases when
the navigation closes.

The light brownish grey coloured specimen from
quarries in the Province of Ekaterinoslav is an example
of the granite of that part of Southern Russia.   This
stone is chiefly employed for foundation work.

On the west flanks of the Caucasus in the government of Kutais, granitic intrusions occur which are
chiefly quarried near Kursebi and Gelati, villages about
40 miles north-east of the town of Kutais.   The granite,
as the specimen indicates, is medium-grained, and
grey in colour.   It is said to be good building material,
and is largely used in all kinds of structures, especially
for engineering works.   It has been employed for facing
the entrances to the railway tunnel which pierces the
Mesques Mountains at the Pass of Suram.   An obelisk
of this stone has been erected at the entrance of the
western end of the tunnel.

SWEDEN.   Large areas of granite occur in many
parts of Sweden and form extensive masses in the
provinces of Kronoberg and Göteborg, where the
principal quarries are situated.

The variety of colour displayed in the specimens
from Sweden is a striking feature.   The granite known
as " Red Oriental" has a singularly rich tint.   An
example of this handsome stone can be seen in Cambridge, forming the façade of the business premises
of Messrs Macintosh on Market Hill, in conjunction
with the Norwegian Larvikite previously mentioned.

Attention is directed to the specimen of "Black
Granite," a trade name given to the pyroxenic rock,
classed by petrologists with the gabbros.   This rock,
which is quarried at Herrestad near Kärda, in the
province of Jönköping is almost unique, as regards its
colour, texture, and density.   It takes a high polish,
and when it appears in monumental and decorative
work is often mistaken for black marble.   The supply

is said to be inexhaustible, for the outcrop of the rock covers an area of fully 200 acres, and since the covering of soil does not exceed a few feet, it is easily quarried. Besides being largely employed in Sweden for all kinds of structural work, it is exported to Germany, France, and Great Britain. Large blocks are sent direct to Aberdeen, where it is worked up and polished, and re-exported to all parts of the world, principally America. An example of it can be seen in the interior of the Ritz Hotel in London.

Besides the fine-grained rock just mentioned, out-crops of coarser "Black Granite" are quarried on Herrestad Manor, and employed for decorative work. The specimen from the Bjoerkelund Hill quarries is typical of the medium-grained variety, and the cube from the Sjoehag Hill quarries represents the coarse-grained rocks. It will be observed the latter possesses, in a small degree, the iridescence of the Larvikites of Norway.

SWITZERLAND. The huge masses of protogine granite which often form the central cores of the rugged mountains of the Swiss Alps are materials useful for building construction, but their almost in-accessible position makes them practically unavailable. Nature has however provided another source. A large part of Switzerland, notably the valley of the Rhone, is studded with gigantic boulders of Alpine granite, transported from the mountains by ice and deposited where they now rest, during the glacial period. Hundreds of these scattered "erratic blocks" have become the scene of numerous diminutive quarries which produce handsome well-chiselled blocks of grey granitic rock for building purposes in all parts of Switzerland.

The grey specimen in the collection is from one of the many workings in the remarkable group of "erratic blocks" near Monthey, overlooking the valley of the Rhone, a few miles below St Maurice.

The light grey coarse-grained specimen from the

Biasca quarries in Canton Ticino is an example of the muscovite-biotite-gneiss which is used in that region for building. It is also used there for gate posts and supports for training the vines.

NORTH AFRICA. EGYPT. Archaeological research shews that the use of Egyptian granite, as a material for construction, is of great antiquity. It was employed in constructing the Great Pyramid of Gizeh, the tomb of Khufu or Cheops. The floor of the king's chamber in that Pyramid is composed of huge blocks of granite, laid with such precision that not only are the joints scarcely perceptible, but the under faces and edges of the stones are so sharp and polished that it is impossible to detect how they were lifted and placed in contact with each other, since no marks of force or of any purchase having been applied can be perceived; indeed some persons imagine that it was not until after they were fixed in their respective places that the outward surfaces of the stones were smoothed down and polished[1]. Part of the outer covering of No. 2 Pyramid of Gizeh, the tomb of Khafra, and that of No. 3 Pyramid, the tomb of Menkaura, were composed of granite[2]. The obelisk, now the sole landmark distinguishing the site of Heliopolis, the ancient city of On, famous for its grandeur and for the learning of its inhabitants, is composed of granite.

Cleopatra's Needle, which now adorns the Thames Embankment in London; the obelisk in the Place de la Concorde in Paris; and that in the Central Park in New York are all made of the same material. Archaeological Museums in most parts of the world contain illustrations of ancient work in granite from Egypt.

The specimen cubes from the quarries in Upper Egypt are typical examples of the granite employed in these ancient buildings, and the wonderful state of preservation of the ancient monuments is a striking

---

[1] *Pyramids of Gizeh*, Col. Vyse, p. 11.
[2] *Ibid.* W. M. Flinders Petrie, p. 32.

evidence of the wisdom of the architects of that period
in selecting a stone of such strength and durability.
In the neighbourhood of Aswan, on the site of the
old quarries, the marks of the picks and chisels can
still be traced on the granite. Allowance must however be made for climatic influence. The well preserved
condition of the ancient monuments and buildings is
not so much due to the inherent quality of the stone,
as to the atmosphere. The obelisk of Luxor, which
stood for centuries in Egypt without being perceptibly
affected by the climate, is now blanched and filled with
small cracks, after only 40 years' exposure in Paris;
and the obelisk in the Central Park of New York has
shed many small fragments since its erection there[1].

The Aswan quarries were the scene of the earliest
quarrying operations of which we have any historical
record. From the fourth dynasty (more than five
thousand, perhaps nearly six thousand years ago) onward
to Ptolemaic times, the Aswan quarries were worked at
intervals to furnish stone for the more ornamental
portions of temples and other structures. The quarrying industry probably reached its maximum in the
eighteenth dynasty, when Senmut, the architect employed by Queen Hatshepsut, completed the great obelisk
of Karnak, nearly one hundred feet high, in seven
months. Senmut was a great organiser of labour, and
to him are doubtless due the quarry roads which still
exist. From about the dawn of the Christian era to our
day, the quarries have remained unnoticed, the more
accessible limestone of Lower Egypt having been preferred to the beautiful but costly granite. With the
construction of the Nile reservoir in 1898 began a new
epoch in the quarrying industry, a hard and heavy stone
being necessary for this class of work, while the highly
ornamental character of the granite, its proximity to
the dam, the absence of weathered overburden, and the
comparative ease with which it can be worked, have

[1] "Building Stones of Wisconsin," E. R. Buckley, *Bul.
Wisconsin Survey*, No. IV. p. 17.

been additional points in favour of its adoption. The whole of the ashler facing and much of the interior rubble of the dam are of this rock[1].

Adolph Erman in his interesting book on Egypt states, "Some of the granite blocks in the temple of King Khafra, not far from the great Sphinx, measure fourteen feet in length, and those under the architrave in the sanctuary of the crocodile god Sobk in the Fayum are more than twenty-six feet long. Among the Theban obelisks there is one of a height of 107 feet, while a papyrus speaks of an obelisk from the Aswan quarries which measured 120 cubits (nearly 200 feet)." Mr Hawkshaw, after visiting these quarries, writes: "In a quarry to the east of Aswan this pink syenite (granite) may be seen and obtained in homogeneous masses of almost any size. A block said to be (for it is partially covered with rubbish) 95 feet × 12 feet × 12 feet, squared on three sides, and still attached to the rock on the fourth, may still be seen in this ancient quarry. Among the ancient monuments of Egypt masses of it weighing many tons (as much as 800 tons in one case) may be seen unaffected by the action of the weather after an exposure of from 2000 to 3000 years[2]."

The foregoing remarks may have created the impression that granite was more extensively used for building purposes in ancient Egypt than was actually the case.

No doubt the many obelisks and sarcophagi, with which we are so familar in many of the capitals of Europe, are made of granite, but very few of the Theban temples and other colossal buildings were built entirely of this stone, although the bases of columns, thresholds, jambs and lintels of doors were usually of granite[3].

---

[1] *Aswan Cataract of the Nile*, Dr Bull, p. 74.
[2] *Quart. Journal Geo. Soc.* Vol. XXIII. J. C. Hawkshaw p. 118.
[3] *What Rome was built with*, M. W. Porter, p. 62.

Perrot and Chipiez write, that there is but one building in Egypt the body of which is of granite, and that is the above-named temple of Khafra at Gizeh[1].

It is instructive to note that the group of specimens of Aswan granite in the collection, presented by the Egyptian Government came from the Syene quarries, at Aswan. It was the name of these quarries which gave rise to the geological term "syenite," said to be first used by Pliny, to describe that class of stone, and afterwards adopted as a scientific designation for a somewhat different kind of rock. Occurring in greatest abundance is a warm red porphyritic rock, which varies slightly in shade and texture. There are three specimens in the collection typical of the variations.

This group contains a beautiful cube of a fine-grained grey granite. Its use in buildings is rare in comparison with that of the well-known red porphyritic rock. The statue of the Sphinx in the Vatican at Rome is composed of grey granite from Aswan, and some of the columns at the entrance of the portico of St Peter's at Rome[2] are of the same rock.

Both varieties of Aswan granite must have been extensively employed by the Romans for beautifying their city. In Corsi's treatise of the decorative stones of Rome, he mentions the existence there of 12 obelisks and 714 columns of the red variety, and 1787 columns of the grey[3].

CENTRAL AFRICA. The huge masses of granitic rocks known to exist in the equatorial regions of Africa have as yet been little used for constructive purposes, though with the advance of civilisation the demand will doubtless be much increased.

That these igneous intrusions will form useful building material, when wanted, is evident from the fact that in a few isolated instances it has been successfully utilized.

[1] *Art in Ancient Egypt.*
[2] *Building and Ornamental Stones,* E. Hull, p. 55.
[3] *What Rome was built with,* M. W. Porter, p. 63.

The specimen from Lakoma on Lake Nyasa is a good example of granite from Central Africa. The stone is quarried on the Island, and has lately been employed for building the exterior of the Cathedral erected by the Universities' Mission at Lakoma.

GOLD COAST, WEST AFRICA. The coast on either side of Accra is composed chiefly of sedimentary rocks. "Further inland are granites and gneisses; and in the Ashantee region, and along the river Volta, are fine black amphibolite-schists abounding with garnets, locally of rather large size[1]." Several specimens in the collection represent these rocks, all of which are occasionally employed for building. The example from quarries at Akusi deserves special notice on account of the large number of garnets in the rock. This stone is extensively used locally for building; the Police Barracks, Court House, and Hospital, at Abuski, being examples.

The close-grained grey granite from the Abbonti-akoon quarries, near Tarqua, represents the building stone of that locality.

It may be well to remark, however, that only the Europeans in this Protectorate employ stone for constructive purposes, the houses of the natives being almost invariably built of a kind of earth, or clay, locally known as "swish."

WEST EQUATORIAL AFRICA also possesses granite which is waiting to be developed for building as soon as there is a demand for it, but except for very special purposes stone is not so used in this region.

The specimen from quarries near Abeokuta is an example of the granite of Lagos. This stone has been employed for the Harbour Works at the port of Lagos. The Church at Abeokuta is also built of it.

---

[1] "Geological notes on W. Africa," O. Lenz, *Geol. Magazine* Vol. VI. 1879, p. 173.

ANGOLA, WEST AFRICA. In the Benguela government of this province there exist masses of granitic rocks which are quarried and used for constructive purposes.

The specimens, which vary from a coarse, black and white, to a pinkish grey, medium-grained rock, were procured from quarries at Lengue in the neighbourhood of the town of Benguela. Besides being used locally for house building this granite has been extensively employed in the making of the Lobito-Katanga Railway, which passes close to the quarries.

PORTUGUESE EAST AFRICA. A muscovite-biotite-gneiss has lately been discovered and opened up near Beira, and in close proximity to the railway which connects Delagoa Bay with Rhodesia. This rock is known commercially as "Moçambique Granite." There are several specimens in the collection of various grades of texture, and although the rock does not admit of a good polish, owing to its foliated structure, it is reported to be a useful stone for engineering work and general building. The abutments of a new bridge spanning Chiveve Creek are constructed of it.

SOUTH AFRICA. Although the southern territories of Africa are more abundantly supplied with granite than those of the north, they are in striking contrast as regards the utilization of the stone for building purposes. As mentioned when referring to Egyptian granite, that of North Africa has been used as a building material from very ancient times, whereas the granitic rocks of South Africa are only beginning to be taken notice of by the architect and builder, indeed, even their existence, let alone their economic value, is a matter of very recent history.

TRANSVAAL. Notwithstanding that water-worn and weather-worn boulders of granite are scattered over a great part of South Africa, and vast masses occur

in situ in the Transvaal, and in Rhodesia, until very
lately no attempt has been made to utilize it for
structural purposes, and even now the colonist as a rule
constructs his house with imported materials.

Since the discovery of gold and diamonds in South
Africa, buildings of a more substantial nature have
sprung up in the towns adjacent to the mines, and
granite quarries are becoming numerous in districts
near Johannesburg. Architects and builders are now
alive to the fact that they have a material at their
doors which will answer every purpose.

The foundations of many and the superstructures of
some of the mills erected for the reduction of the ore
mined on the Witwatersrand are built of granite, as
well as a few of the important buildings recently
erected in and around Johannesburg, and it is believed
by some architects and others that granite will soon
be the chief building material in the Transvaal.

The specimens of Transvaal granite shew that there
is no lack of variety either in colour or texture. The
attractive red specimen from the Pyramid quarries,
near Pretoria, is an example of the syenite existing
and utilized in the northern districts of the Transvaal.
It has been used for the construction of the Railway
Station, also the Municipal Buildings of Pretoria. The
dark grey specimen from the Waterval quarries, near
Pretoria, is another example.

The rock from the Newlands quarry near Johannesburg, which is more of the nature of a gabbro
than a true granite, is used for bridge-building locally.
That from the Wittkopje quarries, a close even-
grained stone, is chiefly employed for engine beds
and for superstructures of the gold-refining mills in
and around Johannesburg. The greenish grey specimen from quarries on the Hyde Park Estate, and
the buff-grey example from the Lake quarries at
Craighall, both suburbs of Johannesburg, represent
the stone that is used for building and maintaining
the numerous road bridges in the district. It is also

employed for constructing the dams that are now being built on many of the small rivers for irrigation purposes.

The granites of the CAPE OF GOOD HOPE are usually grey in colour, resembling in many respects our Cornish rocks. The Paarl Mountain with its well-known group of smooth and naked crags at the summit, yields a biotite-granite which is quarried for building purposes[1]. The specimen from the Allens quarries is an example. This stone is extensively employed in Cape Town for structural work. The Parliament Houses are built of it. It was also used for building the General Post Office in 1893. The foundation stone of the Imperial Institute in London consists of a block of Paarl granite.

South of the Paarl area in the French Hoek valley, there are several masses of granite and quartz-porphyry. The specimen from the Higgo quarries represents the former. It will be observed that this granite is more coarsely porphyritic than the Paarl stone. The colossal memorial to Cecil Rhodes, recently erected near his residence at Groote Schuur, is constructed of this stone. The specimen is well worth examining, the unusually large felspar crystals appear very distinctly.

NATAL. Granite suitable for structural purposes is not so plentiful in this Province as in other parts of South Africa, and what there is has not been much developed. A mass of rock intruded near the coast, a few miles south of Durban, is practically the only granite that is regularly worked in Natal. The specimen from the Park Rynie quarries is a typical example. It indicates that this rock possesses a warm red tint, not unlike the Red Peterhead granite of Scotland. The group of offices in Durban, known as "Butchard's Buildings," is built of this stone.

[1] *Geology of Cape Colony*, Rogers and Du Toit, p. 29.

RHODESIA. The granite intrusions scattered over this territory are many and varied, and suitably situated for quarrying, but at present very few have been developed. The granitic mass forming the Matopo Hills may be taken as the best known, and is occasionally worked for industrial purposes. The specimen from the Rifle Butts Kopje quarry is a good example. This granite hill has recently sprung into fame on account of it being the resting-place of the mortal remains of Cecil John Rhodes; his body was interred, according to his express desire, in a grave hewn out of the solid rock. The pedestal supporting his statue at Bulawayo is constructed of the same rock, and presents a very handsome appearance, the large pink felspars contrasting well with the green matrix.

A grey biotite-granite exists in the Wankie district near the river Gwai, also a pink rock near Manzinyama in the Gwanda district, but as they are not yet utilized for building, specimens do not appear in the collection. Mashonaland is rich in granite but at present there are no quarries.

INDIA. Notwithstanding that the Empire of India abounds in plutonic rocks, most of which are well suited for structural purposes, until lately it has depended largely on Great Britain and other European countries for its supply of granite for engineering and building purposes[1].

It is true there is evidence that native granite has been used in former times, as in the case of the wonderful Temples of Ellora which were hewn out of solid granite[2]; but such instances are rare.

BOMBAY. In 1904 the Chairman of the Bombay Trust, being aware that granite existed in the southern portion of the Presidency, set on foot enquiries with a

---

[1] This statement refers to true granite. Much of the so-called granite of India is a granitoid gneiss which will be mentioned later.

[2] *Architectural Review*, A. W. Cross, p. 13.

view to ascertain if it could be utilized for the New Docks at present under construction at the harbour of Bombay. The investigation was satisfactory, and the specimen cubes shewn from Deccan are samples of the granite now being used in the docks[1].

MYSORE. The uplands of this State are rich in gneissic, granitoid, and porphyritic rocks, all useful for constructive purposes.

The numerous specimens in the collection shew that there is no lack of variety in either the colour or texture of these rocks; but except in the towns of Mysore and Bangalore, these handsome building stones are rarely utilized. Examples of many of them are seen in the New Palace at Mysore. The granite quarries on the slopes of Chamundi Hills near Mysore, are the best known in that State, and the town of Mysore is largely supplied from them. The Black Trap or Turuvekere Stone of Mysore, quarried in the Tumkur district, is a handsome building stone, which, as the specimen indicates, is an extremely fine-grained stone of dark blue-black colour. It has been extensively used in the New Palace at Mysore, and in Tippu's Mausoleum at Seringapatam.

A polished slab of quartz-felspar-gneiss can be seen in the durbar hall of the Rajah's palace at Tanjore which measures 18 ft. × 16 ft. × 2 ft. 1½ in. A small temple in the north-west corner of the Pagoda Court at Tanjore, which is a "perfect gem of carved work," is entirely composed of this stone. The elaborate patterns are as sharp as when they left the sculptor's hands[2].

On Mahendragiri, in the district of Ganjam, there is a good example of what was a common practice with regard to the construction of these

[1] See *Papers on Indian Granite*, published by Bombay Harbour Trust, a copy of which is in the Sedgwick Museum Library.

[2] *Geology of India*, V. Ball, p. 536.

temples. On the top of the hill is an unfinished temple, built with huge blocks of porphyritic gneiss, which, on their exposed faces, are rough and uncut. The practice appears to have been not to attempt any ornamental work until all the stones of the building were in position and then to pare them, so to speak, into shape.

MALABAR COAST. In the maritime districts of Malabar, gneissose rocks exist which are used for building purposes, and known in commerce as "Indian Granite." In South Malabar, Mr P. Lake describes three types of gneiss[1]. First, the quartose gneiss, secondly, the garnetiferous gneiss, and thirdly the felspathic gneiss in which felspar forms the principal constituent. The specimen from near Cochin on the Malabar Coast seems to be most closely associated with the second type mentioned, except that there are no garnets in the rock; the quarries however from which it comes are fully eighty miles from the district of South Malabar where the rocks which Mr Lake refers to are found.

CEYLON. A feature of the Island of Ceylon is the abundance of gneiss overlain by extensive beds of dolomitic limestone[2]. These gneisses were employed for structural work in very early times; for though bricks were largely used by the ancient inhabitants of Ceylon in the construction of their Dāgobas, the platforms on which they rest, the altars, the elephant stables, and the bathing tanks for these animals, were usually built of dressed gneiss. Ruins of these can now be seen in various parts of the Island, the most important being in the neighbourhood of Anurādhapura, the ancient capital of Ceylon. Maha Saeya Dāgoba, built about B.C. 243[3] (on the summit of Mihintale, a hill 1000 feet high, about ten miles from

[1] *Memoirs of Geological Survey of India*, Vol. XXIV. Pt. 3, p. 11.
[2] *Encycl. Britannica*, 9th Ed. Vol. V. p. 361.
[3] Parker's *Ancient Ceylon*, Chap. IX. p. 261.

Anurādhapura) and a work of Devanam-piya-Tissa, is approached by a wide staircase, said to consist of 2400 steps, which are composed of huge blocks of gneiss.

Coming to more modern times, the Dutch, who occupied Ceylon in the 17th century, constructed their fortifications of the local gneiss. These forts, or walls of defence, were erected on a gigantic scale, and required an immense quantity of material to build. The Fort of Galle covers fully 350 acres, and is surrounded by a high massive wall. The Church, Public Offices, and principal residences are situated within. Dates of its erection, ranging from 1667 to 1697, appear on the stone over many of the gates of the fort. Close to it are gneiss quarries, from which many of the specimens in the collection were procured.

Owing to the schistosity of this "Stratified Granite," an appellation freely adopted by quarrymen in America to describe igneous rocks of schistose structure, the gneisses are not now so popular in Ceylon as a material for ordinary building, especially as there is an abundance of "Laterite," which is much easier to work[1]. For heavy engineering undertakings and walling the hard massive gneisses of Ceylon are however still often used, and at present large quantities are being employed for the construction of the breakwater forming the harbour of Colombo. The numerous specimens exhibited are typical examples of the various classes of stone used for this important piece of marine structural engineering, and, it will be observed, they differ considerably both in colour and texture. The presence of garnet in the rock is very apparent. The specimens from quarries in the neighbourhood of Galle are distinguished by their very delicate shades of colour.

STRAITS SETTLEMENTS. The Island of Singapore is the chief district from which granite is procured in the Malay Peninsula.

[1] See p. 241.

Quarries on the Bukit Timah Hills in the centre of the Island produce a biotite-granite useful for engineering, as well as for ordinary building purposes. The Tanjong Pagar Dock Board's new wharf is being constructed of this granite. The specimens indicate that the rock is unusually light grey in colour and of medium texture.

The Mount Faber Range, a line of hills stretching along the south-west coast of the Island, produces a coarse-grained granite, light buff in colour. It has recently been extensively employed in the construction of the New Municipal Water Works of Singapore. Pulo Obin, an island about 4½ miles long by 1 mile broad, between Singapore and the mainland, is composed entirely of igneous intrusions, and yields a useful granite, similar in appearance and texture to the rock of the Bukit Timah Hills. It is at present being used for the construction of the New Sea Wall and Breakwater at Singapore.

CHINA. This great empire possesses granitic rocks yielding a useful and handsome material used for the construction of government and other important structures, but the cost of quarrying and dressing precludes it from being generally adopted for ordinary building, though there is evidence that it was so used in very early times.

Granite was employed for building the Great Wall of China, erected by Che-Hwang-te, "the first universal emperor." This wall was commenced in B.C. 214, but only a few traces of the original structure remains. Capt. Younghusband writes, "The Great Wall north of Peking is a magnificent structure built of immense blocks of granite. It is some 40 to 50 feet in height, and wide enough at the top to drive two carriages abreast on. Winding up and down the steep hill-side, over the summits and across the valley far away in the distance, the credulous European tourist who comes out here to see it, imagines that it extends thus for hundreds and thousands of miles, but scarcely one hundred miles from Peking it has diminished down to a miserable mud wall[1]."

In the Province of Kwang-tung there are several quarries yielding granite, none of which are extensive or important. Most of them are situated near the town of Kowloon, and are worked spasmodically as the demand requires. This is chiefly for Hong Kong, where it is used for general building. The basement and understructure of the Shanghai Bank, also that of the General Post Office, are typical examples. Several specimens in the collection are from this district. It is interesting to note the similarity of one of the specimens from Kowloon to the celebrated Baveno granite quarried on the shore of Lake Maggiore in Italy; the same delicate pink tint characterises both.

The specimen from quarries near Soochow illustrates the granitic rocks of the Province of Kiang-su. This stone is much esteemed by architects, being used for the better class buildings of Shanghai and other towns in China. There are also specimens from granite quarries in the neighbourhood of Ning-po, in the Province of Che-kiang. This rock is not considered to be as durable as the Soochow granite, and is employed for an inferior class of buildings.

Chinese granite has occasionally been exported to foreign countries for building purposes. It is stated that the first stone house erected in San Francisco, California, was built of granite brought from China[2].

JAPAN. The specimen of grey granite from quarries near Kobe is a good example of the granitic rocks of Japan, and is said to be practically the only kind of stone for constructive work used in that country, where it is estimated that 99 per cent. of the buildings are composed of wood and paper. Even in the great cities the dwelling houses are almost entirely con-

---

[1] *Royal Geogr. Proc.* 1888, F. E. Younghusband, p. 489.
[2] *Stones for Building and Decoration*, G. P. Merrill, p. 232.

structed of timber.   Only about half a dozen buildings
in the whole country consist exclusively of granite, and
a few are built of the red bricks of Japan and faced
with grey granite.   This has not always been the case,
because the foundations of all the more ancient temples
throughout the country are of large blocks of granite;
these and the long flights of steps still exist to prove
the durability of this material.   In the Old Castle of
Ôzaka, in the Province of Setza, there is an enormous
piece of granite measuring thirteen paces in length,
and about nine feet in height[1].

The specimen from Rokkosan quarries near Kobe is
an example of the granite used for building the head
office and branches of the Bank of Japan.

AUSTRALIA. New South Wales. Granitic rocks
suitable for architectural and engineering purposes
occur in several parts of this State,. varying in colour
from a rich red, to black and white, or light grey.   The
specimen from quarries on the small Island of Gabo,
near Cape How, is typical of the red granites.   This
decorative stone has been extensively employed for
general building, and all kinds of ornamental and
monumental work, in New South Wales, as well as in
the other states of Australia.   The columns of the
Treasury Buildings in Sydney are Gabo granite[2].   The
specimen from the Barren Jack quarries is another
example of the red granites of this state.   The felspar
it will be noticed has a slight green tinge.   The stone
has been employed for many important engineering
undertakings; for instance it has been used almost
exclusively in the construction of the Barren Jack
Reservoir, some of the blocks weighing 15 tons[3].

The specimen of black and white mottled rock in the

[1] *Encycl. Britannica*, 9th Ed. Vol. XIII. p. 573.

[2] A specimen of Gabo Island granite also appears in the
collection from Victoria, which has a much more brilliant shade
of colour than that sent from New South Wales.

[3] *Building and Ornamental Stones of New South Wales*, R. T.
Baker, p. 20.

collection is an example of the granite quarried at the foot of the Currook Billy range of hills, near Moruya, in Dampier Co. It has been utilized in many of the principal buildings of Sydney. The columns of the colonnade of the General Post Office are constructed with it. The pedestal of the statue of Queen Victoria, Sydney, is composed of Moruya granite. The quarries are situated conveniently near the coast, and the stone is conveyed by water to Sydney.

A handsome porphyritic rock is quarried near Tenterfield in Clive Co., on the western flanks of the New England range of hills. The large flesh-coloured felspar crystals in this stone, scattered throughout a light grey matrix, produce a pleasing effect. It resembles in many respects, the Shap granite of England.

The specimen labelled "Australian Syenite" is a good example of the rock that is known commercially in the State as "Bowral Trachyte," a designation now acknowledged to be erroneous from a petrological point of view[1]. It occurs as an intrusive boss, forming a bold headland known as the Gib Mountain, about 85 miles from Sydney, close to the town of Bowral. The rock varies in colour, from a dark olive-green to a light greenish grey, and, as the specimen indicates, takes a good polish. Being a hard, fine-grained, crystalline rock, it is somewhat difficult to dress, but being found to be very durable, and handsome in appearance, it is rapidly coming into favour with Sydney architects and engineers. The Equitable Life Assurance Society's Offices in George Street, Sydney, is an example, the building being entirely constructed of it, and it was employed in forming the piers of Hawkesbury Bridge. It has been used in the foundations of many of the largest buildings in Sydney.

QUEENSLAND is rich in plutonic igneous rocks; for the great chain of hills which stretches through the

[1] *Mineral Resources of N. S. Wales*, E. F. Pitman, p. 444.

eastern portion of Australia, from Cape York, the most northerly promontory of Queensland, to Bass Straits in the south, is composed largely of granites and other igneous rocks. Nearer the eastern coast granitic rocks appear even in greater force than in the dividing range, and the voyager rarely loses sight of them from Moreton Bay to Cape York.

Up to the present time the supply of granite for industrial purposes has been procured from intrusions on the coast, more than from those further west. As the details in the catalogue shew, the chief centres of the industry are near Townsville in North Queensland, and the district a little to the west of Brisbane.

The specimen from the quarries on Magnetic Island near Townsville deserves special notice. It is a good example of the bright coloured biotite-granite of that district, and furnishes a handsome material for decorative purposes as well as for ordinary building.

VICTORIA rests throughout on a bed of granite, and this is exposed in many parts by the denudation of the overlying strata. The principal mountain chain of the State, known as the Victoria Highlands, is largely composed of granite. On the eastern side of the Lodden valley, in the Harcourt Range, which forms part of this chain, a granite is quarried which is extensively employed for building construction. The quarries being in close proximity to Bendigo, the well-known gold-mining centre, the granite is in demand locally, and railway facilities to Melbourne being good, large quantities are transported to that city for structural purposes.

Stretching eastward and traversing Gippsland, these igneous intrusions frequently occur, and are finally exposed on the small Island of Gabo near Cape Howe, the dividing point of Victoria and New South Wales. An examination of the specimens from Victoria will shew that the granite of Gabo Island differs widely both in colour and texture from that present in the Harcourt mountains. The Harcourt granite, it will be observed, is a light grey, coarse-grained rock, whilst the Gabo Island stone is fine-grained granite of a rich crimson colour. Its bright red colour surpasses in brilliancy the celebrated Peterhead granite of Scotland. The fine even texture of this rock renders it peculiarly suitable for ornamental architectural work, and it is extensively employed in New South Wales as well as Victoria. There are many examples of its use in Sydney[1].

SOUTH AUSTRALIA. There is no lack of plutonic rocks in this state. They are developed near the New South Wales border to the west of Broken Hill, and to the west of Spencer Gulf they widen out to form the great sheet of granitic rocks of the Gawler Ranges. They extend from Reid's Lookout on the east, to the south of Lake Gairdner, and continue until hidden by the Tertiary limestone of the Nullabor Plains[2]. They are said to yield useful building material, judging from the evidence at present available. A comparatively small area however of these vast granitic tracts has been developed for economic purposes, only the rocks within an easy distance of Adelaide having been as yet quarried.

The specimen from Murray Bridge, a coarse-grained chocolate-coloured rock, compares favourably in appearance with granites from other parts of Australia, and indeed from any part of the world.

WESTERN AUSTRALIA. The greater part of this State is covered by Archean rocks, described by H. P. Woodward as occurring in six parallel belts, which run north and south with a slight trend to the north-east. The third belt extends from the southern coast to the Murchison river, and is about 100 miles wide. It is known as the first granite belt[3].

[1] A specimen of Gabo Island granite also appears in the collection representing New South Wales (see note on page 56).

[2] *Australasia*, J. W. Gregory, p. 490.

[3] *Ibid.* p. 525.

The specimen from the Kellerberrin quarries is a light pinkish grey slightly porphyritic rock. These quarries are situated on the Eastern Goldfields Railway, 145 miles from Perth and 807 feet above sea level. This granite has been used for foundations in several of the public buildings in Perth, but up to the present not in any great quantity. There is another specimen from quarries at Meckering, 101 miles from Perth, which is situated on the Eastern Goldfields Railway at an altitude of 636 feet above sea level. This is a fine-grained rock of even texture, light grey in colour, and takes a good polish. It has been used in the lower portions of the Western Australian Museum and Fine Art Gallery, also in several of the banking and insurance offices in Perth[1].

NEW ZEALAND. The specimen of grey granite from quarries at the Mohau in Coromandel County is typical of the plutonic intrusions existing and utilized for building in the North Island of this Dominion. The rock is exposed on the shores of Hauraki Gulf, close to the water's edge, where it is quarried, and transported to Auckland for all kinds of structural work.

The discovery and development of this granite is quite recent, and has been a boon to the Auckland district, since there is a lack of good building material in it. The rock has been used for building the Bank of Australia in Wellington, for the Bank of New Zealand, and for the General Post Office in Auckland. It has been employed for the construction of the Sedden Monument in Wellington (1909).

The specimen of hard dark blue-grey diorite, known as "Ruapuke Granite," represents the granitic in-trusions that are quarried and used for structural purposes in the South Island. The name is taken from Ruapuke Island off the south coast, at the east entrance of Foveaux Strait, which consists entirely

[1] Communicated by the Government geologist, A. Gibb Maitland, Esq.

of igneous intrusions. There are extensive quarries on the Island, but the chief source of supply is from the beach fringing the harbour of Bluff, where the rock is found as boulders, some of considerable size, comparable with the "erratic blocks" of Switzerland. The cost of splitting up these boulders is less expensive than quarrying, and only a small royalty is paid to the Government for a licence. Ruapuke granite is extensively employed in Dunedin, and other towns in the South Island, for ordinary building as well as decorative work. It was used for the foundations and for the polished columns of the Government Life Insurance Buildings.

UNITED STATES. With so vast a territory to deal with as North America, the surface of which is crowded with intrusive rocks, most of which are eminently adapted for constructive purposes, it is not practicable, in the circumscribed space available for this collection, to find room for a specimen of every variety of granite quarried and utilized by our American friends for building purposes. It is hoped however that the specimens collected will fairly represent this important building material at present produced in the United States.

CONNECTICUT. The granites of this State are usually fine-grained, and light grey in colour, characteristics which, as a rule, distinguish them from the granites of the Atlantic States[1]. The specimen from quarries at Waterford is a typical example. It will be observed that it is remarkably fine-grained, and so light in colour that in Connecticut it is often known commercially as "White Granite." Besides being employed for ordinary structural undertakings, it is largely used for monumental work. The Soldiers and Sailors Monument, Syracuse, New York, is an illustration of the latter, and the Green Point Savings Bank, New York, is an example of the former.

[1] *Stones for Building and Decoration*, G. P. Merrill, p. 235

Although it has just been stated that the granites of this State are chiefly fine-grained, light grey rocks, there are exceptions. A specimen will be found in the collection from the Stoney Creek quarries in the township of Bradford, which is a coarse rock and shews handsome large pink orthoclase crystals. This granite is extensively employed for columns and other decorative work in New York and elsewhere. The large polished pillars and slabs lining the interior of the main entrance to the South Union Railway Station in Boston, consist of this granite. It was also used for building the Columbia University, New York.

GEORGIA. This State is said to contain inexhaustible quantities of granite of the finest quality, all of economic value, although not yet extensively developed.

The muscovite-granite from the Stone Mountain quarries near Atlanta, has however recently taken a prominent position in the industrial world, both as a material for road making and for building construction. The granite of Stone Mountain is said to be the largest mass suitable for building in the world. The specimen is light grey in colour and possesses a fine close-grained texture. Examples of this granite can be seen in Chicago, where it has been used for the May Memorial Chapel.

MAINE. The specimens from the State of Maine take precedence, as regards numbers, in the collection of granites from America, that State being the chief source of supply of granite in the United States. So far back as the year 1886 there were some seventy-four granite quarries working in Maine[1], and doubtless the number has largely increased since then. The extensive coast-line of this State, and the proximity of many of the quarries to the sea-board, have given an impetus to the granite industry not enjoyed by other

[1] "Building and Ornamental Stones," G. P. Merrill, *Report of U.S. National Museum*, 1886, p. 413.

States which are less conveniently placed for water carriage.

Many of the quarries are situated on the islands which fringe the coast of Maine. The specimens from the Palmer quarries on the west, and the Sands quarries on the south shore of Vinalhaven Island, in Penobscot Bay, are good examples of the pinkish grey granites worked on that Island. Hurricane Island adjacent, is practically all granite, and, as the specimen indicates, yields a stone similar in colour and texture to that of Vinalhaven. The quarries of these islands are known collectively as the Fox Island quarries, and their rock as the Fox Island Granite[1]. The quarrying of this granite occupies a prominent position in the building stone industry of the United States. The Sands and Palmer quarries together furnished all the stone for the New Custom House in New York. The New Post Office, Washington, and the Masonic Temple, Philadelphia, are examples of the stone from the Sands quarries. The eight imposing columns in the Cathedral of St John the Divine in New York are of granite from the Palmer quarries on Vinalhaven Island; and the United States Post Office, St Louis, Missouri, and the County Court House, Boston, are good examples of the stone from quarries on Hurricane Island.

High Island, so called because it is the highest of the group of the islands in the Muscle Ridge Plantation, about nine miles south of Rockland, yields a useful stone. The specimens indicate that there is more than one variety on the Island, and that they resemble the Vinalhaven granites in colour, although they are finer in the grain. Both varieties are extensively employed for structural purposes in Washington and Philadelphia. The Stephenson Memorial Monument in Washington is an example of its use.

Spruce Head Island, near to High Island, also yields a useful granite, which is quarried for building purposes,

[1] "Granites of Maine," T. N. Dale, *Bul. U.S. Geol. Survey*, No. 313, p. 129.

and occasionally for paving blocks. The specimen shews that this is a quartz-monzonite, without the pink tint usually present in the stone of Vinalhaven and adjacent islands. It is well known as a building material in the Western States of America. The Carnegie Library at Alleghany, Pennsylvania, is built of it.

The pinkish-grey cube from quarries on Hardwood Island in the township of Jonesport, is typical of the granite worked there, which is known commercially as "Moose-a-bec red[1]." This stone takes a fine polish, and is chiefly used for internal decorative work. The staircase in the main entrance to Suffolk County Court House, Boston, and the columns in the Cathedral of Newark, New Jersey, are examples of its use.

Granite worked in quarries on the mainland of this State also holds an important position as a building material. The specimen from the Waldoboro quarries, near the village of the same name, situated on the Boston and Maine Railway, is typical of the stone worked in the district. It will be observed that it is much finer in grain than those granites mentioned above; this stone is said to be unsuitable for polished work. The United States Naval Academy group of buildings at Annapolis, Maryland, and the Chemical National Bank, New York, are examples.

The Hallowell granite, from Kennebec County, which is also grey in colour and fine in grain, as the specimen indicates, is suitable for general building operations, and is in demand for statuary work. The quarries from which the specimen cube was taken, are about two miles north-west of the town of Hallowell, on the southern slopes of Lithgow Hill. It is estimated that about seven-eighths of the product of this quarry is used for general building purposes, and the remainder for carved work[2]. The Hall of Records, New York, including the statuary it contains, is an example of its use.

[1] "Granites of Maine," T. N. Dale, *Bul. U.S. Geol. Survey*, p. 172.
[2] *Ibid.* p. 119.

The light grey fine-grained granite from Long Cove quarries at the mouth of St George's river, about thirteen miles south-west of Rockland, is almost exclusively used for monumental work, although there are a few instances of its having been used for ordinary building.

A light pinkish grey stone, much coarser in the grain than the usual grey granites of Maine, exists in Washington County. The example from the Jonesborough quarries is a representative specimen. It is sometimes referred to commercially as "Jonesboro' Red Granite." The stone takes a high polish and is extensively employed for decorative work. It was used in building the Western Savings Bank, Philadelphia.

In the same county, a diabase, or possibly gabbro, is quarried close to the shores of Pleasant river at Addison, and is in general use in Maine as well as in other parts of the United States. It is also exported freely to Canada, and employed for structural undertakings in Montreal and Quebec. This rock is known commercially in America as "Black Granite," a name which is attached to other gabbros quarried in the States, more or less resembling it[1]. As the specimen indicates, it is a fine-grained dark blue-grey stone.

MARYLAND. The specimen of biotite-granite from the Woodstock quarries, Baltimore County, is typical of the stone that is considered to be the best building material in this State. It is used extensively for structural purposes in Baltimore and Washington, and in many towns of the Western States.

MASSACHUSETTS. Although the State of Maine takes precedence as regards the amount of granite produced for building purposes, other States claim priority for the inauguration of the industry.

In the State of Massachusetts, granite from the boulders on the Quincy Common, and from Chelmsford,

[1] *Stones for Building and Decoration*, G. P. Merrill, p. 284.

began to be used in and about Boston as early as 1737, although the industry did not assume any importance until 1825[1]. The granites of this State, as the numerous specimens indicate, are chiefly coarse-grained, and grey in colour, some having a pink tint.

The group of quarries which yield a stone known as "Quincy Granite," are situated on the flanks of the range of low hills south of Boston, called the Blue Hills, in the townships of Quincy and Milton. There are two distinct qualities of stone from this group of quarries. One of them is specially useful for monumental purposes, and is said to take a high polish. This variety is produced in several shades, which are known commercially as "medium," "dark" and "extra dark[2]." The specimen in the collection is believed to represent the "medium." The Bunker Hill Monument at Charlestown was erected with this stone, and this gave an impetus to the uses of Quincy granite. The other variety of Quincy granite is suitable only for ordinary building purposes.

The specimens from Rockport are typical of the hornblende-granite quarried in this State, and said to hold first rank. It has been in request for building purposes for many years, the development of the quarries dating back to 1830. They are situated in the township of Rockport, which occupies the eastern and northern part of the peninsula known as Cape Ann, on the Atlantic coast. This is almost exclusively composed of granitic intrusions.

Rockport granite suitable for structural purposes is divided into two sorts or varieties; namely, a grey rock sometimes tinted with green, which is commercially known as "Rockport Grey," and another distinguished as "Rockport Green." The specimens in the collection are typical of the grey rock shewing both varieties of tints. The grey stone comes from a quarry

---

[1] *Stones for Building and Decoration*, G. P. Merrill, p. 246.
[2] "Granites of Massachusetts," T. N. Dale, *Bul. U.S. Geol. Survey*, No. 354, p. 94.

situated close to the town of Rockport, while the specimen with the greenish tinge represents the stone which is quarried at the Gloucester quarries, about three miles from the grey-stone quarry. Both of these light grey granites have been extensively used for building in New York, Boston, and New Orleans, and they have been exported to the West Indies for structural purposes. The Boston Post Office was built from a combination of the grey and greenish tinted stones, but the grey was alone employed in the Baltimore Post Office, and the National City Bank, New York.

Another group of quarries in this State are those of Milford, in Worcester County. The prevailing colour of the stone from these quarries is a light pinkish grey, the light flesh-coloured felspar producing this pink tint. This feature has caused the stone to be known commercially as " Pink Granite." Black mica is always present. The rock is chiefly coarse in the grain. Milford granite has recently been employed in the new building of the United States National Museum at Washington, 1906. The numerous pillars of the Pennsylvania Railway Station, New York, are composed of Milford granite[1], as well as the base, and the dome of the station. The coarse-grained pink and black specimen in the collection, from the Milford quarries, represents this group.

NEW HAMPSHIRE. The granites of this State have enjoyed a wide reputation over a lengthened period for all kinds of constructive purposes, indeed, although it ranks only fifth in the list of granite yielding States in the industrial world of America, New Hampshire is popularly known as the "Granite State[2]." For many years these rocks have been extensively employed for structural work in Boston and New York; they also appear in the buildings of Chicago.

[1] "Granites of Massachusetts," T. N. Dale, *Bul. U.S. Geol. Survey*, No. 354, p. 80.
[2] "Building and Ornamental Stones," G. P. Merrill, *Report of U.S. Nat. Museum*, 1886, p. 422.

The quarries may be described as forming three groups. One is in the neighbourhood of Concord, in the Merrimac river basin, in the southern part of the State. Another is near Milford in Hillsboro County and in the Merrimac river basin; while the third is on the south side of the White Mountain region, in the east-central part of the State, near Conway[1].

The Concord intrusions yield a muscovite-biotite rock, bluish grey in colour, of medium fineness, and somewhat porphyritic in texture. The specimen from the quarry on Rattlesnake Hill is a typical example. This quarry has been in operation for a great number of years (it was working in 1812), and the product has been mainly employed as a building stone. The Congressional Library, Washington; the Standard Oil Buildings, New York; and the Blackstone Library, Chicago, are representative examples of its use. One remarkable feature of this stone is the wonderful ease with which it can be worked; it is recorded that blocks can be split out with a hammer almost as easily as if it were wood[2]. The Milford quarries lie within a radius of about four miles of Milford village, and yield a light grey, medium-grained stone. There are, however, quarries in this group affording a very fine-grained granite which is practically a monumental stone. The specimen from Lovejoy quarries is an example of the medium-grained variety which is almost exclusively used for building purposes. The Majestic Theatre, Chicago, is an example.

NEW JERSEY. The quarries yielding granitic rocks in this State suitable for building purposes are not numerous, and the products are chiefly employed for railway construction. There is one variety however found in the eastern border of the Highlands of this State, and quarried near Pompton Junction, which is

[1] "Granites of N. Hampshire," T. N. Dale, *Bul. U.S. Geol. Survey*, No. 354, p. 144.
[2] *Stones for Building and Decoration*, G. P. Merrill, p. 254.

considered to be one of the most ornamental granites in the United States, and a glance at the specimen serves to confirm the statement. Up to the present this stone has not been extensively employed, the toughness of its texture rendering it difficult and expensive to work, only when the architect desires something of special colour or quality is it brought into requisition. The texture of the rock is often quite variable, ranging from a fine to a very coarse-grained stone. The specimen in the collection can be classed among the latter. There is also a corresponding variation in the proportion of pink, white, and green colours[1]. An example of the use of this handsome granite will be found on the platform forming the approach to the New Building of the United States National Museum at Washington.

It is interesting to note that besides the display of this New Jersey granite in the New Museum Buildings, granite from Milford, Mass., is used for the basement of the platform, with a superstructure of the white granite from Bethel, Vermont, both of which rocks are represented in this collection.

NORTH CAROLINA. The granite areas of this State produce stone chiefly of a light grey colour and of various degrees of fineness, the majority of which makes a useful building material. The most important intrusion in this area is that which forms Mount Airy, a solid hill of granite which rises 128 feet above the level of the railway at its base. A surface of about forty acres is exposed on the flanks of the mountain. The specimen exhibited from these quarries testifies to the colour and degree of fineness of this valuable stone which is extensively used for structural purposes. Another example of the granite in this State, in good repute for building, is from quarries near Salisbury.

SOUTH CAROLINA. The grey, close-grained specimen from the Anderson quarries near Winnsborough in

[1] "Geol. Survey of N. Jersey," *Annual Report*, 1908, p. 65.

Fairfield County, is the only representative in the collection of granitic rocks of this State.

A pinkish granite occurs in the same county, also a light grey rock near Columbia, Richmond County[1], but it is said that the Winnsborough quarries are the chief, if not the only source of granitic building material in this State, and the demand is only local. The stone employed in the Post Office, Florence, Darlington County, is an example.

PENNSYLVANIA. With the exception of some isolated masses of gneiss, granitic rocks are not known in this State.

"The southern gneissic district, described in the geological reports of Pennsylvania as ranging from the Delaware river at Trenton to the Susquehanna, south of the State line and lying south of the limestone valley of Montgomery, is the district in which are situated nearly all the quarries of gneiss in the State, and those furnishing most of the material are in the vicinity of Philadelphia[2]." The specimen from quarries at Holmesburg, a suburb of Philadelphia, is a good example. The stone is extensively employed in that city for structural purposes, especially for foundations. The Armoury of the First Troop of Cavalry built in 1901, St Paul's Church, Broad Street, in 1905, and the Cumberland County Court House, Bridgeton, New Jersey, in 1909, are constructed of it.

A rock quarried near Gettysburg, in Adam County, is used for building, and is known commercially as "Gettysburg Granite," although, strictly speaking, it belongs to the diabases. The specimen is a good example of this hard, tough rock which is useful for general building, also for ornamental work.

RHODE ISLAND. This State yields granites of various shades of colour, some varieties being exceedingly fine-

---

[1] *Stone for Building and Decoration*, G. P. Merrill, p. 261.
[2] "Building and Ornamental Stones," G. P. Merrill, *Report of Nat. Museum*, 1886, p. 424.

grained. These outcrop in dyke-like masses from 50 to 100 feet thick[1].

The centre of the granite industry is in and about the town of Westerly, which is at the extreme edge of the State, where most of the quarries are situated. Some of them are on an east and west ridge, about a mile from the town. Others are about a mile southeast of Westerly. The granites from these quarries are divided into three varieties, "Westerly White Statuary," "Blue Westerly," and "Red Westerly." The first of these is a monumental or statuary stone, and therefore is not represented in this collection of building stones.

"Blue Westerly," although largely used for monumental purposes, is also occasionally employed for building, but owing to its increasing scarceness, and ready sale for monumental work, both in America and Europe, it is now less used for building than formerly. It is a uniform bluish grey rock, and varies in fineness. The specimen from the Newell quarries is exceedingly fine, while that from the New England quarries is somewhat coarser. Newell granite was used for constructing the Monument of Senator Sherman, erected in Mansfield, Ohio; the obelisk to General Wallace, Crawfordsville, Indiana, is another example. The coarser stone from the New England quarries was employed for building the offices of the Mutual Life Insurance Co., Philadelphia, also the Masonic Temple, New York.

The third variety, "Red Westerly," is a biotite-granite of reddish grey colour, and as the specimen indicates, is much coarser than those just mentioned. It is used almost exclusively for building operations. The Washington Life Insurance Buildings, the American Bank, New York, and the Colonial Trust Buildings, Pittsburg, are examples of its use.

---

[1] "Granites of Rhode Island," T. N. Dale, *Bul. U.S. Geol. Survey*, No. 354, p. 190.

TEXAS. The red biotite-granite of Burnett County, although up to the present not extensively quarried, makes a useful material for structural purposes. It occurs in coarse and fine varieties. The specimen in the collection, from the Granite Mountain quarries, is an example of the coarse rock which is chiefly used for foundations and ordinary building. The State Capitol Building at Austin is constructed of it, as are also the Court Houses in the counties of Galveston, Harris, and Wise.

VERMONT. Numerous isolated intrusions of granitic rocks occur in this State. In Washington County there are three groups of quarries, the most important being that in the Barre district, which is considered to be the chief granite producing centre of Vermont[1].

"Barre Granite" is a grey biotite rock of various shades, and is known as "Barre Dark," "Barre Medium," and "Barre Light." The specimen in the collection is an example of the "Medium" variety. Most of the quarries are situated either on the summit, or on the slopes of Millstone Hill, about three miles south-east of Barre. A railway skirting the hill communicates with that town, and thence with other parts of the State. The granite is useful both for monumental work and ordinary building purposes. An example in Barre is the Robert Burns Statue and pedestal.

In the northern part of Windsor County is the small but important White Granite area of Bethel[2], which yields a rock known commercially as "Bethel White Granite." The specimen in the collection deserves special attention. It is exceptional as regards colour, being so nearly white that it might easily be mistaken for marble. It is in demand for all sorts of decorative work in buildings, as well as for monuments and statuary. It was used in the construction of the New

[1] "The Granites of Vermont," T. N. Dale, *Bul. U.S. Geol. Survey*, No. 404, p. 50.
[2] *Ibid.* p. 9.

National Museum at Washington. The decorative work over the entrance of the American Bank Note Building in New York is another example.

WISCONSIN is rich in granite, but its geographical position has to a certain degree impeded the development of the quarries.

Recent statistics shew that this State furnishes thirteen granites which differ considerably in texture and colour[1], varying from a brilliant red to a dark brown; while some are exceedingly fine-grained, and others coarsely porphyritic.

The specimen from the Montello quarries, which are said to have been the pioneers in the development of the Wisconsin granite industry, represents the close-grained brownish red rocks of the State. They are situated near the centre of Marquette County at an elevation of about 90 feet above the general level of the adjacent country. Montello granite is said to be the strongest true granite ever tested, and the details which appear in the catalogue, furnished by the Geological Survey of Wisconsin (verified by the U.S. Arsenal at Springfield, Mass.), seem to confirm this assertion. Montello granite is chiefly in demand for monumental work. Many examples can be seen in the cemeteries of Wisconsin and adjoining States. The sarcophagi for General and Mrs Grant at Riverside, New York, were hewn from this rock. It has however also been used in the construction of many important buildings, of which the *Herald* Offices, the Kirk Block, and the Stone Buildings, Chicago, are examples.

The specimen from the Granite Heights quarries represents the bright red variety of rock obtained in this State. The quarries, situated about ten miles north of Wausau, were among the first to be worked in Wisconsin, and the stone is commercially known as "Wausau Granite." It is very abundant and extends

[1] "Building Stones of Wisconsin," E. R. Buckley, *Bul. Wis. Survey*, No. IV. p. 89.

continuously in a broad belt, on either side of the
Wisconsin river, from Granite Heights as far as Pine
river, a distance of about six miles[1]. The Marathon
County Court House in Wausau, is built of it. A specimen of Wausau granite in the collection, which has
an even more brilliant colour than the last named,
comes from another quarry in the same area.

Quarries which have but lately been opened in the
Granite Heights area of the Wausau district, yield a
rock which differs widely both in colour and texture
from those just described. The coarse green and black
specimen from Parcher quarry is an example. This
rock has been so recently employed for industrial purposes, that no record is as yet forthcoming of its
chemical or physical characters, or of its use for any
particular work.

The specimen from the Pike River quarries, near
Amberg, is an example of the fine-grained grey
granite this State produces, which it will be observed
resembles in many respects some of the grey granites
of Scotland. This stone is mainly employed for monumental work, but sometimes for building, the Citizen
National Bank, Green Bay, Wis., being entirely constructed of it.

The dark brown, nearly black, specimen from the
Berlin quarries in Green Lake County, represents a rock
which is quarried extensively in that district, and although classed among the granitic rocks of Wisconsin
is better known as "Berlin Rhyolite." This rock when
polished has a dense black background, through which
numerous porphyritic crystals of pink felspar are
scattered. Physical tests shew that it is exceedingly
hard and tough, and is said to be one of the most
durable stones in the market. It is extensively employed for monumental work, and is also used for
building purposes. The Science Hall, Madison, Wis.,
also the Bartlett Building, Chicago, are built with it.

[1] "Building Stones of Wisconsin," E. R. Buckley, *Bul. Wis.
Survey*, No. IV. p. 131.

About twelve miles north-west of Berlin in Wau-shara County are numerous small mound-like outcrops of granite rising above the general level of the country. These are quarried, yielding a pink and black, medium-grained stone. As the colour of this rock is more subdued than those already described, it is not so popular for monumental work, but it is well adapted for all purposes of construction.

The two small cubes of pink and black coarse-grained rock represent a granite from the eastern por-tion of Waupaca County, known as "Waupaca Granite." The quarries are situated on a hill about 100 feet above the general level of the country, five miles north of the town of Waupaca. A number of boulders, some of considerable size, are found on the east side of the hill, which are also split up and used for building. This granite has been extensively employed in Chicago and elsewhere for both construction and ornamental work. The Omaha Bee Building, the gateway to Lake View Cemetery, Minneapolis, and the State Soldiers Monu-ment at Chickamauga, Tenn., are examples.

CANADA. In the township of Standstead, Quebec Province, grey granite occurs in abundance, occupying in one place an area of fully six square miles. As is shewn by the specimen cube, this stone admits of a fine polish, and it is said to be easily worked: characteristics which recommend it as a useful building material, especially for decorative purposes. It has been largely used in the city of Sherbrooke. The handsome wall surrounding the grounds of the Parlia-ment Buildings at Quebec is built of this granite.

The same Province yields a handsome dark brownish red rock which is quarried at Staynerville. This granite has only recently been developed for industrial purposes; but the appearance of the specimen warrants the asser-tion that it has already gained repute among architects and sculptors in Montreal. It has been used for building the New Prison House, Montreal.

Many of the islands comprising the celebrated group in the river St Lawrence, popularly known as the Thousand Islands, are chiefly composed of granitic rocks and yield a useful building material. The prevailing colour of these rocks is a warm red, and they vary considerably in texture. The specimen from Leek Island, it will be observed, is coarse-grained, whilst that from Forsyth Island is classed among the fine granites. They are extensively employed in Montreal for monumental purposes as well as for ordinary building.

The specimen from the Bay of Fundy is an example of the red granite procured from the group of quarries near St George in Charlotte County, New Brunswick. These quarries are situated on the slopes of a range of hills about two and a half miles from the village of St George, which stands at the head of the Salt Water Navigation on the river Magaguadavic. The power derived from the falls on this river is utilized for dressing the stone and preparing it for the market. The St George granite, as the specimen indicates, possesses a bright red colour, resembling in tint and texture the Red Correnie granite of Aberdeen. The quarries have been worked for many years, and the granite has been employed for both ordinary and decorative work in Canada and the United States. The Museum of Natural History in Central Park, New York, is built entirely of this stone; it was also used for making the pedestal of the Sir John A. MacDonald Monument, in Kingston, Ontario.

SOUTH AMERICA. In the Province of Córdoba, in the Argentine Republic, there are numerous igneous intrusions which yield a granite that is used for building purposes both in the town of Córdoba and in the surrounding district. The prevailing colour of the stone is grey, some of the rocks having a pink tint. The specimen from La Falda quarries is an example of the pink variety, and that from Capilla del Monte quarries represents the grey granite.

IGNEOUS ROCKS (VOLCANIC).

The ejectmenta of volcanoes, in the form of consolidated dust, mud, scoria, or lava, which yield a stone useful to the builder, whether cellular or solid, are of very different character physically, even when similar in chemical composition. These variations are due to the manner and rapidity of cooling, or to internal changes that may arise from long continued pressure, or to chemical re-assortment of their component ingredients. A cursory glance at the specimens arranged under this head will confirm the statement, and convince the observer how widely these volcanic products differ in lithological character. A few words in reference to their economic value may be useful.

BRITISH ISLES. Since the stone obtained from this division of igneous rocks in the British Isles is hard and tough, it is difficult to dress and carve, and thus of little importance to the builder. Although occasionally employed for rubble work in the immediate neighbourhood of the quarries, its chief economic value is for paving and road making, hence specimens are purposely excluded from this collection for the reason explained in the introduction. There are however one or two exceptions, and brief reference will now be made to these.

The cleaved ash beds of the volcanic series of Borrowdale in Cumberland, yield a stone which is extensively employed for building in and around Keswick. Owing to the hard and tough nature of this rock it is seldom tool-dressed, the operation being tedious and expensive in ordinary building, but is usually only rough-hewn and built in broken courses, the quoins and sills of the houses being composed of dressed sandstone. This stone is however frequently employed by local sculptors for fashioning the Celtic memorial crosses so often seen in the cemeteries of Lakeland.

On the banks of the Tweed, under the shadow of the Eildon Hills, near Melrose, there is a trachytic rock said to be of Old Red Sandstone age[1], which has lately been quarried for building purposes. The nature of the rock, as the specimen indicates, renders it unsuitable where decorative work is required, but it makes a useful building material for ordinary structures.

In the counties of Kinross and Clackmannan some of the Scottish dolerites are quarried. This rock is known in building and industrial circles as " Alloa Granite," and, although very tough and hard, it is sometimes employed for structural purposes. The stone, besides being used locally, is transported to the port of Alloa, whence it is shipped to the Thames and to other centres, for building. The columns in the house of Mr Barnato in Park Lane, London, are of "Alloa Granite."

Among the felsitic rocks of Ireland, one of the most interesting, from a petrological point of view, is the tridymite-bearing liparite of Tardree. It consists of porphyritic crystals of glassy sanidine, plagioclase, and smoky quartz, in a pale yellow or white trachytic or felsitic matrix[2]. The rock is cut into by a series of quarries extending from south to north along the flanks of the Tardree mountain in Co. Antrim, and is thence transported to Belfast and to other centres where it is employed for building purposes. It is chiefly utilized for sills, quoins, and jambs, indeed for all purposes where granite is usually employed, except

---

[1] *Petrology for Students*, A. Harker, p. 175.
[2] *British Petrology*, J. J. Harris Teall, p. 347.

where polished decorative work is required, for which its nature is unsuitable. The stone when freshly quarried is soft and saturated with "quarry water," and of a dark brownish grey colour, but when exposed to the atmosphere the water soon evaporates, leaving the stone much lighter in colour, as shewn by the specimen in the collection[1].

CONTINENTAL EUROPE. The lavas as well as the fragmental products of volcanic action, occupy a more important position in the building industry than those of the British Isles, as the numerous specimens in the collection will testify, and many of them are extensively employed for structural purposes.

FRANCE. In Brittany, not far from the town of Brest, a rock is quarried which is known commercially as "granite," but is classed, by petrologists, among the lamprophyres. It belongs to the Kersantite type of this group, named from Kersanton in the Department of Finistère[2]. The specimen shewn was procured from a quarry near this place. It is a dark blue-grey fine-grained holocrystalline rock, and is used for building and for road making.

GERMANY. The bare plateau of the Eifel in Rhenish Prussia, with its numerous extinct volcanoes, yields a dark coloured porous trachyte-tuff (Eifel Tuffstein), which is utilized for structural purposes on either side of the Rhine. The specimen from the Brohl quarries is a good example. Several buildings of importance in Cologne are constructed of it.

The basalts of the Rhine, which, in more than one place between Bonn and Coblenz, contribute to the scenic charm of the river, furnish a material which is invaluable for heavy constructive work and for engineering undertakings. Some of these rocks are

[1] *Trans. Royal Dublin Society*, Vol. VI. G. A. J. Cole, p. 86.
[2] *Petrology for Students*, A. Harker, p. 141.

columnar in structure, and this physical characteristic
has been turned to account by experts in civil
engineering. The prismatic columns can be easily
detached in the quarry without injuring the smooth
surfaces of the hexagonal or square pillars; they are
thus practically self-dressed and ready for re-adjust-
ment, forming, when put together again, a solid
structure, equal in strength and solidity to a building
composed of ashlar masonry. Immense quantities of
these extremely hard prismatic columns have been
employed for constructing sea walls and coast defence
works in Holland, as well as river bank work in Germany.
Several sea walls in England are built with this Rhenish
basalt, for instance those at Blackpool, Clacton-on-Sea,
Southend-on-Sea, as well as one of the docks at
Hartlepool, and sea defence works at Hastings.

A brief description of the *modus operandi* when
constructive work is undertaken with these basaltic
columns may perhaps be of interest. The rock, it
should be said, is separated, in commercial nomenclature, into two divisions, Columnar Basalt, and Basalt
Sinkstone, the latter being again sub-divided into two
classes, Heavy Sinkstone, and Light Sinkstone. The
Heavy Sinkstone consists of large masses of basalt,
generally columnar in form, and is chiefly used in
Holland for maritime engineering work, being thrown
into the water round the piers and at the foot of sea
walls to break the force of the waves. It is also used
for sinking rafts to form the foundation for building
piers out into the sea. The subsoil of Holland is so soft
that it is impossible to make a foundation by simply
throwing down loose stone, so that large rafts are
made of willows and floated into the required position.
They are then sunk by throwing about a hundred tons
of Sinkstone on the raft, the large surface of which
prevents it from sinking into the sand. On the top
of this foundation the piers are built. The Columnar
Basalt blocks are used for sea walling, and are placed
as nearly as possible in the same position as they

occupied when in the quarry. They are not dressed in any way, and owing to their form, interlock so well, that sea groynes can be built of these columns without any mortar or cement. The light Sinkstone is used for filling up at the back of a sea wall built in the manner just mentioned. The name Sinkstone has reference to its use for sinking purposes as described above. The specimens in the Museum are from the Weilberg quarry at Heisterbach near Königswinter. In order that the uses of this basalt, as a material for construction, might be better understood and appreciated the owners kindly sent a full series of columns, square, pentangular and hexagonal, also two massive blocks of Sinkstone, instead of the usual specimen cubes as specified. These interesting examples have been placed in the outer court of the Museum, where they can be inspected.

ITALY. It is not necessary to remind the student of geology that a large proportion of the rocks of the central and southern provinces of Italy are formed of the ejectamenta of volcanoes. This eruptive matter has yielded building material which was largely used by the Romans, and is still extensively employed for structural purposes. There are several specimens of these rocks in the collection, and it may be interesting to refer to a few of them.

The specimen of Volcanic Tuff from Rome represents the fragmental rocks employed for the construction of that ancient city. Indeed the use of this stone for building can be traced to a date anterior to the founding of Rome, if, as is believed, it was used in the ancient Etruscan city of Veii[1]. As may be seen from the specimen of this tuff, it is more compact than the generality of rocks composed of volcanic scoriae. This stone, which is known in Italy as "Tufa Litoide" must not be confounded with the granular tuff of Rome known as "Tufa Granulare," in which the catacombs

---

[1] *What Rome was built with*, M. W. Porter, p. 15.

of Rome are excavated, which is softer than the Tufa Litoide, and not suitable for building. There is another and even softer tuff known as " Pozzolana," which when ground up and mixed with lime, is used for building purposes as mortar.

" Peperino " is also composed of volcanic ashes, the grey ground mass of which is studded with black particles of scoriae resembling peppercorns, hence the name. There are several varieties of this rock, all of which are harder than the Tufa Litoide. The specimen shewn was taken from the quarries at Marino, and resembles the Peperino that was formerly found near Lago d'Albano, and was known in the time of the Romans as " Lapis Albanus." Both Peperino and Tufa Litoide were largely employed in the construction of ancient Rome.

Alternate layers of Soft Tufa and Peperino may be seen in the walls of the so-called Temple of Cybele on the Palatine.

It is instructive to note the manner in which the Roman builders placed their different materials according to the weight they had to carry. While Tufa was frequently used for the main walls, Peperino (e.g. in the Servian wall on the Aventine) or Travertine[1] (e.g. in the forum of Augustus and the temple of Fortuna Virilis) was often inserted at points of special pressure, such as piers and arches. The Colosseum is a particularly elaborate example of this mixed construction with three degrees of pressure supported by different materials.

It is also interesting to follow chronologically the use of the three rocks just referred to. At first Tufa only was used in the opus quadratum, as we see in the so-called wall of Romulus. Next, the harder Peperino began to be worked; it is used, though sparingly, in the great Servian wall, and during the later Republic appears to have been largely employed for exterior

---

[1] See p. 237.

walls or at points where there was heavy pressure, while other parts were built of Tufa. Thirdly, Travertine appears to have been introduced about the 2nd century B.C. but was employed at first for merely ornamental purposes, very much as marble was under the Empire. After about the middle of the 1st century A.D., Travertine began to be largely used for the solid mass of walls as in the Temple of Vespasian and in the Colosseum[1].

In the region round about Naples there is also a group of volcanic ejectamenta, which are quarried and employed for structural purposes in that city and the surrounding neighbourhood. This group is represented by several specimens in the collection. The most important are the trachytic lavas which are principally found in the Phlegræan Fields, a district to the west of Naples. From the craters of volcanoes, now extinct, flowed the lava streams that now furnish building material for Naples. Large blocks of this stone were used for constructing the Municipal Buildings of Naples.

"Piperno"[2] from the Arecco quarries, easily identified by the flame-like streaks of darker lava in the mass, and the heavier grey sanidine-trachyte from Pozzuoli are used for constructing the palatial edifices of Naples, as well as for ordinary building. The Post Office, and the National Museum of Naples, are examples of the use of the former.

The leucite lavas of Vesuvius, known to the Italians as "Pietarsa," are also utilized for building. Many of the basements of the principal palaces and public offices in Naples are built of this rock.

The soft fragmental tuffs found near Naples vary more in colour and texture than those near Rome. The rock from Fontanelli, as the specimen indicates, is composed of fine, even-grained, yellow consolidated dust, whilst that from Fiano quarries near Caserta is a

---

[1] *Encycl. Britannica*, 9th Ed. Vol. xx. p. 809.
[2] *Piperno* must not be confounded with *Peperino*, the grey tuff of Rome, which has lately been described.

coarse grey scoriaceous stone with black inclusions. These materials are used chiefly for internal work. The stone from Fiano was employed for constructing the mausoleum of Baron Schillizze on the Hill of Posilipo.

AFRICA. Among the granitic rocks of Saldanha Bay in the CAPE OF GOOD HOPE dykes of quartz-porphyry are common[1]; these are quarried and the stone utilized for constructive purposes in Cape Town and in the surrounding neighbourhood. As the specimen shews, this rock admits of a good polish, and is frequently employed as a decorative material as well as for ordinary building.

TURKEY in ASIA. Some of the numerous islands of the Grecian Archipelago, as well as that portion of the coast of Asia Minor which is washed by the Ægean Sea, are composed of volcanic lavas, chiefly of an andesitic type. These were employed in very ancient times for building, and are still used for that purpose on the islands where they occur, also in and about Smyrna and other towns in Asia Minor. The numerous specimens of this group in the collection shew that the rocks differ considerably in colour and texture, although closely related petrologically.

The specimen of red rock from the quarries near Dikili, in the Mytilene Channel, the port of Pergamos, is typical of the quartz-andesites, or dacites, and the grey Dikili cube from the same quarries, represents the hornblende-andesites[2]. The ancient fortifications of Pergamos were chiefly built with these rocks.

The greenish-grey specimen from the quarries at the foot of Mount Pagus, just outside Smyrna, and the purple-coloured cube from the Bariakili quarries, near the supposed site of ancient Smyrna, opposite the modern town, are also examples of these andesites.

[1] *Geology of South Africa*, Hatch and Corstorphine, p. 45.
[2] *Petrology for Students*, A. Harker, p. 185.

Both of these make excellent material for building, and are largely employed for that purpose in Smyrna.

The light purple-coloured specimen of andesite-tuff from quarries on the Island of Mytilene, known to the Greeks as Lesbos, is typical of the stone employed for building in the town of Mytilene, the capital of the Island, and in Antissa and other adjacent towns.

The peninsula of Karabouran, which forms the bay of Smyrna, is almost entirely of volcanic origin, and furnishes stone useful for structural purposes. The quarry at Lithri on the peninsula, the ancient Erithroe, yields a grey andesite, which, as the specimen indicates, is denser than most of the rocks mentioned above. This stone was used for building the walls of the ancient town. The specimen from the quarries near Alatazata, a small town on the south-west coast of the peninsula, is typical of the light yellow rhyolitic tuffs of this district, which are employed for building. It has recently been extensively used for restoring Alatazata, which was partly destroyed in 1882.

Near Fokies, sometimes called Focha, the ancient seaport of Phocaea, situated in a small inlet on the north side of the Gulf of Smyrna, are quarries which furnish a rhyolitic tuff used for building purposes. The specimens indicate that this stone is paler in colour than the other rhyolites of this region; they are of a very light cream colour, and are useful for internal decorative work.

Among the specimens of volcanic rocks from Turkey in Asia, there is one of a dark grey-brown basaltic stone from Palestine. This is typical of the rocks of volcanic origin which occur in the country west of the river Jordan in the mountains near the valley of Jezreel[1]. The specimen is from the Houran quarries not far from the Lake of Gennesaret; the natives of this district use the stone for building their houses.

INDIA. In and around Bombay the building stone

[1] *Encycl. Britannica*, 9th Ed. Vol. XVIII. p. 171.

most in use belongs to the great series of volcanic
sheets which extends over a large portion of the
territory of the Bombay Presidency, and is known in
geological nomenclature as " Deccan Trap." This rock
is some form of dolerite or basalt, but it varies considerably in character in the different beds. Some beds
are excessively hard, compact, and homogeneous, the
crystalline structure being so minute as to be detected
with difficulty. Others are soft and earthy, evidently
from partial decomposition[1]. The rocks selected for
structural purposes belong to the former class, and are
in constant demand for building. As the specimens
indicate, the rock even in the same quarry varies
considerably in colour. The stone is known to the
builder as " blue " or " yellow basalt," according to the
colour, or as " Kurla Stone," the name being derived
from the quarries about ten miles from Bombay where
it is worked. The "Bombay Basalt" is of a uniform
blue-grey colour, and is a harder and tougher rock
than the Kurla Stone.

MYSORE. A pale green felsite occurs in this State as
a dyke about one mile long and five yards wide. The
rock which is quarried at Belagola about six miles from
the town of Mysore yields an ornamental stone and
is chiefly employed for internal work. The specimen
indicates that it is soft in texture and suitable for
carving.

A considerable number of specimens representing
the porphyries of this State will be found in the
collection. They are too numerous to describe individually, but taken as a group they are well worth
a close inspection. The variety of types including
orthophyre, bostonite, and quartz-porphyries, presents
an interesting field to the petrologist, while the display
of colour cannot fail to attract the ordinary observer.
The rocks usually occur in dykes, some of which are
several miles long. Nearly all these porphyries have

---

[1] *Geology of India*, R. D. Oldham, p. 257.

been employed in the construction of both the New and
Old Palaces of Mysore, as well as for other public buildings in Southern India. There are several specimens
from quarries in the district of Seringapatam.

SIAM. The conical hills, which in the south-west
of this kingdom surround the town of Pechaburi, sixty
miles south-west of Bangkok, consist of trachytic lavas
and scoria. These are excavated by the natives to form
elaborately fitted temples; the material is also utilized
for building purposes. The specimens in the collection
shew that they differ considerably in colour, varying
from a pinkish brown to a light green. They are close
in texture, and make a useful building material, which
is extensively employed in Bangkok.

CHINA. The dark blue-grey porphyrite from the
Kowloon district is usually known in China by the name
of "Blue Stone." This rock is very hard and tough and
is chiefly employed for the basements of buildings, the
superstructure generally being of brick. The specimen
is an example of the stone used for the basement of
the new Law Courts at Hong Kong.

Throughout Southern China, which comprises that
mountainous tract of country stretching from Shanghai
on the north-east to the coast of Tonkin on the south-
west, are extensive areas of plutonic intrusive rocks,
together with some volcanic eruptives. The former
yield granites, represented by several specimens, and
the latter produce a tufaceous stone which is extensively
employed for building in the province of Che-kiang and
surrounding districts.

The specimens from quarries near Ning-po are good
examples of this volcanic tuff. It will be noticed that
they differ considerably both in colour and texture.
One is a pale green stone possessing a fairly high
specific gravity, another is pink, whilst a third is
light grey, flecked with dark pieces of scoria. These
rocks are known as Ning-po Green, Ning-po Red, and
Ning-po White Stone, respectively. The Green Stone is

considered to be the most durable and therefore the
most valuable for constructive purposes, and is used for
better-class buildings. Next comes the Red Stone, which
is chiefly employed for Chinese buildings, and lastly the
White Stone, which is only occasionally used for structural work and exclusively by the Chinese; it is also
utilized for flagging.

QUEENSLAND. On many of the downs stretching
inwards from the coast-line of this State there are
numerous craters now extinct, but surrounded by sheets
of lava and masses of volcanic ash. These deposits yield
a tuff which is employed for structural purposes in
and around Brisbane, and is generally known among
builders as "Brisbane porphyry." The specimens shew
that this rock is generally of a delicate pink tint altered
to a light buff colour in some cases.

NEW ZEALAND. There are several lines of volcanic craters across the North Island. One of these
traverses the county of Auckland, and yields a basaltic
rock much used for building and engineering work.
The specimen in the collection is from the Paerata
quarries, which are situated close to the New Zealand
main line of railway, twenty-eight miles from the
town of Auckland. The rock is known commercially
as "Paerata Bluestone," and has been extensively employed for constructing the many small bridges and
culverts on the railway between Drury and Pokena.

Huge boulders of the basaltic rock, spread over
the plains of Central Otago, South Island, furnish an
excellent building material, which is known as "Kokonga
Stone," and can be obtained with no other cost than
that of splitting and conveyance. It is largely employed
in all parts of the Island for structural purposes,
although its toughness makes it unsuitable for decorative
work, for which purpose a softer stone is generally employed. The specimen in the collection represents the
stone used for building the New Railway Station of

Dunedin, which is entirely constructed of it, except the facings, which are of "Oamaru Limestone[1]."

Banks Peninsula, on the east of the South Island, and the coast line of Otago Harbour, are largely composed of volcanic rocks. The specimen from the quarries near the seaport town of Timaru, a little south of the peninsula, represents the basalt used for building in that locality. It is chiefly employed for foundations, although sometimes also for general construction, as for instance at St Mary's Church, Timaru.

The greenish grey specimen from the Port Chalmers quarries, about seven miles from Dunedin, represents the fragmental rocks used for building in that district. It is specially useful for foundations, and for steps, but careful selection is needful, the stone being rather uneven in quality, and some beds are said to weather badly. The Law Courts of Dunedin are built of it, with Oamaru Limestone for the facings.

WEST INDIES. St Christopher. The two specimens of basaltic rock from St Kitts, an island of volcanic origin, are typical of the rocks used by the inhabitants for building. They are known on the Island as "St Kitts Fire Stone" and "St Kitts Hard Stone." Both of them are generally tool-dressed, and are said to be fairly durable, not requiring any coating of cement or mortar to protect them from the destructive influence of the weather. The specimens shew that the Fire Stone is the softer of the two rocks. The "Public Buildings," the only structure of any architectural importance on the Island, is built of this material. The specimens are from a quarry near Basseterre, the chief town of the Island, but both varieties occur in other localities.

Grenada. The specimen of basalt tuff from this Island, one of the Windward group of the West Indies, illustrates the chief material in use for building.

---

[1] See page 225.

The rugged central mountain range of the Island is of volcanic origin, and on its slopes, as well as spread over some parts of the lowlands, the basalt tuff is found in great abundance. Since it occurs practically on the surface, no quarrying is required. In some instances it is collected and built in as rubble, at other times it is smoothed and dressed. Many of the chief buildings in St George, the capital of the Island, are constructed of this stone.

St Lucia. This Island, the most northerly of the Windward group, is of volcanic origin. The central mass consists of two huge cone-shaped rocks composed of late volcanic accumulations, which rise out of the sea to a height of fully 3000 feet, and are chiefly made up of andesitic lavas and scoria. The northern portion of the Island is largely composed of basalt, which however is not much quarried, a sufficient supply being obtained from boulders.

The specimen of greenish grey tuff, and the cube of light purple hornblende-andesite, are typical of the material which the inhabitants use for internal structural work, and are known by the name of " Cut Stone." The dark blue-grey specimen is an example of basalt which is employed for rubble-walling and rough exterior work. This is chiefly obtained from the loose blocks scattered about on the surface of the land. The specimen of dark greenish grey andesite represents the class of stone which is generally utilized for steps, flooring, and sometimes for outside paving.

MADEIRA. This Island is almost wholly composed of basaltic rocks. Every precipice shews sheets of lava, interstratified with tuffs, scoria and bright brick-coloured earths[1].

The cube of light purple basalt is an example of the material which is chiefly used on the Island for struc-

[1] *Quart. Journal Geol. Society*, Vol. xxxviii. 1882, J. Starkie Gardner, p. 277.

tural purposes. Practically every building of importance
has been constructed of it, among them being the
Governor's Palace, the Cathedral, the Infantry Barracks
and the churches of all the parishes on the Island. The
specimen indicates that it is hard and tough, and somewhat difficult to dress and carve.

A basalt of less density and greater softness is
found at Caxäo, about five miles west of Funchal, the
capital of the Island. This rock is used, together with
the harder basalt, for the sills, quoins, and other
dressed work required in building. The colours of
the two rocks, as the specimens shew, combine har-
moniously.

The tufaceous rocks are also used for structural
purposes, but are only suitable for certain classes of
internal work. They resist heat remarkably well, and
are chiefly employed for constructing ovens, and for
the flues of chimneys and stoves. The specimens
from the Quinta Granda quarries at Cabo Giraõ, six
miles west of Funchal, are good examples, which indicate that this variety of stone owing to its vesicular
nature is useless for decorative work. It will be
noticed that there are two specimens in the collection
from these quarries, one of a dark reddish brown,
and the other of a yellow colour. The dark-coloured
stone is considered to be superior to the yellow in
resistance to heat. The quarries are near the coast,
and the material is brought round to Funchal by boat.
This fire-resisting rock is also quarried at Caniçal,
near Porto de St Lourenço, on the eastern side of
the Island. It may be interesting to note that the
Portuguese colony in the Sandwich Islands imports
a quantity of this stone for structural purposes.

Porto Santo, the other island of the Madeira group,
is also almost entirely composed of volcanic rocks, of
which the trachytic variety bears a much greater ratio
to the basaltic, than in Madeira. These trachytes, as
the specimen indicates, are nearly white, and when
quarried are utilized for monumental and decorative

work. Considerable quantities are transported annually to Funchal for that purpose.

CANARY ISLANDS. These Islands are wholly volcanic, consisting partly of trachytes, but more largely of basalt[1], both of which are employed for building purposes.

The specimens of lava from Teneriffe display a great variety of colour. The example from Tajao has a delicate pink tint rarely met with in rocks of volcanic origin. The specimen from Las Palmas in Grand Canary is a typical example of the tuff used for structural purposes on that Island. It will be observed that this rock contains fragmentary pieces of dark scoria. These black particles do not harmonise with the light straw colour of the matrix, and give the stone a somewhat unsightly appearance. In order to hide this, as well as to protect the stone, which is of a soft and porous nature, most of the buildings constructed of it are coated with plaster or cement.

ST HELENA. The building stone of the Island of St Helena is compressed tufaceous rock, coloured red by oxide of iron, and containing augite[2]. As the specimens indicate, these rocks vary in compactness, some being much more hardened than others by heat and pressure. Many of the principal buildings in Jamestown are constructed of this description of rock, but most of them are faced with a thin coating of plaster or cement, which protects the stone and renders the building more durable.

The specimen of bright red colour and cellular texture, represents a class of stone that is usually known as " Red Stone," and is said to possess peculiar properties for resisting the action of fire. For this reason it is selected by the inhabitants of the Island for building and lining their baking ovens.

[1] *Volcanoes*, T. G. Bonney, p. 266.
[2] *St Helena*, J. C. Melliss, p. 77.

MAURITIUS. On this volcanic island nearly all the rocks are lavas, chiefly greyish in colour, some being basalts. The latter is the only variety used in the Colony for constructive purposes. The specimen cube is a good example of the quality and colour usually selected. The Post Office and other buildings of importance in Port Louis, the capital of the Colony, are constructed of stone from the quarry which supplied the specimen.

SOUTH AMERICA. The specimens from Arequipa in Peru are typical of the volcanic rocks which are used for building in that region. The immense sheets of these andesitic lavas and tuffs which surround the city of Arequipa, and extend as far as Jambo, are believed to have been formed by the eruptions from the now extinct volcano Misti, a perfect cone 20,300 ft. above the level of the sea, situated about twenty miles north-east of Arequipa[1]. As the specimens shew, the colour and texture of these deposits vary considerably. This tufaceous stone is easily worked, and owing to its property of non-conducting heat it is popular for the building of dwelling houses. The stone is known in Peru as "Sillar." The Cathedral and the Ladies' College in Arequipa are built of it.

[1] "Die Vulkangruppe von Arequipa," F. H. Hatch, *Tschermak's Min. u. Pet. Mitth.* Vol. VII. p. 328

METAMORPHIC ROCKS.

Brief reference will now be made to a few specimens of metamorphic rocks used in building, which for convenience have been placed in the cabinets set apart for examples of igneous rocks. They will be found immediately following the specimens of the volcanic division of the igneous systems.

As is well known, this class includes rocks both of igneous and of sedimentary origin.

In some instances it is doubtful whether the specimen represents an altered igneous or an altered sedimentary rock.

As mentioned in the introduction, the stone from this division of rocks is not often utilized for building purposes, the tough nature of its texture rendering it more useful for paving and for the making and maintenance of roads. There are however a few exceptions, and these must have a position in the collection, since it attempts to be illustrative of every kind of stone employed for building, even if some may not be very suitable for the purpose.

BRITISH ISLES. At Polyfant near Launceston in Cornwall, is a famous quarry of a serpentinous rock, popularly known as "Polyfant Stone," which has been worked and utilized from very early times. The greater part of the Saxon and Norman arches in the eastern portion of the county are built of this stone. As the specimen shews, its structure is somewhat schistose, and at the same time brecciated. It is of a dark greyish green colour with brown ferruginous spots, which are caused, as shewn by the chemical analysis, by the presence of hydrated oxide of iron. The rock, although compact and hard in appearance, is very easily wrought, and can be readily cut with a chisel or even with a knife. This property causes it to be sought after for highly finished decorative work, and it is much used in the county for that purpose, especially in ecclesiastical architecture. The rock is also very infusible, and is sometimes employed for making crucibles for melting copper and tin. Mr Collins designates this stone as a "Pot Stone[1]."

The pretty little town of Salcombe, nestling on the sunny banks of the estuary of the same name in South Devon, is built on a much crushed green schist which is quarried in the neighbourhood, and freely used for building. As the specimen shews, this rock is best suited for rubble masonry, the schistose nature of the stone making smooth dressing difficult.

A fine grit, known locally as "St Catherine's Stone," is present on the west banks of Loch Fyne in Argyllshire, near Inverary, and is quarried for building purposes. The colour, as the specimen indicates, is blue-grey, and the stone being soft in texture yields readily to the mason's chisel, and is suitable for all kinds of carved work as well as for ordinary building. Inverary Castle, the seat of the Duke of Argyll, was constructed with it.

CENTRAL AFRICA. A serpentinous rock, known as Soap Stone, is quarried at Limbue on the east coast of Lake Nyasa, and has been extensively employed in the construction of the Anglican Cathedral which has been recently erected on the Island of Lakoma. Owing to the soft nature of this rock, it is only suitable for internal construction, and, being very readily carved, it is largely used for decorative work. The pulpit, font, choir stalls, screen, and other ornamental structures in the Cathedral are made entirely

[1] *Geological Magazine*, 1886, J. H. Collins, p. 366.

of this stone. As the specimen shews the rock is slightly veined and somewhat resembles alabaster in colour and appearance.

GOLD COAST, WEST AFRICA. The specimens of quartzite, from quarries near Dodowa, are typical of the stone that is used in this Protectorate for the construction of aqueducts, culverts and other engineering works connected with irrigation. It is also employed for general building in and near Dodowa. The cube from the Aburi quarries is another example.

TRANSVAAL. The specimen of dark brownish grey foliated rock from a quarry in Pretoria represents the stone which is largely used in that district for walling and foundation work. The schistosity of this rock, combined with the hardness of its grains, renders it unsuitable for decorative building. The foundations of the Law Courts in Pretoria are constructed with it.

INDIA. In the State of Mysore a valuable building material, especially for internal work, is quarried from a large dyke of rock classed with the picrites, and known as "Grey Potstone." The specimen cube is from the Manhalli quarries in the Heggadadevankote district. This stone has been extensively employed for the re-construction of the Palace at Mysore, as well as for building several Hindu temples in the Mysore State. Potstone, as the name implies, is also used largely by the natives for making bowls and platters of divers sizes and shapes for various culinary and other domestic purposes[1].

In the same State a quartzite containing pale green mica is quarried at Belavadi in the district of Kadur, and is used for internal decorative work. The specimen is one of the most attractive in the series, and is well worthy of inspection. Mysore also possesses a crystalline

---

[1] *Memoirs of Geological Survey of India*, Vol. xxv. p. 205.

limestone which makes a handsome building stone. It will be seen on examining the specimen from near Huliyar, in the district of Tumkur, that this rock is prettily veined and can be classed with the marbles.

CEYLON. A greyish-white dolomite is found near Kandy in the Central Province of this Island, which is quarried, and occasionally used for building in Kandy and other towns in the Province. The specimen shews that this stone is highly crystalline, rendering it somewhat difficult to carve.

PERSIA. Sun-dried and occasionally burnt bricks, formed from the pale coloured alluvial loam which extends over a large portion of this country, are almost the only materials employed by the Persians for structural work. Important buildings in Ispahan, Teheran, and other large towns, are however sometimes made more substantial by the introduction of stone to form a coping for basements, and also for pillars and steps.

The mountain and hill ranges of Persia comprise a considerable variety of geological formations, the prevailing rocks being of Cretaceous age. Older rocks however occur, consisting chiefly of granites and metamorphosed sedimentary deposits, forming a great band which extends from Lake Urumiah in the north to a point nearly due west of Ispahan; and the same crystalline masses appear in the ranges between Ispahan and Kashan. It is these formations which furnish the stone for the purposes just mentioned. The specimen from the quarries on the slopes of Kuh-i-Sufi Mountain, close to Ispahan, is a typical example of the metamorphic rocks also used for this class of work.

SEDIMENTARY ROCKS.

The chemical and physical characteristics of the consolidated sediments are so various, that in very few instances is the stone produced in one quarry identical in colour, texture, and other obvious characters, with that worked in another and perhaps closely adjoining quarry, indeed it frequently occurs that the same quarry will yield several distinct qualities of stone. Bearing in mind these facts, it is manifest that, in a collection such as this, it is not practicable to exhibit specimens representing every quality and shade of colour employed in building. It is hoped however that the specimens selected are fairly representative of the more important divisions and subdivisions recognised by geologists.

CAMBRIAN AND SILURIAN.

The Cambrian and Silurian systems of GREAT BRITAIN are but small contributors to building material, especially when roofing slates are excluded. A few examples, however, belonging to these formations, appear in the collection.

The specimens of Hartshill Freestone are typical examples of the tract of Cambrian rocks which is worked at Atherstone in Warwickshire, and on the summit of Caldecote Hill near Nuneaton. They consist usually of siliceous grit, cemented into a quartzite, with occasional partings of purple or grey shale. It will be observed, on examining the specimens, that their colour ranges from a light pink to a rich purple. This rock is very hard, the crushing strain it resists exceeding that of many of the toughest granites. This stone was used for building Hartshill Abbey (1125).

Useful building material is procured in Shropshire from the Caradoc or Bala series of the Lower Silurian rocks, now better known in geological nomenclature as the Ordovician system.

The deposits are developed along a line of country extending from Acton Burnell on the north, to Horderley on the south, and skirting the east flank of the Church Stretton Hills, where the quarries are situated.

The chief supply of building material is derived from the Chatwall group. It is sometimes distinguished as "Soudley Sandstone," taking its name from the famous quarry at Soudley near the picturesque village of Hope Bowdler. The lower stratum of this deposit, as the specimen indicates, consists of a fine-grained greenish yellow sandstone. The upper stratum yields a stone which develops distinct ferruginous bands. These bands or streaks are generally of a deep purple colour, and give the buildings which are constructed with it quite an ornamental appearance. The parish church of Hope Bowdler is an example of a building erected with the upper or banded rock, but the stone from the lower bed is also used. Sir R. Murchison, when referring to the Soudley Sandstones, writes, "In some quarries they are striped in ribbon-fashion, with dark red and light green layers. They are extensively quarried for use, and are as easily worked as many freestones of the younger formations[1]."

The basement beds of the Caradoc series are exposed on the flanks of the Hoar Edge Ridge which yields a building stone known as "Hoar Edge Grit." It will be observed from the specimen, that this rock is much coarser than the Soudley Sandstone of the Chatwall group. The Parish Church of Church Stretton is built of it.

[1] *Siluria*, R. Murchison, p. 66.

This series of rocks, as has just been mentioned, extends as far south as Horderley. Quarries are worked there which yield a useful building material, known as "Horderley Stone." The specimens shew that as the beds extend southward the stone darkens in colour, the purple shade, which exists only as bands in the Soudley Sandstone, being the predominating colour in the Horderley Stone. It is frequently employed for structural purposes in Ludlow, and in Church Stretton.

In the Ludlow group of the Upper Silurian system, there occurs in Shropshire a fine-grained, and thin-bedded, yellowish stone which is quarried for building purposes at Downton Castle near Ludlow and is known as "Downton Sandstone." The bed is about 80 feet thick and considerable quantities of the stone are used for structural purposes in the neighbourhood of Ludlow, and throughout the counties of Shropshire and Hereford. This rock resembles in appearance and texture the Soudley Stone of the Caradoc series.

In districts adjacent to the quarries which yield roofing material from the Cambrian and Silurian systems, the "Slate Stone" is frequently utilized for structural purposes; such for instance is the well-known Green Slate Rock from the Honister quarries, on Honister Crag, in the heart of the beautiful English Lake district, far above the famous Pass of the same name. They command a magnificent panoramic view, and it is interesting to note that communication between the Slate Company's offices in Keswick and the quarries, is maintained by the use of carrier pigeons[1]. This Slate Stone is extensively employed in Keswick and the neighbourhood for building, especially where dressed stone is required. The basement of the War Memorial in the City of York is a good example of Honister Slate Stone and the War Memorial at Penrith is a representative of the same set of rocks

[1] These little messengers carry to the quarry the orders brought in by the morning post.

quarried at Elterwater, near Ambleside in Westmorland. This stone is finer in grain than the Honister rock. The Tilberthwaite cube is another example from that county.

The Blue Slate Rocks of North and South Wales are used for building purposes in their respective localities, and are also in constant demand for steps, walls of cisterns, and other structures, for which the use of a material impervious to moisture is essential. The specimens from the Penrhyn quarries near Bangor are examples of these Slate Stones.

The cube of Blue Flagstone from Settle represents the building stone from the Upper Silurian formation in Yorkshire, which is known by geologists as the "Horton Flags." Professor Hughes writes, "This series consists of a sandy Mud-stone, splitting into flags of enormous size. There is a rough cleavage, but when this approaches within fifteen degrees or so of the bedding planes, the rock splits along the beds, and the cleavage is practically obliterated. When the cleavage predominates, the rock splits along it into rough slabs, which are found useful for building or walling[1]." It is also often employed for constructing cisterns.

IRELAND. Rocks of Lower Cambrian age occur in the north-west of Co. Wexford, and yield a slatey stone, which is quarried and used locally for building, the National Bank, Newtownbarry, being constructed of it. The specimen from Glaslacken quarries, near Newtownbarry, is an example of this stone which is also very suitable for flagging, and is extensively employed for that purpose.

In COLONIAL and FOREIGN countries, rocks belonging to the Cambrian and Silurian systems, suitable for building, are not abundant.

[1] *Proceedings of Yorkshire Geol. Society*, Vol. XVI Pt. 1, p. 58.

DENMARK. Bornholm, an island in the Baltic, besides furnishing a great variety of granitic rocks for building purposes[1], yields a useful sandstone belonging to the Cambrian system, which is employed extensively in Copenhagen, and other parts of Denmark, for all kinds of structural work.

The rock is usually classed among the "Freestones," although the specimen shews distinct lamination. It was used in constructing the Column of Liberty in Copenhagen.

RUSSIA. Silurian rocks are present in horizontal undisturbed beds between the Baltic and the flanks of the Ural Mountains, but over most of this extensive area they are covered up and concealed from view by later formations. In the Baltic provinces however they are exposed along the southern margin of the Gulf of Finland, where they consist chiefly of marls and limestones. The latter are quarried in the neighbourhood of Reval and used for building purposes. The specimen indicates that the rock is exceedingly fine-grained and capable of taking a fairly good polish. It is chiefly employed for internal structural work.

SWEDEN. The Lower Silurian system of Sweden consists of shales and beds of grey and red limestone, the latter being largely used as building material. It is present in the provinces of Örebo, Östergothland, and Christianstadt, where there are numerous quarries. The specimen from the Lanna quarries near Nerike, a district in Örebo, is typical of the grey variety, which was employed for building St Stephan's Church, Stockholm, in 1903.

The red variety is represented by the specimen from Borgholm quarries on the Island of Öland in the Baltic. The limestone forms a sterile plain extending across the whole length of the Island from north to south, sloping towards Kalmar Sound on the west, and towards

[1] See p. 31.

the Baltic sea on the east. It is quarried in large quantities for building purposes and is shipped to Stockholm, and all parts of the Baltic.

SOUTH AFRICA. In the collection there are several specimens of "Table Mountain Sandstone," a rock employed for constructive work in the CAPE OF GOOD HOPE, which belongs to a group of the Palaeozoic rocks bearing the same name. This forms part of a system existing in South Africa, correlated with the Lower and Upper Silurian and Carboniferous formations of Europe, to which the term "Cape System" has been applied[1]. The Table Mountain group belongs to the lowest division of this system, therefore the specimens find a place in the cabinets allocated to the Cambrian and Silurian rocks[2].

The series occurs in the extreme west of the Cape, forming the precipitous wall and the extensive plateau of Table Mountain, the well-known landmark on the coast of the Cape of Good Hope. A small outlier also forms the upper portion of the Lion's Head, another familiar beacon to the voyager approaching the South African coast. Again it crops out on the coast near Simon's Town, and the series stretches eastward as far as Algoa Bay, although much of it is hidden by sand dunes. The rock, as the specimens indicate, is more of the character of a quartzite than a typical sandstone. It will be noticed that the colour varies from a light yellow to a bluish grey. The stone is light in colour when first quarried but turns darker by weathering.

The specimens from the quarries on the slopes of Table Mountain are good examples of the rock forming this lofty plateau, and used for building in Cape Town. The stone is exceedingly hard, and is not well adapted for highly decorative architectural work,

[1] *Geology of South Africa*, Hatch and Corstorphine, 1st Ed., p. 57.

[2] Some geologists group these rocks with the Devonian of Europe.

being difficult and expensive to carve. Owing to the quantity of useless stone that has to be removed in quarrying the best beds of rock, it is not employed so much as one might expect from its wide distribution[1]. The external walls of the new Cathedral, Cape Town, are built of it (1906). The specimens from Simon's Town are taken from quarries in that district where large quantities of stone are being excavated and utilized for the construction of extensive government harbour works in Simon's Bay, a class of work for which this stone is eminently adapted.

The specimen of sandstone from the quarries near Buiskop in the TRANSVAAL, is a typical example of the "Waterberg Sandstone" from the formation existing in South Africa bearing that name, which is believed by some geologists to be on the same horizon as the Table Mountain Sandstone of the Cape, just mentioned[2], whilst others correlate it with the Matsáp beds of the Cape[3].

This formation consists of a great thickness of sandstones and grits, separated from those of Pretoria by a marked unconformity, and it is the former which are utilized for building. The Waterberg Sandstone covers an enormous extent of country in the Northern Transvaal, forming the whole extent of the Palala Plateau. As the specimen indicates, the rock is of a dark terra-cotta colour, and possesses an even and fine grain. It makes a handsome building material for decorative as well as for ordinary structural work. The Government House and other important buildings in Pretoria are constructed of this stone. It is also used in Johannesburg, where Eckstein's Buildings were erected with it.

---

[1] *Geology of Cape Colony*, A. W. Rogers, p. 117.
[2] *Geology of South Africa*, Hatch and Corstorphine, 1st Ed., p. 183.
[3] *Geology of Cape Colony*, A. W. Rogers, p. 78.

INDIA. In the Allahábád division of the North-West Provinces, a fine-grained sandstone is quarried which is held in great reputation as a building material both in that region and in the adjoining province of Benares. It is also transported by water down the Ganges to Calcutta, where it is extensively employed in constructive work. It belongs to the Káimur group of the Vindhyan system of India, which is correlated with the Silurian of Europe. This great system ranks third in superficial extent among the rocks of the Peninsula[1], and it is much in evidence between the valley of the Sone, and that of the Ganges. The stone is fine-grained, greyish, yellowish, or reddish white, sometimes speckled brown. It will be observed that the colour which predominates in the specimens is yellow, although one example has a pink tint.

These handsome rocks have yielded the material for the masterpieces of Indian art from the time of Asoka to the present day. Amongst the buildings of Vindhyan sandstone may be mentioned the Buddhist stupas of Barhut, Sanchi, and Sarnath, the exquisite temples of Kajraha, the palaces of Gwalior, Delhi, Agra, Fatehpur-Sikri, Amber, Dig, and the magnificent Jumna Masjids of Delhi, Agra, and Lahore. The beds vary to such an extent that it is possible to obtain monoliths of Egyptian magnitude, or flags of the thinness of slates. Such a variety of excellent material is available that, in certain parts of India, public buildings as well as private houses from the flooring to the rafters and ceilings are built entirely of this stone, and Vindhyan sandstone is sometimes even used for telegraph posts[2]. The High Court and Foreign Offices in Calcutta, the Queen's College in Benares, and the Muir Central College in Allahábád are also examples of buildings constructed with this stone.

---

[1] *Geology of India*, R. D. Oldham, p. 93.
[2] *A Summary of the Geology of India*, E. W. Vredenburg, p. 27.

A fine-grained limestone is quarried conveniently near the shore of the Irawadi river in Upper Burma, and is transported by water to Rangoon, where it is largely employed for building purposes. The existence of the older Palaeozoic beds in extra-peninsula India is not yet very clearly made out, especially in the country east of the Irawadi valley, but fossils recently found seem to be sufficient to prove that the rock is Silurian[1]. The specimen shews that it is a very dense rock, of a rich purple colour.

SIAM. In the neighbourhood of Bangkok there are limestones probably of Pre-Cambrian age, which are quarried, and furnish a grey and white streaked stone, commercially known as " Siam Marble." As the specimen indicates, the rock admits of a good polish, and appears to be specially suitable for decorative work, but as it is employed in Bangkok for general structural purposes, it is included in this collection.

SOUTH AUSTRALIA. The few mountain ranges of this State are generally composed of Palaeozoic rocks, those to the south-east, between the Murray Valley and the Yorke Peninsula, consisting chiefly of Cambrian and Silurian strata. On the flanks of these hills are quarries yielding building stone, which is extensively employed in Adelaide and the surrounding neighbourhood. The specimens shewn demonstrate that some of these rocks are distinctly arenaceous, and may be classed as sandstones, one being very coarse, while others are fine-grained mudstones, displaying marked planes of lamination.

UNITED STATES. The well-known Potsdam series of the Upper Cambrian which occurs in the areas bordering on the great American Lakes[2] yields a sandstone which is used for building purposes throughout the Northern States. Although it is said to be one

[1] *Geology of India*, R. D. Oldham, p. 118.
[2] *Geology*, Chamberlain and Salisbury, Vol. II. p. 219.

of the most durable of building stones in the United States, it is not very popular among architects and builders as its extreme hardness renders it difficult to work. When first quarried the stone is fairly soft, but it becomes intensely hard when it has been exposed to the atmosphere.

The rock is practically a quartzite, being composed entirely of quartz grains cemented into a compact mass by the deposition of siliceous matter. Physically the rock varies slightly in the different localities where it is quarried, as will be seen from the following details.

NEW YORK. The two specimens of Potsdam stone, from quarries in this State, both situated in St Lawrence County, exhibit two distinct shades of colour, one being a bright red while the other is brown and slightly banded. Stone from these quarries has been extensively employed for building in the City of New York, and in Potsdam, notwithstanding the toughness of its texture. The Houses of Parliament in Ottawa, Canada, are also built of it.

MINNESOTA. A great quartz-sandstone formation comparable to the New York Potsdam is found in Wisconsin and Minnesota, lying below the St Croix beds, but separated from them by a great unconformity[1].

In Minnesota these beds consist of large quantities of sandstone found over a wide area along the Kettle river in Pine County, where it is quarried for building purposes, and is known commercially as " Kettle River Sandstone." It is a beautiful stone of a uniform light salmon colour, even in grain, and, as the specimens indicate, varying a little in fineness, the cube from Sandstone quarries being slightly coarser in the grain than the one from the Banning quarries. In past

[1] " Report on the Lower Paleozoic," N. H. Winchell, *International Congress of Geologists of America*, 1888.

years it was not much used on account of the remote-
ness of the locality, but now it is being brought into
the eastern markets of the United States and is rapidly
growing in favour. Kettle River Sandstone is said to
resist the action of frost in a marked degree.

WISCONSIN. The light fawn-coloured, fine-grained
specimen from Cofax quarries, Dunn County, is a
good example of the Potsdam Sandstone existing in
this State, and is found near the southern limit of
the series. The Cofax Sandstone varies from a very
coarse-granular rock to one which is exceedingly fine-
grained. The specimen in the collection is an example
of the latter, which makes an excellent building stone,
especially when carved and dressed work is required.
It has been used in the construction of several public
buildings in St Paul, Minnesota, as well as in other
towns in that State and in Wisconsin.

The Lower Magnesian division of the Silurian
system, which rests conformably on the Potsdam Sandstone[1] in this State, yields a limestone which is employed
for building, of which a good example is to be seen
in the cream-coloured, fine-grained specimen from the
quarries near Fountain City. These are situated
near the summit of the bluffs on the banks of the
Mississippi river, about a quarter of a mile east of the
town. The bluffs are of a considerable height and somewhat steep, which to a certain extent militates against
the economical working of the quarries, so that the
use of the stone is chiefly confined to the locality.

SOUTH AMERICA. In the Province of Buenos
Ayres a hard quartzite occurs which, although classed
among the Pre-Cambrian rocks, for convenience sake
is placed here. There is an outcrop of this in the
neighbourhood of Mar del Plata, in which town it is
used in considerable quantities for building purposes.

[1] "Building Stones of Wisconsin," E. R. Buckley, *Bul. Wis.
Survey*, No. IV. p. 256.

As the specimens shew, the stone is exceedingly hard, consequently very difficult to work, and therefore chiefly used for structural purposes when a building of a very substantial nature is required, or when no other material is available. It is only employed locally, the cost of transport, coupled with the expense of working it, precluding its use in Buenos Ayres and the other great cities in the Republic.

The houses of Buenos Ayres itself are almost exclusively constructed of bricks made from argillaceous deposits found over nearly the entire area of the vast alluvial plains of the Argentine. These bricks are of a soft and friable nature, and most of the buildings constructed with them are covered with cement. Where decorative work is required Italian marble is generally employed.

DEVONIAN AND OLD RED SANDSTONE.

Though the tough flags, grits, and limestones of the Devonian system occupy a comparatively insignificant position among building materials, one or two of these rocks are of some importance.

In the BRITISH ISLES, the marine deposits, which occupy a large area in Devonshire, include limestones of various shades of colour, most of which are beautifully veined. They are popularly known as "Devonshire Marbles," but being extensively developed in South Devon they are largely employed as ordinary building material in addition to their use for decorative purposes, and thus claim a position in this collection. Most of the buildings in Torquay are constructed with this handsome stone. St Peter's Church at Ilfracombe and the Naval College at Dartmouth are also built of it. The specimens in the collection are from extensive quarries practically in the town of Torquay. The Upper Devonian of North Devon is represented by the cube of Pickwell Down Sandstone from the South Down quarries, Morthoe. St Sabinus Church, Woolacombe, is being built with this stone (1910).

The Old Red Sandstone, believed by some geologists to be a lacustrine deposit, yields a material which is useful for building. The specimens from quarries near Hereford are typical examples. One of these cubes, which can easily be distinguished from the rest by its deep chocolate-red colour, represents a stone which is much softer than the ordinary rock and is generally

known as "False Bedded Old Red." In Pembrokeshire
Old Red Sandstone is utilized as building material
when the strata can be conveniently quarried. The
stone was first turned to account for structural
purposes when a bed of it was pierced by the Great
Western Railway Co. near Templeton, and the stone
excavated was used for the construction of bridges on
other sections of the line. The rock has since been
quarried for ordinary building purposes, and is employed
in Tenby and other towns in South Wales. It is known
locally as "Templeton Stone."

Stretching eastward across the Bristol Channel,
this system occupies a limited area in Worcester and
Gloucester. The specimen from the Mitcheldean
quarries in Gloucestershire is a typical example, and
is known as Red Wilderness Stone. It is a hard,
reddish brown, micaceous sandstone, and is suitable for
sills, steps, or flagging, as well as for ordinary structural
work. The pavement and steps of the new Cathedral
in Liverpool are formed of this stone (1908). It was
also used in the building of Harrow School.

The Upper Old Red Sandstone, occurring near Peel,
on the west coast of the Isle of Man, yields a building
stone which is employed for ecclesiastical and other
purposes. The Cathedral Church is built of it, and
although, as the specimen indicates, the colour of the
stone is somewhat sombre, it is generally used throughout the Island for building. Peel Castle on St Patrick's
Island is also constructed of it. The rock is
generally known as Peel Sandstone, and is said to be
the only "Freestone" in the Isle of Man[1].

The specimen, from Doddington Hill quarries near
Wooler in Northumberland, represents a deposit of
arenaceous rock exposed on the Scottish border and
occurring locally at the foot of the Cheviot Hills.
These beds are, by some, classed with the Old Red
Sandstone, whilst others assign them to the Lower

[1] "Geology of Isle of Man," G. W. Lamplugh, *Memoirs of
Geol. Survey of U. K.* 1901, p. 563.

Tuedian at the base of the Carboniferous system. Sir A. Geikie has shewn that the lowest strata merge insensibly into the upper part of the Old Red Sandstone[1]. The quarries are situated in a picturesque district of North Northumberland, sheltered by the rugged Cheviots. The stone is popular for structural purposes on both sides of the Scottish border, and considerable quantities are sent to Edinburgh and even further afield. Dunblane Cathedral has recently been restored with Doddington stone. The Observatory and the Wesleyan Memorial Hall in Edinburgh are built of it.

SCOTLAND. Crossing the border the formation just described as being present in Northumberland also occurs in the adjoining county of Berwickshire, and which Sir A. Geikie includes in what he terms "the Lake Cheviot area of the Upper Old Red[2]." This series stretches with some interruptions from Berwick to Chirnside, and then onwards past the county town of Greenlaw as far as Teviotdale. Where the rock is exposed near Greenlaw it yields a useful building stone, the two specimens, one from Swinton, and the other from Whitsome Newton quarries, both being good examples. We may note that although these quarries are close to each other, the rocks differ widely in colour. The Swinton stone is the more popular of the two, its delicate pink tint being highly esteemed by architects. Considerable quantities are sent to Edinburgh for building villas in and about that city. The Whitsome Newton stone is light buff in colour and slightly coarser in grain. It was used in the restoration of Dunkeld Cathedral.

Another belt of Old Red Sandstone, rather more than a mile in breadth, occurs in Haddingtonshire, beginning at the sea a little to the south of Dunbar,

---

[1] *Geology of Northumberland*, G. A. Lebour, p. 43.
[2] Some geologists now group all these rocks with the Carboniferous.

and stretching along the base of the Lammermuir Hills. This rock is worked at Broomhouse near Dunbar where there is an important quarry which supplies that town, as well as the surrounding district, with a useful and handsome building material. The specimen shews that this stone has an unusually brilliant light red tint for a rock belonging to this system. The Parish Church of Dunbar, a handsome building erected on an eminence just outside the town, and a well-known landmark to seamen, is a good example of a building constructed of this stone. It was erected nearly a hundred years ago and notwithstanding its exposed position the carved work, of which there is a considerable amount, is in perfect preservation.

The Old Red Sandstone occupies an extensive area in the basins of the Forth and Clyde, yielding red, or chocolate, as well as grey and yellow sandstones, which form useful building material. The specimen from Hamilton in Lanarkshire is a typical example of this rock, which was used for building the Nautical College in Leith and the Roman Catholic Church in Lanark. The bright red specimen from the Auchincarroch quarries is a good example from Dumbartonshire. That from Aberdalgie represents the rock which is quarried for building in Perthshire.

Others are from Forfarshire, where the system is largely developed, fine cliff sections being displayed along the coast-line from the mouth of the Tay to Stonehaven[1]. The cubes from the Carmyllie quarries near Arbroath are good examples. This stone has been extensively used for engineering work, and was employed for building the piers of the Forth Bridge (1885).

In Aberdeenshire the rocks of this system are exposed, and quarried for building purposes. An example is seen in the specimen from the Dalgaty quarries, near Turriff, where the stone is largely used; Dalgaty Castle is constructed with it.

---

[1] *Text Book of Geology*, A. Geikie, 4th Ed. Vol. II. p. 1008.

More specimens may be seen from quarries on the shores of the Moray Firth, the district which formed the happy hunting ground of Hugh Miller, whose name will always be associated with this formation. The specimen from the Tarradale quarries is an example. The stone was used for building Inverness Castle.

Still further north is found the celebrated Caithness Flagstone, which is one of the most important economic products of this system. It is used locally for general building, but it is in special repute more as a material for flagging, and for staircases, both external and internal, for which purposes it has long been employed not only in the British Isles but all over the world. Baron Liebig's great establishment on the river Plate, in South America, for the manufacture of his well-known meat extract, is floored throughout with Caithness flags. Sir Roderick Murchison, in his third edition of *Siluria* published in 1858, writes, "The Flagstones of Caithness, which were first described by me in the year 1827 under the name of Bituminous Schists (*Trans. Geol. Soc.* ii. sec. vol. ii. p. 213) are in many places impregnated with bitumen, chiefly resulting from the vast quantity of fishes embedded in them. The most durable and best qualities, as flagstones, are derived from an admixture of this bitumen, with finely laminated, siliceous, calcareous, and argillaceous particles, the whole forming a natural cement more impervious to moisture than any stone with which I am acquainted." A good practical illustration of this useful stone, employed for steps, can be seen in the staircase leading from the ground floor to the first floor of the Sedgwick Museum, Cambridge, connecting the economic and palaeontological sections. This example, as well as other specimens, shew that by no means all the rocks of the Old Red Sandstone are "red."

Leaving the mainland, the system extends northward to the Orkney Islands, where an outcrop occurs at Fersenes near Wick, which supplies the islanders with a useful building stone. This rock, as the specimen shews, is much lighter in colour than the Old Red occurring further south, and possesses the character of a very fine sandstone.

BELGIUM yields several limestones from the Devonian system which are useful for building purposes. The chief quarries are situated on the banks of the Meuse in the Province of Hainault. The specimens shew that some of the varieties are streaked with slender veins of calcite, and as they admit of a fair polish, they are frequently employed as marbles.

The lenticular limestone from quarries near Frasnes-les-Couvin (*Calcaire de Frasne*) belonging to the lower stage of the Upper Devonian series[1], is of a bluish grey colour when first quarried, but appears to possess the characteristic of whitening with age. It is said to be much esteemed by architects and builders in Belgium and France.

DENMARK. Besides the numerous examples of granite in the collection from the Island of Bornholm, there is a specimen cube of the rock which represents the Devonian system in that Island, and furnishes a building stone which is transported to Copenhagen and other towns in Denmark for all kinds of structural work. It will be noticed that the specimen shews planes of lamination, not usually seen among the Devonian building stones. This stone was used in the construction of the Column of Liberty in Copenhagen.

GERMANY. The Devonian rocks existing in the hilly districts of the Eifel include several types of sandstone, differing considerably in fineness and colour, which are largely employed for building purposes in the regions bordering on the Rhine. The specimens indicate that some of the varieties contain mica.

UNITED STATES. The specimen of "Warsaw Bluestone" from the State of New York is a typical

[1] *Stratigraphical Geology*, A. J. Jukes-Browne, p. 176.

8—2

example of the "Bluestone" existing in that State, which in American geological nomenclature belongs to the Hamilton period of the Devonian system. This rock is chiefly found in thin beds, and splits up readily into slabs which are extensively employed for steps, sills, and similar purposes. In some quarries however the beds are massive, and the joints allow blocks of very large size to be obtained. This Bluestone, which is the popular name given to the rock owing to its colour, is largely worked, but the district of production is confined to comparatively narrow limits west of the Hudson river, and mainly to Albany, Green, and Ulster counties[1]. The Roman Catholic Church at Balater, New York, is built of this stone.

CANADA. The Devonian limestone deposits of Ontario yield a useful building stone which is extensively employed for structural purposes in that province. The specimen from the Beamsville quarries in Clinton township, Lincoln County, is a typical example. The quarries are situated on the top of the "Mountain" a little more than a mile from Beamsville. Two beds are exposed in the quarry, the upper one being seven feet, and the lower eight feet in thickness. An examination of the specimen shews that the rock is somewhat crystalline in character, and of a rich cream colour. It is very dense and hard, and is more suitable for heavy bridge masonry than for ordinary house building. This stone has been used for the last additional masonry on the celebrated Victoria Bridge, Montreal, and many of the large bridges of the Grand Trunk Railway in Canada are constructed with it.

[1] *Stones for Building and Decoration*, G. P. Merrill, p. 323.

CARBONIFEROUS.

In the BRITISH ISLES this system is usually divided into the Carboniferous Limestone (or Lower Carboniferous) series, the Millstone Grit, and the Coal Measures, the last two divisions comprising the Upper Carboniferous rocks. The system is of the greatest economic importance, for besides yielding coal it produces much valuable building material.

When the CARBONIFEROUS LIMESTONE series is shorn of its marbles, its value for furnishing building material is greatly diminished[1]. Owing to the high polish displayed on the face of the cubes, several of the specimens look as if they should be classed among the marbles. These however are not, as a rule, employed for decorative work, but are used for ordinary building construction. The specimens have been polished on one face in order to shew up more distinctly the grain of the rocks, and in some instances the abundant fossils which they contain. The cube of crinoidal limestone from Hopton Wood, in Derbyshire, is a striking example. Hopton Wood Stone has been employed for constructing many important buildings, including the Guildhall (1789), and the Imperial Institute (1881), in London; it was also used for enlarging Chatsworth (1820), the Duke of Devonshire's residence in Derbyshire.

The Mendip Hills in Somersetshire yield a compact

[1] As mentioned in the introduction, marbles are excluded from this collection. Examples of these will be found in another part of the Museum.

stone from the Carboniferous Limestone series which is used locally for building, and is popularly known as " Stink Stone." This name owes its origin to the fact that the rock, when newly quarried, emits a strong fetid odour.

Yorkshire contributes its share from the same series. Along the eastern flanks of the terraced Ingleborough, and on the lower slopes of picturesque Ribblesdale, are several quarries. Some of these yield a rock that is better adapted for road making than for building, but there are others, producing a stone which is extensively employed for structural work. The specimens demonstrate that there is much variation in the colour of these beds, the same quarry, for instance, yields " Ribblesdale White" and " Ribblesdale Blue."

The rocks known as the Yoredale Beds from the upper part of the Lower Carboniferous series consist of limestones with intercalated bands of shale and sandstone. They occur in the North Riding of Yorkshire and traverse the Yare or Yore valley, from which the name is derived, a district now better known as Wensleydale. Both the limestones and sandstones of these beds are used for building purposes. The limestone, which is frequently highly fossiliferous, is employed for constructing the numerous walls which intersect the rich pasture ground of that district. The sandstones form a useful building material with which the dairy farmers of the Dale construct their dwellings, and the agricultural towns of Leyburn, Aysgarth and Hawes are also chiefly built of this rock. The specimen from the Sedbusk quarry near Hawes is a typical example of it. Another bed is exposed at Stag's Fell in the same parish as Sedbusk. It will be seen, on examining the specimens, that the Stag's Fell stone is darker in colour than that from Sedbusk, owing chiefly to the numerous small ferruginous specks which are present in the stone.

The Lower Carboniferous rocks lose their calcareous

character as they are traced northward, so that in Northumberland the series is largely composed of sandstones. To the upper division of these strata Professor Lebour has given the name "Bernician," from *Bernicia*, the Roman name for Northumberland[1], and the lower is generally known as the Tuedian group.

An historical and antiquarian interest surrounds the Bernician rock represented by the specimen from the Black Pasture quarries in South Northumberland. The stone from these quarries was used by the Romans to build the piers of the bridge that spanned the North Tyne, and formed a part of their great wall between the North Sea and the Solway (A.D. 120). Dr Bruce, in his *Handbook to the Roman Wall*, p. 75, writes, "Next we come to the most remarkable feature on the whole line of the wall, the remains of the bridge over the North Tyne. The stone is from the Blackpasture quarry.... The peculiar feathered tooling of the facing stones will be noticed.... There have been three water piers. It has been ascertained, by partial excavation, that one of them lies immediately under the eastern bank of that river. Two others are, when the waters are low and placid, to be seen in the bed of the stream. Blocks of masonry, which have resisted the roll of this impetuous river for more than seventeen centuries, are a sight worth seeing, even at the expence of being immersed in cold water to the full extent of the lower extremities." In modern times Black Pasture stone has been largely used for the restoration of Durham Cathedral; it was also employed for building the Mitchell Library, Glasgow.

The series extends into Central Northumberland where there are outcrops, yielding building material, at Prudham and Woodburn. These quarries have supplied stone for important buildings both in the north of England and in the Lothians of Scotland. The General Post Office (1870) and the Central Railway

---

[1] *Geological Magazine*, 1875, p. 543.

Station (1850) in Newcastle, are built of Prudham
Freestone. It was also sent to London for additions
to the University College Buildings (1882-3). Woodburn Freestone was used for building the *Scotsman*
Office in Edinburgh (1901-2), and for extensions to
the Waverley Station of the North British Railway in
that city (1895).

The specimens from the Denwick quarries, near
Alnwick, are typical examples of the rock belonging to
this series nearer the coast. It will be observed that
the colour of the stone varies somewhat in the same
quarry, and that the rock is fine-grained throughout,
rendering it useful for all kinds of structural work.
It has been largely employed by the Dukes of Northumberland for the external decorative work of Alnwick
Castle, including the famous stone sculptures on the
battlements. A great deal of this stone has been used
in Newcastle-upon-Tyne for building. Another example
of this group of Bernician rocks is that from Spylaw
quarry, near Bilton, where the beds vary considerably
in texture; the specimen in the collection represents
a coarse variety. At Glanton, a little further west,
another outcrop occurs, and is quarried for local
purposes, the Parish Hall being built with it.

The specimen of Pasture Hill Stone, also belonging
to this group, deserves attention. It is of a rich
cream shade, being much lighter in colour than the
other examples of this series in the collection. The
quarries, which have only recently been developed, are
situated near Chathill, a few miles north of Alnwick.
A considerable quantity of this stone is being sent to
Edinburgh, where it is said to be finding favour among
architects on account of its colour.

The Upper Tuedian or Fell Sandstone group of the
Lower Carboniferous rocks existing in North Northumberland yields a massive grit which forms a useful
building material. The specimen from the King's
quarry, situated about one and a half miles south of
Berwick-upon-Tweed, is a good example of this stone.

It will be seen by the specimen that this rock is of a delicate pink tint, and it is said to be suitable for all kinds of architectural work. The stone has been extensively employed by the War Department in the construction of military buildings at Berwick, and it is also used for repairing the old walls and ramparts which surround that ancient border town.

In SOUTH WALES the Carboniferous Limestone is of great thickness, and consists almost entirely of a grey, black, or brown compact rock, most of which is useful for building.

Good sections are exposed in Pembrokeshire, and the pretty little town of Tenby is practically built on the rock, which is quarried and used for construction. The limestone is also found in Caldy Island, where it is worked and transported to many parts of South Wales for building. In some places the rock is very fossiliferous, and when this is the case it is frequently employed as marble, as it admits of a good polish.

The specimen from Tenby shews that the rock is often veined with white calcite. This makes it unpopular with some architects and builders, and the stone quarried further inland, which belongs to the same series, but is free from these veins, is often preferred. The specimen from the Williamston quarries is an example of this variety which is also slightly more crystalline than the Tenby rock.

The ISLAND OF ANGLESEY yields a useful building stone belonging to the Carboniferous Limestone series which, as the specimens shew, is generally close-grained and compact, and admits of a fairly good polish. The blue-grey rock quarried near Beaumaris is usually known as " Penmon Marble Stone," although it is employed as a rule for ordinary building. Owing to the dense and hard nature of this Anglesey rock it is well suited for heavy constructive engineering work. The Mersey Dock and Harbour Board have recently acquired quarries for the supply of this stone to their extensive operations on that river.

ISLE OF MAN. " Besides affording the chief source of lime for the whole Island, and being to some extent used for road making, the dark grey flaggy Lower or Castletown Limestone beds at the southernmost part of the Island supply the local building stone, both dressed and in the rough[1]." The excellent state of preservation of Rushen Abbey at Castletown bears testimony to the durability of this stone. There are several quarries near Castletown, the cube from the Scarlett quarries being a representative example.

Above the Castletown beds, and often forming knolls, is the Poolvash Limestone, sometimes known as the Black Marble of Poolvash. As the specimen indicates this stone is too soft to take a good natural polish, although sometimes when smoothed it is covered over with a black varnish, and in this way utilized for decorative building. Poolvash Limestone has been extensively employed for staircases and similar structures. The steps of St Paul's Cathedral in London are said to have been originally constructed with this stone.

SCOTLAND. The Scottish Calciferous Sandstone, which belongs to the lower part of the Carboniferous Limestone series, yields a material of great value to the builder. The outcrops in the neighbourhood of Edinburgh are the most important sources of supply.

Craigleith Stone, belonging to the upper division of this group, has had a good reputation among architects and builders for very many years. The original quarries, about three miles to the west of Edinburgh, are nearly exhausted after having furnished an immense quantity of building material to that city. It was long the principal stone used for both public and private

[1] " Geology of the Isle of Man," G. W. Lamplugh, *Memoirs of Geol. Survey of U. K.* 1901, p. 513.

buildings in the Scottish Metropolis, and was also transported to London and to the Continent. The Bank of England in London (1770), as well as the British Museum (1828) and parts of Buckingham Palace (1825–35) were built of it. The specimen indicates that it has a delicate grey-drab colour, and is close-grained, so that it is suitable for decorative work.

It will be observed that there is another cube in the collection labelled " Craigleith Stone," although this is from the Barnton Park quarries a few miles distant from Craigleith. As this stone is in many respects similar to the original Craigleith Stone, and belongs to the same group, it is known by the same name although it is not identical with it in colour or texture. The Barnton Park quarry produces three distinct classes of stone. The lowest bed yields the so-called Craigleith Stone just referred to, which was used for building the Imperial Institute, London (1880). Next is the Blue Liver Rock, a dark blue-grey stone which is used for building as well as for polishing glass. The top bed is composed of a light fawn-coloured rock interspersed with dark markings due to carbonaceous matter. It is known among builders as "Common Rock." As the specimen indicates, it is a soft stone, and the markings render it unsuitable for other than second class structural work.

Another building stone, well known in Edinburgh, belonging to this group, and known as "Hailes Sandstone," is represented in the collection. It will be seen that the same quarry yields three distinct varieties, blue, pink and white. They all make good building material, and are extensively used for structural purposes in and around Edinburgh. The Free Church Assembly Hall, Edinburgh, erected in 1846, is an example of the " Blue Hailes," and the Royal Infirmary, completed in 1875, is built of the " Pink Hailes Stone."

The Calciferous Sandstone group extends eastward into Haddingtonshire, and outcrops occur in the neighbourhood of Prestonpans, where the stone is quarried for building purposes. The specimen from Tranent quarries indicates that this rock has an extremely delicate, light cream tint, being almost white, a characteristic which renders it very popular among architects for decorative work.

The Lower Carboniferous rocks extend into Stirlingshire where the Upper Limestone group is extensively quarried. The specimens shew that the rock is finer in grain and lighter in colour than those just referred to. The light drab stone from the Plean quarries which is commercially known as " Plean White Freestone," is a typical example. It was employed for erecting the Bass Rock Lighthouse.

In Renfrewshire the same group yields the well-known Giffneuk or Gifnock Rock, which is largely employed for structural purposes in and around Glasgow, the University having been constructed with it (1865), besides many of the other public buildings. The light pinkish drab tint of this rock is looked upon with favour by Glasgow architects and builders.

Crossing over to the north side of the Forth, we find beds of Calciferous Sandstone which are quarried in the south of Fifeshire, and used for building in Edinburgh and the surrounding district. It will be noticed that the prevailing colour of the stone is light yellow with small brown ferruginous specks. The specimen from Humbie quarries, near Aberdour, is an example of the stone which was employed for constructing the Exchange Buildings, Glasgow.

A light cream-coloured, fine-grained rock, belonging to the upper beds of the Calciferous Sandstone, is quarried at Cullaloe, also near Aberdour. This stone was employed for building Fettes College (1870), and St Mary's Cathedral (1879), Edinbugh, and is at present being used for extensive alterations and additions to St Giles' Cathedral. The restoration of the nave of Iona Cathedral was also executed with this stone.

A little further north, in the same county, there is

an outcrop of these strata, near Fordell. The specimen shews that this rock is of a light drab colour, and resembles in many respects the Giffneuk stone of Renfrewshire. Although the quarries have only recently been developed, it is said that they were worked previously, but had been abandoned for a number of years. The stone was used in 1908 for building Dundee College, it has also been exported to Russia and employed for structural purposes in St Petersburg.

There is only one example of the use of Scotch Sandstone known to occur in Cambridge. The bridge over the Cam at King's College, built in 1819, is constructed with Calciferous Sandstone from Fifeshire[1].

IRELAND. Although the Carboniferous Limestone underlies almost the whole of the plain of central Ireland, and practically stretches from shore to shore, it is only occasionally that the limestone crops out on the surface, and can readily be quarried. Where this occurs the rock is used for building as well as for other industrial purposes. This is specially the case in King's Co., and in the counties of Kilkenny, Carlow and Kildare. The specimens exhibited from these districts indicate that the rock is usually very compact, and in some cases admits of a fine polish.

The cube from the Surra quarries, near Birr, is an example of the rock quarried in King's Co., which is a blue-grey fine crystalline stone and makes a useful building material. It is used in Birr Castle, the seat of the Earl of Rosse.

In Co. Kildare the beds of this age yield stone of various colours, the prevailing shades being black and drab. The specimen from Donaghcumper quarries represents the former variety. The deep black colour of this stone, the compactness of its texture, and the

[1] *Architectural History of Cambridge*, Willis and Clark, Vol. I. p. 573.

high polish which it takes, suggest that it would make a useful marble, but the owner of this valuable deposit states that, up to the present, it has not been utilized for that purpose. Christ Church Cathedral, Dublin, was built of it. The specimen from the Carrick quarries near Edenderry, is an example of the drab-coloured rock quarried in this county. It also admits of a high polish and can be usefully employed for decorative purposes.

The same group in Co. Kilkenny yields a stone suitable for building, which is more variegated in colour than the rocks just described. The black and grey mottled fossiliferous specimen from the Three Castles quarries is an example, and, as will be observed, it takes a good polish. Many buildings in Dublin are constructed with this stone.

The handsome dark blue-grey close-grained specimen from the Royal Oak quarries is representative of this group of rocks in Co. Carlow. It has been exten-sively employed for constructive engineering work on the Great Southern and Western Railway of Ireland.

The formation extends further south and is exposed in the counties of Limerick and Cork. The numerous specimens from Co. Limerick shew that the litho-logical character of the rock is similar to that of the stones from central Ireland, and the same may be said regarding the beds existing in Co. Cork.

The rock suitable for building which belongs to this group in the north of Ireland, is distinctly arenaceous in character. The specimen from the Drumkeelan quarries in Co. Donegal is a good example. This stone, known commercially as "Mountcharles Sandstone," was employed for building Letterkenny Cathedral, and for the extension of Dublin Museum.

In the Ballycastle Coalfield, in the north-east of Co. Antrim, red and yellow sandstones occur, belonging to the Lower Carboniferous series[1], which

---

[1] *Quart. Journal Geol. Society*, Vol. XXIII. E. Hull, p. 625.

are quarried and used for building in Belfast and other towns in the north of Ireland. Rocks of both colours are frequently found in the same exposure, and various shades between these two are often seen in the same quarry. The specimens from the Drumaroon quarries are examples of the red and yellow varieties. This stone was employed for building the Old Court House, Belfast; also the Parish Church, Ballymoney, and two churches in Ballycastle, besides many others in the same county.

In England the MILLSTONE GRIT yields building stone over a wider area than does the Carboniferous Limestone series. These rocks, as the name indicates, are largely used in the manufacture of millstones and grindstones, but they are also extensively employed for all kinds of structural work. In many cases their density and their imperviousness to moisture, render them peculiarly adapted for engineering operations, such as dock walls, marine piers, or reservoirs.

The long narrow outcrop of the Millstone Grit in Derbyshire, which traverses the picturesque dale of Matlock and extends to the high table-land of the Peak, yields a valuable building stone, which is quarried extensively in the neighbourhood of Matlock, Darley Dale and Grindelford. As may be seen by the numerous specimens from these localities, the stone varies considerably not only in colour but also in texture, although most of it is very coarse, being composed of large quartz grains and fragments of orthoclase, cemented together with siliceous matter. Indeed, so coarse are some of the beds, that the rock might be called a conglomerate. The specimen from Grindelford is an example of the rock specially suitable for engineering work. The stone from these quarries is at present being employed for the construction of large storage reservoirs for the supply of water to Sheffield, Derby, and other towns in the Midlands.

The cube from the Stancliffe quarries, Darley Dale, is an example of the more finely-grained rocks in the group.

This is known commercially as "Darley Dale Stone." It was used in building St George's Hall, Liverpool (1842), also for constructing the Thames Embankment, London (1870). Some of the beds in the Darley Dale district yield a pink, fine-grained stone. This is represented by the specimen from the Hall Dale quarries. It was used for building the extensive premises of the Wholesale Co-operative Society in Manchester (1886). A rock of a similar colour is obtained from the Cromford quarries near Matlock, but, as the specimen shews, the stone is much coarser in grain than that from Darley Dale. The same bed crops out a little further north at Whatstandwell, the specimen from Duke's quarries being an example. This stone was used for building Euston Railway Station in London, and was also employed in erecting the gaols at Birmingham in 1849, and Leicester in 1828.

The specimens from quarries at Tansley, near Matlock, are examples of the stone more generally employed for the manufacture of millstones than for building purposes, although there are many instances of this stone being used for structural work. Several of the Board Schools in Sheffield are built of it.

In Lancashire the Millstone Grit suitable for building is worked chiefly in the north of the county, and is generally of a finer texture than that of Derbyshire. The specimen from the Longridge quarries, near Preston, is an example of the stone used for building the New Town Hall at Lancaster (1908). Manchester Cathedral, built in the 15th century, is said to be constructed of Millstone Grit[1].

The same formation stretches into the West Riding of Yorkshire. The beds are exposed near Pateley Bridge, as well as further up the beautiful valley of Nidderdale. The cube from the Scar quarries represents this stone, of which large quantities are being employed for the construction and maintenance of the Bradford Water Works at the head of the valley.

[1] *England's Chronicle in Stone*, J. F. Hunniwell, p. 198.

One of the specimens from the Hanging Stone quarries about a mile south-east of Ilkley, is a typical illustration of very coarse Millstone Grit. This characteristic prevents many of the rocks of this group from being used for decorative work, the coarseness of the grain and hardness of the stone rendering it difficult and expensive to carve. Nevertheless it has been sometimes employed for highly ornamental buildings. This can be noticed, and its weathering qualities observed, in Kirkstall Abbey near Leeds, which was built in the 12th century with the original "Bramley Fall Millstone Grit" from quarries long since exhausted. Large quantities of stone of a somewhat similar character are quarried in the Leeds district, and sold as "Bramley Fall Stone," which has become a general term for that class of stone, wherever it may be quarried. In Cambridge, Bramley Fall Stone was employed for constructing the river wall within the grounds of King's College.

Several specimens of so-called "Bramley Fall Stone" are in the collection; that from the Horsforth quarries is an example of the rock used for building part of Euston Railway Station in London. It was also employed for constructing Millwall Docks, London, and buildings at Brunswick Dock (1843). The specimen from Pool Bank quarries near Otley is an example of the stone used for building the Roman Catholic Church, Leeds (1904).

Referring again to the specimens from the Hanging Stone quarries, near Ilkley, it will be noticed this outcrop yields stone of several degrees of fineness, from a very coarse to an extremely fine-grained one, all of the same colour. The medium-grained stone was employed for erecting the well-known Hydropathic Establishment at Ben Rhydding (1844), as well as for many other important buildings in and about Ilkley. The coarser qualities have been largely used by the Urban District Council, which now owns the quarries, for constructing and repairing bridges in the neighbourhood.

The Millstone Grit, as it is followed northward, seems to lose its coarse gritty character. The numerous specimens belonging to this group from quarries in the counties of Durham and Northumberland are much finer in their texture than the corresponding Yorkshire stones. These rocks make excellent building material and are much used in the north of England.

Barnard Castle, in the county of Durham, was built in the 12th century with the stone quarried near the town bearing the same name. This rock was also used for building the Bowes Museum in Barnard Castle, and the Municipal Buildings, York (1889). In this county the rock belonging to the Millstone Grit varies considerably in colour. The stone just referred to is dark brown, whilst that quarried at Heworth near Felling, on the south bank of the Tyne, sometimes has a blue-grey shade. This stone was used for building the Central Arcade, Newcastle (1831), and the river Wear Commissioners' Offices, Sunderland (1905). The same quarry also yields a light brown stone, and they are distinguished commercially as "Heworth Blue" and "Heworth Brown" stone. The light brown specimen shews that this variety is speckled with dark ferruginous spots, but they are so small that they do not interfere with the general appearance of the stone. The County Council Offices in Durham are built of it.

In Northumberland, the Kenton quarries, situated about two miles north-west of Newcastle, yield a Millstone Grit varying in character. The specimens indicate that one section of the bed is light drab in colour and of a medium grain, while the other is much finer and brown. This well-known deposit has been worked for a great number of years, and many of the principal buildings in and around Newcastle are constructed with it. The New Town Hall (1856) is a good example.

At Stocksfield, a village on the south side of the Tyne, about midway between Newcastle and Hexham, are outcrops of Millstone Grit which have been extensively quarried. The specimen from Bearl quarry is an example. Besides furnishing building material for many of the principal houses in the locality, which is a favourite residential district, it has been used in the manufacture of millstones.

On the north banks of the Coquet, in Mid Northumberland, under the shadow of the famous ruins of Warkworth Castle, the group is exposed and quarried for building. The bed, which runs into the sea near Alnmouth, produces, as the specimens indicate, stone of two distinct shades of colour. One is brownish-pink, and the other bright yellow. The pink rock is said to resist fire remarkably well, and it has been exported to Spain for lining smelting furnaces.

SCOTLAND. Rocks of Upper Carboniferous age are not so fully developed in Scotland as in England, and the Millstone Grit suitable for building purposes is by no means abundant. The specimen from Braehead quarries, in Linlithgowshire, is an example of it. Owing to the coarse grain of this stone it is not very suitable for decorative work, but makes a useful building material for ordinary purposes.

The group is also present in Stirlingshire, and is represented by the straw-coloured specimen from the Old Brighton quarries, near Polmont Station. This group of sandstones is generally known in Scotland as "Moorstone Rock[1]."

IRELAND. The Millstone Grits of this country are uniformly of a closer texture than those existing in England. The specimen from the Doonagore quarries in Co. Clare is an example of it. This rock is known in commerce as "Shamrock Stone," and besides being much used for building, it is in demand for steps and flagging. The steps leading into the Cork Exhibition, held in 1906, were composed of this material. As the specimen indicates, the rock is exceedingly fine in the grain and it

---

[1] *Stratigraphical Geology*, A. J. Jukes-Browne, p. 265.

is said that it can be dressed to a very smooth surface and assumes a soft grey appearance.

The COAL MEASURES of England yield rocks which are extensively quarried, and are of great value as a building material. A region where coal exists and is commercially worked, usually becomes a centre of industry with a large population, thus causing a good demand for building material. Quarries are therefore very numerous wherever stone is accessible.

The specimens in the collection from the Coal Measures indicate that many of the rocks belonging to this group are highly laminated. The cube from the Thornton quarries, near Bradford, in Yorkshire, is a good example, and is known commercially as "Self-faced Flagstone," tool-dressing being unnecessary for most purposes. The specimen from the Howley Park quarries, near Morley, in the same county, represents a rock from this group which is free from lamination, and has more the structure of a "Freestone." This stone was used for building the Colosseum, Leeds (1883).

It is from Yorkshire that a great proportion of the building material from this group is produced. Large quantities are sent to London, and other centres of industry; indeed so widely is the Yorkshire Sandstone known, that the name "York Stone" has become a generic term[1]. The drab fine-grained specimen from the Tuck Royd quarries, near Halifax, is from one of the many quarries that have supplied the London market. The Post Office Savings Bank (1900), the New War Office (1906), and the extension to the British Museum (1907), were built with it.

The Crosland Hill quarries, near Huddersfield, supplied stone for building the Town Hall (1877) and the Exchange (1871), Manchester; as well as the Liverpool Exchange (1868). The cube shews that the rock from this bed is slightly micaceous, as is common with

[1] *Masons' Practical Guide*, R. Robson, p. 67.

many of the building stones of this group. The specimen of Bolton Wood Stone represents the material that was used for building the Town Halls of Bradford (1873) and Leeds (1858). The Arthwith quarries, near Bradford, yield two stones differing widely in colour. As the specimens indicate, one is fawn-coloured and the other brown. The former is known commercially as "Hard Blue," and the latter "Hard Brown" York Stone. The Hard Blue was employed for building the Admiralty Docks, Portsmouth (1900), and the Royal Infirmary, Newcastle-upon-Tyne (1906). Woolwich Arsenal Buildings (1900), and the Dock Offices, Leeds (1905), are built of the Hard Brown variety.

The coalfields of South Wales, Lancashire, Durham, and Northumberland, also contribute their share of building stones. As the specimens shew, the sandstones from the Coal Measures are chiefly very fine in grain, and of a greyish or yellow colour[1].

It would be going beyond the intended scope of these notes to attempt to mention in detail the numerous examples of stone collected from these scattered coal-bearing regions and enumerated on pages 288 to 290. Keeping in view however the important position these rocks occupy as building materials in the districts where they exist, they must not be altogether passed by.

The Pennant beds of Gloucestershire, Monmouth, and South Wales, yield a hard blue-grey fine felspathic sandstone which is much in evidence in the buildings of Bristol, Cardiff, and the neighbouring towns. The Forest of Dean specimens are good examples, which, it will be observed, vary considerably both in colour and texture, ranging from a medium-grained drab, micaceous sandstone, to a dark blue-grey, close-grained rock. These stones have been extensively employed in London for building purposes, as well as in the provinces. The New Sessions House, Old Bailey, is an example of its

---

[1] *Geology of Northumberland*, G. A. Lebour, p. 23.

use in London; the New Dock, Avonmouth, and the Military Barracks, Salisbury, are provincial buildings constructed with it. Attention may be drawn to the pink specimen from Huckford quarries, near Winterbourne, locally known as " Red Pennant Stone," which also belongs to this group, and is sometimes preferred to the blue by architects and surveyors. It is said to be harder, and equally durable.

The purple-brown fine-grained sandstone of Tintern, which was used for the construction of the Abbey, is also well known in Cardiff as a building material. The Abbey, founded in 1131, and rebuilt in the 13th century, is a good example of the weathering properties of this stone, for the building is almost entire except the roof.

In North Wales the Coal Measures, which form an irregular band along the south side of the Dee traversing the east of Denbigh as far as Oswestry, yield a light drab or yellow, fine-grained sandstone, which is quarried for building purposes, as well as for making grindstones. The stone is worked at Broughton, near Wrexham, the specimen from the Bryn Teg quarries being an example. Wrexham Parish Church is built of it. The bed is again exposed at Ruabon, producing a useful building material known as Cefn Stone, which was employed for Ruabon Church (13th century), the Free Library and Museum (1859), and the Walker Gallery (1877), Liverpool, as well as the New University College, Bangor (1908).

The Yorkshire Sandstones from this group have already been mentioned on page 132. In the adjoining county of Lancashire, outcrops of these rocks occur near Burnley, which are extensively worked for building purposes. The light greenish grey specimen from Worthsthorne is a good example.

Passing farther north we come to the Durham and Northumberland coalfields, where the quarries, if not so numerous as in Yorkshire, are of equal importance. The Middle Coal Measures include sandstones, which besides being used for building purposes, are employed for making the celebrated Newcastle grindstones. The bed is known locally as " Grindstone Post." The specimen from the Windy Nook quarries, near Gateshead, in the county of Durham, is a typical example of this rock, which was employed for constructing the Exchange Buildings, Newcastle, in 1860.

In most of these beds there is a tendency to segregation of the colouring matter in the form of concentric bands of various tints, sometimes giving rise, after long exposure to the air, to singularly variegated effects. This is well illustrated in many of the buildings of Newcastle and Gateshead[1].

Some of the beds of the northern counties yield a stone that is capable of resisting fire in a marked degree. The specimen from the Burradon quarries in Northumberland, is an example, and is known locally as "Burradon Firestone." That from the Penshaw quarries, Co. Durham, is another, which is largely used for building furnaces. This stone was also employed for constructing the Sunderland Piers (1885).

In the Cumberland coalfield, there are alternations of grey, yellow, and pinkish sandstones, which furnish useful building material. The specimen of Round Close Freestone from the Castle quarries, near Whitehaven, is a good example. The pink tint of this rock is said to find favour among builders and architects. The stone has been used for building the Carnegie Free Library, Cockermouth (1903), the Keswick Branch of the Bank of Liverpool (1900), and many other buildings of importance in Cumberland. The Parish Church of Clifton in Westmorland, built in 1899, was also built with it.

SCOTLAND. The Coal Measures of the Upper Carboniferous rocks of this country are represented in the collection by numerous specimens of sandstones, varying considerably in colour and texture.

[1] *Geology of Northumberland*, G. A. Lebour, p. 23.

Those from the Lanarkshire coalfields consist of grey and whitish medium-grained sandstones, the light drab specimens from the Auchinlea quarries in this county being good examples. It will be observed in one of these that there are minute brown ferruginous specks. These are frequently present in the Sandstones of this group in Scotland. The Auchinlea Stone was used for constructing the Carlisle Citadel Railway Station built in 1880, and for the Western Infirmary in Glasgow (1874).

An outcrop occurs and is quarried in Clackmannan. The pinkish grey specimen from the Devon quarries, near Alloa, is an example of this rock.

The Coal Measures in Kincardineshire yield a stone useful for building, and the specimen from the Sands quarries near the town of Kincardine, represents this group. This stone was used for extensions to the Waverley Railway Station in Edinburgh (1892).

The rocks belonging to this formation in Ayrshire sometimes have a warm drab colour, approaching a pink shade, which is highly esteemed by architects and builders. The specimen of Scares Liver Rock from quarries near Cummoch is a good example. That from the Sevenacres quarries near Irvine represents the light yellow rocks of this group. It will be noticed that this cube is also speckled with ferruginous matter, but the particles are so small that they do not interfere with its general appearance.

CONTINENT OF EUROPE. Here, on the whole, the building stones obtained from the Carboniferous system are not so numerous or varied as those of the British Isles.

BELGIUM. The so-called *Petit Granit*, which is really a crinoidal marble, belongs to this system. It is a dense black limestone containing small fragments of shells, corals, and crinoids. The outcrop of this formation stretches across the country in an east and west direction, and yields an immense quantity of useful building stone, which is extensively employed in Belgium, and is exported to France, Germany, and Holland. The specimen from Soignies in the Province of Hainault is a representative example, and shews that this stone admits of a good polish. It was employed for building the Law Courts in Brussels.

There is another variety of Carboniferous Limestone in Belgium, which is extensively quarried on the banks of the Meuse, and largely used for building in Brussels and other towns. In texture it is more compact than the *Petit Granit*, and, as the specimens shew, it varies considerably in colour, ranging from a dark blue to a light grey. The example from the Méhaigne quarries near Namur represents the stone used for building the East Railway Station, Antwerp.

RUSSIA. The Carboniferous system of this country is important, the lower division yielding limestones, some of which form excellent building material. The beds of this series extend from at least a hundred miles south of Moscow to as far north as Archangel. In the greater part of this area they are hidden by Permian and Triassic deposits, but extensive outcrops appear in the Governments of Tula and Moscow, where the chief quarries are situated.

As the numerous specimens indicate, the rock is, as a rule, light straw or cream-coloured, generally close in texture, and, in some cases, admits of a fairly good polish. This limestone is known locally as "White Moscow Stone," and the town of Moscow is practically built of it[1], besides most of the churches, in that, and the neighbouring Governments. The cube from the Kreise Wenew quarries is a typical example. The fine-grained specimen from the Barybine quarries, near Tula, is an example of the stone largely used for façades and steps in that town, also in Moscow. The same formation extends eastward and appears in the Government of Nizhni Novgorod, and is there

---

[1] *Limestones and Marbles*, S. M. Burnham, p. 215.

quarried for local building. This bed yields a darker and more crystalline stone than the "Moscow White," as the specimen from the quarries near Murom shews. It is specially useful for foundation work.

SWITZERLAND. In this country a rock from the Carboniferous system, suitable for building purposes, has lately been used in considerable quantities for railway undertakings. The outcrop extends up the valley of the Trient from that of the Rhone towards Chamonix. This stone, as the specimen in the museum indicates, is an extremely hard and somewhat crystalline rock, so that it is difficult to work. It has been employed for constructing the bridges on the Martigny and Chamonix Railway.

TURKEY IN ASIA. Rocks belonging to the Carboniferous system occur in the north-west province of Asia Minor, extending along the coast of the Black Sea, which yield coal in limited quantities, as well as sandstones that are used for building purposes. The specimen from quarries near Lefke, on the river Sakaria, is an example. It is a fine-grained stone, and has been employed for constructing the Anatolian Railway. Some of the Stations are built of it.

AUSTRALIA. The main chain of mountains in the east of the Continent stretching from Cape York in the north, to Bass Straits in the south, as mentioned when referring to the granites of Australia[1], is chiefly composed of igneous intrusions, while the Carboniferous, and other sedimentary deposits, now form the slopes of the hills on each side.

QUEENSLAND. Along the eastern flanks of the range the Carboniferous rocks are quarried, and yield useful building material which is transported to Brisbane and other towns on the coast. Some of these limestones have been metamorphosed into marbles, which

[1] See p. 58.

vary greatly both in colour and quality. Most of these marbles are used exclusively for decorative work, but some are employed for ordinary building, and specimens of these are included in the collection.

VICTORIA. The Grampian sandstones, regarded as belonging to the Lower Carboniferous series[1], are quarried on the flanks of the Grampian Hills in Western Victoria, and supply a useful building stone. The quarries are situated in the Stawell district, celebrated for its gold-mining operations. There being good railway communication from that locality to Melbourne, large quantities of the Grampian sandstones are sent thither for structural work. The stone, as the specimen indicates, is of a soft dove colour, and has a fine even grain, so that it is very suitable for decorative work, as well as for ordinary building purposes. It was employed for erecting the Parliament Houses in Melbourne.

UNITED STATES. There are several specimens in the collection belonging to the " sub-Carboniferous age," which is the name given by the geologists of the United States to the lower division of the Carboniferous period.

INDIANA. The most important rock of the series in this State is that from the Bedford Quarries in Lawrence County, which is popularly known as " Bedford Limestone." As the specimen indicates, this stone is light grey in colour, and rather fine-grained. In texture it strongly resembles the European Oolites, being composed of small, even, rounded concretionary grains, bound together with a calcareous cement, which is said to be very tenacious, thus enabling the stone to sustain an abnormally heavy pressure. It is also a true " Freestone," and is therefore greatly esteemed for all kinds of carved work.

[1] *Australasia*, J. W. Gregory, p. 415.

Ohio. The Sandstones of the sub-Carboniferous age in this State hold a very important position as building materials. The specimens include one from quarries near Berea, which is generally known as "Berea Grit." Another example comes from the Euclid quarries in Cuyahoga County, and is termed "Euclid Bluestone." This rock is finer in texture than the Berea Grit although similar in colour. The latter is said to become yellowish in colour on exposure, but this does not affect the quality of the stone[1].

CANADA. New Brunswick. Here Freestones are abundant among the Lower Carboniferous rocks, and have been worked for many years. They are of red, yellowish, and olive tints, often so soft when freshly quarried as to be readily cut, but harden on exposure, and are largely used for building purposes both in Canada and in the United States[2].

The "Millstone Grit," though its name is not in all cases lithologically appropriate, is the chief source of supply. It is exposed in Gloucester County, on the coast of Chaleur Bay, an inlet of the Gulf of St Lawrence. The light greenish grey fine-grained specimens from the Chaleur Bay quarries are representative examples of this stone. It has been used for the construction of many well-known Canadian buildings, including the Royal Victoria Museum, Ottawa; the Parliament Buildings and the City Hall, Toronto; and the Legislative Buildings, Halifax.

There are also numerous quarries on the banks of the Miramichi river, yielding stone of varying colour and texture, which is known commercially as "Miramichi Sandstone." The specimen from a quarry at Indian-Town is an example. This stone has been extensively employed for structural purposes in Montreal, Quebec, and Ottawa. The C.P.R. Telegraph Buildings, Montreal;

[1] "Building and Ornamental Stones," G. P. Merrill, *Report of the U.S. National Museum*, 1886, p. 457.
[2] *Acadian Geology*, J. W. Dawson, p. 249.

the Jeffrey Hale Hospital, Quebec; and the Government Departmental Buildings, Ottawa, are examples. There, is another cube in the collection from this group, which is much darker, and the stone is occasionally used for building in Montreal, but it is not a favourite, on account of its colour, Canadian architects and builders preferring the brighter New Red Sandstones from Great Britain[1]. Near the town of Sackville, in Westmorland County, there are other quarries yielding useful building material from this formation, of various colours, the specimens from Rockport and Woodpoint quarries being good examples.

NOVA SCOTIA. Grey Sandstone, useful for architectural purposes, is found in great abundance in the well-known Pictou coal formation of this Province[2]. It is quarried, for local building, as well as for exportation to the United States. The specimen from the Pictou quarry, situated about one and a half miles from the centre of the town of the same name in Cumberland County, represents this rock, of which large quantities are sent by rail to Pictou Harbour, in Northumberland Strait, thence to be shipped to all parts of America. The Bank Buildings at Pictou were constructed with it.

The Wallace quarries in the same county also yield useful building material from the "Millstone Grit" group. As the specimens shew, the colour and texture of the stone often vary in the same quarry. The beds near the surface yield the light blue-grey, fine-grained rock, sometimes known as "Blue Stone." Those beneath produce the light olive-coloured variety, which, it will be seen, is a coarse-grained rock. Both kinds were used for building the National Museum at Ottawa.

[1] See p. 156.
[2] *Acadian Geology*, J. W. Dawson, p. 344.

PERMIAN.

BRITISH ISLES. The arenaceous rocks of the British Permian system furnish useful building materials. The stone, as the specimens indicate, varies in colour, some of the examples having a rich red shade, others a delicate pink tint, while others again are light yellow or buff. Their various shades, combined with their even texture, render these rocks popular among architects, and they are in request for structural purposes in all parts of the kingdom.

Devonshire is the only county in the south of England where Permian rocks occur that are useful for building. Here the beds which underlie the Trias are composed of sandstones, breccias, and marls. It is the breccias which are utilized by the builder. These are well exposed in the fine cliffs between Babington and the estuary of the Teign, where there are numerous quarries yielding stone which is used for structural purposes in and around Teignmouth, Torquay, and Paignton. The specimen in the collection demonstrates that this stone is a coarse breccia in which fragments of Devonian limestones are abundant. The Parish Church at Paignton is an example of a building constructed with it.

Some deposits of the Permian age are extensively quarried for building in Central England. The specimen from the Nesscliffe quarries near Great Ness, in Shropshire, is a good example. This stone was used for restoring the Abbey Church of Shrewsbury in 1887. In Nottinghamshire rock of this age also occurs, and there are quarries near Mansfield, which yield building material of different colours, the two chief varieties being known as Red and White Mansfield Stone. These are represented by the specimens from the Lindley and Sills quarries. The Red Stone from the Lindley quarries was used for restoring Chichester Cathedral (1861), and for enlarging the royal residence at Sandringham in Norfolk. The White Stone has been employed for restoring Ely Cathedral, and for the buildings of King's College, Cambridge, erected in 1823. The Red stone from the Sills quarries was among those chosen for building the St Pancras Railway Station Hotel (1873), and the New Law Courts in London. The Municipal Buildings at Windsor (1852) are built with Sills White Mansfield Stone.

Permian rocks occur in North Wales and crop out near Wrexham. The specimen shews that the stone from this district is of a darker colour than are most rocks of this age. It was used for restoring Hereford Cathedral.

In Cumberland and Westmorland the well-known "Penrith Sandstone," belonging to this system, forms a useful building stone, varying in colour from bright red to light buff. The specimen from the Remington quarry near Penrith represents the light red variety which has been extensively used by the London and North Western Railway Co. for building bridges on their system. It is also being used for external string-courses in Liverpool Cathedral. The specimens of Lazonby Red and White Stone in the collection, represent the rock that has been quarried for building purposes over a long period on the Fells of Lazonby, a few miles to the west of Penrith, and has been used for all kinds of structural work in Cumberland and elsewhere. Owing to the rather coarse, gritty nature of this stone it is sometimes known among architects as "Lazonby Grit." St Aidan's Church, Carlisle, built in 1900, is an example of the Red variety. The specimen from the Moat quarries near Netherby, is an example of the bright red rock in this group, locally known as Red Moat Stone. It was employed for constructing the National Gallery in Edinburgh (1850) and Bamburgh Castle in Northumberland has lately been restored with it. That from the Half Way Well quarry near Penrith illustrates the dull red variety, while the yellowish stone with small ferruginous specks, quarried on the Heights near Appleby in Westmorland, is a typical example of the light buff rock.

The upper beds of the Permian system east of the Pennine range consist of MAGNESIAN LIMESTONE or " DOLOMITE," which is a name often given to those varieties of limestones in which carbonate of magnesia is present as well as carbonate of lime. Outcrops of this group occur in Nottinghamshire, which are worked, and yield a very fine-grained, rich yellow building stone. The texture of this rock is so compact and the grain is so fine, that the stone is capable of taking a fairly good polish. The specimen of Mansfield Woodhouse Limestone is typical of this group, which overlies the beds that yield the Red and White Mansfield Stone[1].

The example from Bolsover Moor quarries represents the rock which is quarried in Derbyshire. This Magnesian Limestone was selected by the Royal Commission, appointed in the year 1839 to report on and select the most suitable building stone for constructing the new Houses of Parliament in London. The reasons for choosing this stone are summed up in the report of 1839 in the following words: "In conclusion, having weighed, to the best of our judgement, the evidence in favour of the various building stones which have been brought under our consideration, and freely admitting that many sandstones, as well as limestones, possess very great advantages as building materials, we feel bound to state, that for durability,

[1] See p. 143.

as instanced in Southwell Church, etc.; and from the results of experiments . . .; for crystalline character, combined with a close approach to the equivalent proportions of carbonate of lime and carbonate of magnesia; for uniformity of structure, facility and economy in conversion, and for advantage of colour, the Magnesian Limestone, or Dolomite of Bolsover Moor, and its neighbourhood, is, in our opinion, the most fit and proper material to be employed in the proposed new Houses of Parliament[1]."

The decay that has taken place in the fabric of these buildings has caused architects and others to question the practical value of scientific knowledge in the selection of building stones. Acquaintance with the true cause of this unsatisfactory result ought however to be the means of dispelling such doubts, and a brief reference to this may be of interest. The erection of the Parliament Houses had not proceeded far, when it became evident that the supply of stone from Bolsover Moor quarries was quite inadequate to complete the work, neither were blocks obtainable of sufficient size to carry out the designs, therefore when only a small portion of the superstructure, to the top of the basement windows, had been built of this stone, it was necessary to procure a supplemental supply. At first the Mansfield Woodhouse Limestone was selected, but owing to the difficulty in obtaining blocks of the required size, this was also superseded by Anston Stone from a neighbouring quarry in Yorkshire[2]. The services of an expert were offered to supervise the quarrying and selection of the stone, but for some reason the appointment was never made, and stone was quarried and delivered indiscriminately, without regard to the nature of the bed, the lie of the rock in the quarry, or the necessary seasoning of the stone.

As evidence to prove that those in authority were

[1] *Report of the Building Stone Commission*, 15 July, 1839.
[2] "The Building Stones of London," *Quarry*, March, 1901.

w.   10

justified in selecting the stone they did, to supplement that from Bolsover Moor quarry, it is only necessary to mention that the stone from Anston quarry was also used in the erection of the Museum of Practical Geology in London, opened in 1851, and was quarried at the same time as that for the Houses of Parliament, but in the case of the Museum, it was most carefully selected by Sir H. de la Beche, who was then Director General of the Geological Survey, with the result that scarcely a single defective stone found its way into the structure. Although, as the details in the catalogue shew, Bolsover Moor and Anston quarries are in different counties, they are within eight miles of each other. In the year 1847 the flying buttresses of Westminster Abbey were restored with Magnesian Limestone from Anston quarries.

There are many examples, up and down the country, which shew that Magnesian Limestone has stood the ravages of time admirably. Among others, may be mentioned Huddlestone Hall in Yorkshire, which was built of Huddlestone Limestone in the 15th century, and is still in a good state of preservation. Huddlestone Stone was employed for building King's College Chapel in Cambridge, the foundation stone of which was laid by Henry VI in 1446. The stone at first used for building the basement of the Chapel was from Thesdale quarry near Tadcaster in Yorkshire, which had previously supplied material for York Minster[1], but after the work had proceeded for about three years, that from Huddlestone was substituted. On 25 February, 1449, the King obtained a grant from Sir John Langton to work a part of the Huddlestone quarry, lying next to that belonging to the Dean and Chapter of York[2]. The remainder of the basement and a large part of the superstructure were built with this stone. The limit of its use may be clearly seen in the building, running in an oblique line

[1] *King's College*, A. Austen Leigh, p. 19.
[2] *Architectural History of Cambridge*, Willis and Clark, Vol. I. p. 397.

from four or five feet above the ground at the west end, to the spring of the arches of the side windows at the east, the white surface of the stone contrasting with the Weldon Oolite used for the remainder of the Chapel[1]. The light tint is due more to the gradual bleaching and decay of the stone than its natural colour. The specimen of Huddlestone Stone in the collection shews that the normal colour is yellow. The Eleanor Cross in front of Charing Cross Station, London, was erected with Huddlestone Stone.

In reference to King's College Chapel, it should be mentioned that its beautiful groined roof is also composed of Magnesian Limestone, from the Roche Abbey quarries in Yorkshire. The specimen shews that it is similar in colour and texture to the Huddlestone rock. This stone was used for building the Abbey, and the good state of preservation of what remains of the building is evidence of the durability of this rock. Probably one of the reasons which induced the monks to settle on this site was the unlimited supply in the neighbourhood of a building stone beautiful in colour, easily worked, and yet very durable.

The same group crops out near Pontefract, and although, as the specimens indicate, the beds in this district do not yield a stone so fine in the grain as those just mentioned, they produce a useful building material, which has been employed for the construction of many important buildings, including Scarborough Town Hall (1903). This rock is sometimes used for making grindstones. The specimen from Smaws quarries near Tadcaster, is another example of this group from Yorkshire. The rich cream colour of this stone and the fineness of its texture render it suitable for decorative building. It has been extensively employed for restoring and repairing York Minster, and Ripon Cathedral. It was also selected for rebuilding Selby Abbey Church after the disastrous fire in 1906.

[1] See p. 171.

Permian rocks occupy the coast line of the county of Durham, from South Shields to Hartlepool. Where the beds are exposed they yield a light yellow concretionary Magnesian Limestone which is useful for bridge building and similar engineering work. The specimen from the Marsden quarries, situated about three miles from South Shields, represents this rock. It indicates that the stone is not suitable for decorative building, being too hard to carve, although it is somewhat softer when freshly quarried.

GERMANY. The lower division of the Permian system in Germany, better known in that country as the Rothliegende, yields sandstones and conglomerates, the former furnishing useful building material. It occurs most extensively in the region of the Saar and Nahe south of the Hunsrück, and the chief supply of building stone is procured from this district. The yellow and brown, fine-grained, banded specimen from the Medarder quarries near Kreuznach is a representative example of the rock known to German geologists as the Kreuznach beds[1]. The specimens from quarries at Oderheim, in the Grand Duchy of Hesse, are also examples of the same formation.

RUSSIA. The Permian rocks extend eastward from Germany through Russia, as far as the Ural mountains, on the flanks of which they are largely developed. Outcrops also occur in the Government of Moscow, where the rock is dolomitic, and is quarried for building purposes. The light yellow, slightly mottled specimen from the St Koloma quarries represents this stone, which is used in and around Moscow for all kinds of structural work.

SOUTH AFRICA. Several specimens in the collection represent the rocks belonging to the great "Karroo system," a large part of which may be correlated with

[1] *Text Book of Comparative Geology*, Kayser and Lake, p. 170.

the Permian of Europe. The lowest or Dwyka series being chiefly composed of shales and conglomerates, does not yield good building stone, but the group known as the Ecca beds, which lie conformably on the Upper Dwyka shales, largely consists of sandstones which make valuable material for structural purposes. The rock from quarries near Greytown, in NATAL, deserves special attention. It closely resembles some of the best building stones from the Oolites of England, having a rich cream colour and great evenness of texture. These properties have earned for it a wide reputation throughout Natal, and it is sent to all parts of the Province for building. The new Town Hall at Durban (1910) is faced with Greytown Stone.

Rosetta Stone, which takes its name from the district of Rosetta, where most of the quarries are situated, also belongs to the Ecca series, and is much esteemed by architects in Natal. From the several specimens shewn it will be seen that it possesses a pleasing greenish grey colour in a great variety of shades. The Legislative Council Buildings in Pietermaritzburg (1902) are built of this stone from the Downing Farm quarries. Special care however is necessary in selecting this class of stone in the quarry, as some strata develop an unsightly blotched appearance when exposed to the atmosphere. This defect may be seen in the Colonial Offices at Pietermaritzburg, built in 1897–1901.

The Ecca beds are also worked for building purposes close to Pietermaritzburg. The specimen from the Town Bush quarries is a characteristic example, and shews that the rock in that district has not the green shade present in the Rosetta stone. It is much used for structural work in Pietermaritzburg. The specimen from the Effingham quarries, near Avoca, represents the stone of this group that was used for building the Standard Bank, Durban.

The building stones from the Ecca beds in the TRANSVAAL are now being freely developed and employed for structural work in Johannesburg, Pretoria, and other towns that are springing up in that Province. The examples in the collection shew that generally the colour of the rocks from these beds in the Transvaal is distinctly darker than that of the Natal stones. They are more like the rocks of the Carboniferous system in Europe. The specimen from the Kockemoed quarries at Klerksdorp is an example of the stone from this group. The Parish Hall of St Mary's Church in Johannesburg is built of it. The stone from Waring and Gillow's quarries on the East Rand is another example, and was employed for building the Carlton Hotel in Johannesburg (1906). That from the quarries in Pretoria has a pinkish shade, and was used for the foundation and lower structure of the New Law Courts in Pretoria.

The next member of the Karroo system is the Beaufort series, which lies conformably on the Ecca beds. In it sandstone occurs in abundance, and is used locally for building. The beds do not extend over so wide an area as do those of the Ecca series, and there is not the same variety of stone. The yield is chiefly from the western portion of the CAPE OF GOOD HOPE. The Beaufort sandstone, as the specimens indicate, is darker in colour than the Ecca beds, consequently it is not so popular among builders. The stone quarried at Queenstown is however lighter in colour and therefore more esteemed for constructive purposes[1]. It resembles in appearance the Oolites of Europe. The Municipal Buildings at Queenstown are built of it.

The specimen from the Kivelagher quarries near East London represents the stone from this series existing on the coast, which, as the specimen shews, has a light purple shade. It is employed for structural purposes in and around East London, the Fire Station being an example.

TASMANIA. The Upper Palaeozoic rocks of this State belong to the so-called Permo-Carboniferous

[1] *Geology of Cape Colony*, A. W. Rogers, p. 226.

system, the lower members of which are of marine origin and include limestones and grits[1].

These beds occupy a large portion of the eastern half of the Island. The limestones occur in the south-eastern area, including the Islands of Maria and Bruni, and are chiefly worked for lime-burning. The north-eastern territory consists mainly of sandstones, which appear on the flanks of Ben Lomond, and stretch in a southerly direction through the Fingal district, and the counties of Glamorgan and Pembroke, as far as Adventure Bay. These sandstones, which some geologists maintain may belong to the base of the succeeding Mesozoic age[2], yield excellent building material, which is much used locally, and is also exported in considerable quantities to Australia. The cube from the Spring Bay quarries, Pembroke County, a little to the north of Maria Island, is a characteristic example. This stone was employed for building Melbourne Town Hall.

[1] *Australasia*, J. W. Gregory, p. 462.
[2] *Geology of Tasmania*, R. M. Johnston, p. 145.

TRIAS.

BRITISH ISLES. The building stones from the Triassic system often closely resemble those belonging to the Permian, as a glance at the specimens from these two systems will prove. The colour in many instances is identical, and the texture very similar.

The rocks furnishing building material occupy a large area in central England, ranging southwards to Glamorganshire and Somersetshire and northwards through Lancashire, where they outcrop near the bold headland of St Bees, and again appear in the valley of the Eden.

The Lower or Bunter group of this system as the name implies, produces in some places red and white variegated sandstones, and in others a bright red rock.

In the south of Lancashire, the lower part of the Pebble beds of the Bunter yields useful building material. The rock from the top layer of these beds often contains many pebbles and nodules of clay, but that below is a good solid stone which is in request for structural purposes[1]. The specimen from the Mill quarries near Rainhill, is an example. This stone is being used for the internal walls of the Cathedral at present being erected in Liverpool. The Pebble beds are again exposed in the village of Woolton, about five miles east of Liverpool. Pebbles are very numerous in some of the beds, but there are others in which they seldom occur, and these afford a good building stone. The

[1] *Geology of the Country around Liverpool*, C. H. Morton, p. 84.

specimen indicates that the Woolton Stone is harder in texture than the Rainhill Stone, and the details of their crushing strain, which appear in the catalogue, confirm this. The whole of the external walls of Liverpool Cathedral are being built of it. The Wavertree Clock Tower, erected in 1884, was also constructed of it.

The bright red variety of the Bunter is represented by the specimen of St Bees Red Freestone from the Bankend quarries near St Bees Head, in Cumberland. This stone has been used for structural purposes for a long period. It was employed for Windsor Castle, as well as for other important buildings of architectural beauty in all parts of the kingdom. It was also used for constructing the docks at Barrow-in-Furness.

The quarries near Aspatria in the same county, yield a chocolate-coloured sandstone which is used locally for building churches; it has also been transported to Ireland for structural work. The hotel recently erected at Newcastle, Co. Down, is built of it.

A few miles further north there is another exposure of this division of the Trias. The cubes of red and mottled stone from the Shawk quarries, situated between Dalston and Thursby, are examples of it. Besides being used locally for building, this stone was employed for erecting the Rylands Library in Manchester (1899), and for extensions to the Cathedral in that city. It was also used for the construction of the Bute Docks in Cardiff.

The rocks of the Upper or Keuper beds of the Triassic system hold a very important position as furnishers of building material. The usual colour of the rocks is red, although there are many beds which yield a light drab stone, whilst in others the two colours are mixed, producing what is known as "Mottled Stone." The three varieties frequently occur in the same quarry. The specimens from the Hollington quarries near Stoke-on-Trent, in Staffordshire, are examples of this peculiarity. These are known as Red, White and Mottled Hollington Stone, and have been used in the construction of many important buildings. Croxdale Abbey (1128), also Hereford Cathedral (1148), are built of Hollington Red Stone. The Town Hall at Walsall (1904) and the Gladstone Memorial Church at Hawarden (1900) are constructed with Hollington White, and Trinity Church, Burton-on-Trent (1882) are built of the Hollington Mottled Stone.

The specimen from quarries at Penkridge is another example of the mottled variety. This stone was used for restoring the west front of Lichfield Cathedral in 1890. The specimens of Bromsgrove Stone from Worcestershire, and the Grinshill Stone of Shropshire, are also illustrations of the three varieties of rock quarried in these counties. Red Grinshill was employed for building the west tower of the Abbey Church of Shrewsbury, in the 14th century, and Morton Corbet Castle in the 13th century. White Grinshill was used in the construction of St Mary's Church, Shrewsbury, in the 12th century. The light red specimen from the Harmer Hill quarries, Shropshire, represents the stone that was used for the restoration of the Abbey Church of Shrewsbury in 1900.

The outcrops of the Keuper beds in Cheshire yield useful building material. The specimens of Weston Sandstone from Runcorn are examples of this group. The red variety is being used for building the interior of the New Cathedral in Liverpool. St Peter's Church, Oldham, was constructed with it, and it has also been extensively employed in the restoration of Chester Cathedral[1]. The Mottled Stone, better known in Cheshire as "Flecked Sandstone," was also used for building St Peter's Church in Oldham.

The pinkish buff-coloured cube from the Higher Bebington quarries situated in the Wirral division of Cheshire, is an example of the well-known Storeton Stone of the Keuper group, which has been employed for building over a great number of years; its name is

[1] *England's Chronicle of Stone*, J. F. Hunniwell, p. 193.

also associated by geologists with the " Fossil Footprints
of Storeton." Examples of these can be seen in the
Palæontological department of this Museum and are
well worth inspection. The " Footprint Beds" of this
rock are of little use economically, owing to their
irregular liability to break up into slabs, but there
are other beds that yield excellent building stone.
Numerous examples may be seen in Liverpool, Birkenhead, Southport, and some as far away as Rugby. In
Liverpool, the Custom House (1828), St George's
Church (1833), the Wellington Monument (1863) and
the Lime Street Railway Station (1871) are all built
of it. The Town Hall of Birkenhead, built in 1833,
and the Central Library are other examples.

The specimen of purplish red micaceous stone from
the Alderley Edge quarries is another example of rock
worked for building purposes in Cheshire, where it
has been largely used for churches.

Rhaetic beds producing stone for building purposes
occur in the south-west of England. The stone differs
from that of the underlying Keuper, in being more
calcareous, finer in texture, and widely different in
colour. The specimens of White and Green Querella
Stone from the Bridgend quarries in Glamorganshire,
are excellent examples of this beautiful building
material which was employed for restoring Llandaff
Cathedral in 1869. It is stated that the three best
beds are the White Bed, 10 feet, the Green Bed, 9 feet,
and the Grey Bed, 4 feet thick[1].

SCOTLAND. In the county of Elgin, a limited
area of Triassic rock occurs, which yields a useful
building material. It rests unconformably upon the
Upper Old Red Sandstone, and it will be seen from the
specimen shewn that it exhibits the drab and mottled
colour, so common in the building stones from the
Bunter division of this system in England; though

[1] "Geology of S. Wales Coal-field," A. Strahan and T. C.
Cantrill, *Memoirs of Geol. Survey*, 1904, p. 52.

whether this deposit can be included in that division, has not yet been decided by geologists.

Until recently this outcrop was believed to be the only example of the Triassic system in Scotland yielding building stone, but several exposures quarried in the south-west, notably in Dumfriesshire and Ayrshire, which were formerly classed as Permian, are now being identified with the Trias, although the writer of these notes is informed by the Geological Survey of Scotland that the question is still uncertain[1]. Meanwhile, in accordance with the recommendation of the Survey, the specimens are placed with the Trias.

There are numerous quarries in Dumfriesshire which furnish this excellent building material, and the specimens in the collection shew that the prevailing colour of the rock is red, differing in shade according to the locality. It has been employed for structural work in Scotland for many years, and has been sent to London, besides being exported to the United States and Canada[2] where the warm red tint of some varieties is much appreciated.

The specimen from the Corncockle quarries represents the stone that was used for building Jardine Hall, Lockerbie, early in the 19th century. The Closeburn Red Freestone, near Thornhill, was employed for St Enoch's Railway Station, Glasgow (1876). In Cambridge an excellent example will be found in the façade of Messrs Hallack and Bond's business premises on Market Hill, built in 1908.

The bright red specimen from Gatelaw Bridge quarries, also near Thornhill, represents the stone that has been largely exported to America for building purposes. The Post Office, Paisley (1893), is an example of it in Scotland. The Corsehill quarries near Annan, furnish building material of two shades of red. The darker coloured stone was employed for building the Liverpool

---

[1] Communication from Dr J. Horne, Edinburgh, 5 March, 1909.

[2] See p. 141.

Street Station of the Great Eastern Railway in London (1875). There are several buildings in New York constructed with it. The rock from these quarries, and that from Kirtle Bridge, are distinctly micaceous, as the specimens indicate, but this characteristic does not seem to diminish their value as a building stone. As previously mentioned, the majority of the beds in Dumfriesshire yield red-coloured rocks, there being few which furnish a light buff-coloured stone. The specimen from quarries at Kirtle Bridge is an example.

The same series of rocks is exposed in Ayrshire and quarried for building purposes, large quantities being sent to Glasgow. The specimens indicate that the stone from this outcrop is similar in colour and texture to that in Dumfriesshire. The example from the Ballochmyle quarries near Mauchline exhibits a few ferruginous specks, but this feature does not interfere with the utility of the stone. It was selected for constructing the Burns Monument, Kilmarnock, in 1879.

IRELAND. Triassic deposits are found only in the north-eastern part of Ireland, and the rock useful for structural purposes occupies a very limited area in Co. Down. The stone is finer-grained and lighter in colour than that of the average English Trias. One specimen it will be noticed has a distinctly yellow shade. This sandstone is in places overlain by a great sheet of basalt, and numerous sills and dykes of that rock traverse the beds and interfere somewhat with the quarrying[1]. Notwithstanding this the stone is in good demand for building purposes in and around Belfast, and as it yields kindly to the mason's chisel it is popular for decorative architectural work and ornamental sculpturing.

GERMANY. The builders of central and southern Germany are largely dependent on the rocks of the Triassic system for the building materials of their

[1] *Building and Ornamental Stones*, E. Hull, p. 270.

houses. Bunter, Muschelkalk, and Keuper, the three divisions of this system in Germany, all contribute useful stone for structural purposes, and do not lack variety, as the numerous specimens in the collection demonstrate. They are, with the exception of the calcareous Muschelkalk, composed mainly of sandy or argillaceous rocks[1], and cover a large area in the Palatinate, Baden, and Würtemberg.

Of the Bunter group, the Lower beds yield firm sandstones valuable for building material, and the Middle or Main subdivision consists of a more or less coarsely granular quartzose sandstone, also useful for constructive work. The specimens from the Freudenstadt quarries in Würtemberg are examples of these two classes of stone. That representing the Lower division shews in a marked degree the presence of rounded inclusions of dark red clay, the so-called "clay galls," which are everywhere present in these sandstones. The Upper Bunter Sandstone or "Roth" contains beds of pale-coloured quartzitic sandstone, useful for decorative work. The specimen from the Hausen quarries near Leonberg is an example.

It is the Upper division of the Muschelkalk that yields the well-known hard, light drab, encrinital limestone which is in demand for structural purposes all over central and southern Germany. The specimens from Crailsheim in Würtemberg are good examples of this shelly rock. The same division yields a hard, compact blue-grey limestone. The cube from quarries at Rottenburg, on the Neckar, is an example of it.

The Keuper, the uppermost group of the German Trias, contains pale sandstones of various colours and degrees of fineness. Its Lower division, known in Germany as "Kohlenkeuper," or the "Lettenkohl," contains yellow, brown, or greenish grey sandstones, all forming valuable building material. The light grey specimen from quarries at Crailsheim, also the light brown cube from Maulbronn, are examples.

---

[1] *Comparative Geology*, Kayser and Lake, pp. 195, 198, 200.

The Middle or Main division, the thickest of the whole of the Keuper, besides yielding the well-known gypsum-bearing shales, furnishes sandstones of diverse shades of colour, varying from a light drab-grey to a rich brown and dark purple.

The Lower beds of this Main division, known to German geologists as the "Schilfsandstein[1]," yield fine-grained rocks, yellow and brown in colour; some of a warm shade, others of a lighter tint with rich purple streaks. The specimen from the Maulbronn quarries is an instance of the latter, while those from quarries at Heilbronn, a few miles north of Stuttgart, represent the brown rock. The specimen from the Renningen quarries near Magstadt, in Würtemberg, is an example of the yellow rock found at the bottom of the "Schilfsandstein" series.

Above this, forming the Upper or Top beds of the Middle Keuper, are very light drab, nearly white sandstones, known in Germany as "Stubensandstein." They vary considerably in texture, some being fairly fine-grained, while others are very coarse quartzose sandstones. They make useful building material. The specimen from the Hirshau quarries near Rottenburg, is one of the fine variety, and that from the Nürtingen quarries, also in Würtemberg, is an example of the coarse sandstone.

The Rhaetic, which by German geologists is included in the Upper Keuper, is composed mainly of pale thick sandstones and grey shales. The former are quarried for building purposes in Würtemberg. The light brown, fine-grained specimen from the Pfrondarf quarries, near Tübingen, is a good example.

RUSSIA. The specimens from the government of Radom, in Poland, represent the building stones belonging to the Trias formation of Russia. This system is widely spread in the governments of Kielce and Radom,

---

[1] *Comparative Geology*, Kayser and Lake, p. 210.

yielding a fine building stone[1]. As will be seen by the specimens, it varies in colour from a rich dark red to a light yellow. The stone is extensively employed for building in Poland, and a large quantity of it is sent to St Petersburg for the same purpose. The red variety is in request for façades and for staircases.

GOLD COAST, WEST AFRICA. Close to Accra is an outcrop of red, coarse-grained, and somewhat argillaceous sandstone, which is said to resemble that of the Triassic system of Germany[2]. Beds of yellow rock also occur, some varieties being streaked with bright red bands, as illustrated by the specimens in the collection. The Church at Accra was built of the yellow variety, the stone being procured from a neighbouring quarry, which provided one of the specimens in the collection; the other, that with numerous red markings, is from a quarry in Horse Road, near Accra. Several buildings in the town are constructed with this stone.

SOUTH AFRICA. The Stormberg group of the Upper Karroo system of South Africa, the equivalent of the Rhaetic beds of the European Trias[3], yields a high quality of building stone. This series occupies practically the entire area of Basutoland, but as the Basuto does not aspire to a stone-built dwelling, the demand in that region is but limited.

The formation extends into the ORANGE RIVER PROVINCE nearly as far as Bloemfontein, but in many cases the rock is not well situated for quarrying. At Ladybrand however, where it forms the summit of the Platberg, it can be easily reached, and quarries have

[1] *Encycl. Britannica*, 9th ed. Vol. XIX. p. 308.

[2] "Geol. Notes of W. Africa," O. Lenz, *Geological Magazine*, Vol. VI. p. 173.

[3] There is some uncertainty as to the correlation of these beds with European deposits, as some geologists have recently referred them to the Jurassic. (See *Geology of Cape Colony*, Rogers and du Toit, p. 242.)

been developed there. The specimens from these quarries indicate that the stone makes a useful building material, being fine and even in the grain, and yellow in colour. The Schools, and other Government buildings in Ladybrand, are built of it. In South Africa this rock is known as the "Cave Sandstone." There is a specimen in the collection from quarries near Kroonstadt, where an outcrop of this rock occurs. The stone is coarser in the grain than the rock on the Platberg. It was used in the construction of the Standard Bank at Durban.

The "Forest Sandstone" of RHODESIA, which is correlated with the Cave Sandstone of the Cape and the Orange River Provinces, yields good building material. The specimens in the collection deserve special notice owing to their colour. One example is of a rich salmon tint, another of a delicate pink. The Forest Sandstones occupy most of the country north of a line drawn from Buluwayo to Gwelo, and west of the Mashonaland border, and is also extensively developed in the Tuli district. It is probable that it also covers a large area in the southeastern portion of Mashonaland, but as the country is practically unexplored, it is impossible to say more than that the sandstone undoubtedly occurs in the district. This stone is known to have been used for structural purposes for a great number of years. It was utilized even at the remote period when the Sabaeans are supposed to have occupied the country, and the Wankie ruins are constructed of it[1]. It is now employed for building in Buluwayo, the Government Buildings and Market Hall being examples[2].

AUSTRALIA. NEW SOUTH WALES. The Trias formation of this State yields a useful building material known as the "Hawkesbury Sandstone." This

---

[1] There is not, however, universal agreement as to the date or origin of the builders of these ruins.

[2] *Geology of Southern Rhodesia*, F. P. Mennell, p. 18.

rock underlies the whole of the Sydney district[1], and crops out round the harbour, forming the picturesque cliffs that fringe the coast south of Port Jackson. The stone is well adapted for architectural and ornamental work, since it can be easily sawn and carved; after being quarried it tones down to a light straw colour which it retains for an indefinite period. It is composed of small grains of waterworn quartz, with cementing medium of varying composition. The specimens from the Pyrmont and Hunter's Hill quarries, the former of which is practically in Sydney, are good examples of this rock. It has been largely employed for government, ecclesiastical, and mercantile buildings in Sydney, including the Town Hall, General Post Office, Public Library, and University[2]. The stone from the Pyrmont group of quarries varies somewhat in colour, as the specimens indicate. The University of Melbourne was built with the stone from the Hunter's Hill quarries. The Hawkesbury Sandstone covers a much larger area of the State than the Sydney district. It occupies the coastal line stretching south as far as Shoalhaven river. The specimen from the Bundanoon quarries, about 60 miles from Sydney, represents the southern beds of this series.

QUEENSLAND. The Lower Mesozoic rocks in this Colony are divided into the Burrum and Ipswich beds; the former being generally correlated with the Trias of Europe[3]. It extends along the coast from Laguna Bay as far as Blackwater Creek, and stretches inland to the Darling Downs. It yields a useful building material. The specimens indicate that the rock varies considerably in fineness. That from the Yangan quarries represents the fine-grained stone occurring near Warwick, where numerous quarries

---

[1] *Mineral Resources of N. S. Wales*, E. F. Pittman, p. 444.

[2] *Building and Ornamental Stone of N. S. Wales*, R. T. Baker, p. 41.

[3] *Australasia*, J. W. Gregory, p. 363.

exist, all of which furnish stone extensively employed for building in and about Brisbane. The cube from the Helidon quarries, situated seventy miles west of Brisbane, is an example of the coarser rock, which is also used for building. The interior of Brisbane Cathedral is constructed of it.

NEW ZEALAND. The specimen of close-grained, greenish grey sandstone, from quarries at Waikola, on the south-east coast of the South Island, represents the rock belonging to the Trias of that district, and employed for building purposes. The quarries at the present time are only worked intermittently, but large quantities of this stone were used in some of the finest buildings in Dunedin, and it is still in demand for better class work. It was employed for building the Government Insurance Buildings in that city. The pedestal of the Stuart Monument, in the Triangle, Dunedin, is another example.

UNITED STATES. The Triassic formation of America furnishes some useful building material.

MARYLAND. Beds of sandstone belonging to this system occur chiefly in the southern area of this State, the most important quarries being situated near the mouth of Seneca Creek in Montgomery County. The specimen exhibited is from one of them, and it will be seen that this rock possesses a rich reddish brown colour, and undoubtedly makes a handsome building stone. As is the case with most of the rocks belonging to this system in America, the beds occasionally contain numerous small enclosures of clay, which naturally diminish the value of the stone as a building material, causing it to weather badly and exhibit unsightly blotches on the surface when exposed to the atmosphere. Care has therefore to be exercised in the selection of the stone when quarrying. The Smithsonian Institution Building of the National Museum in Washington, erected in 1848–54, was built of it[1].

[1] *Stones for Building and Decoration*, G. P. Merrill, p. 314.

PENNSYLVANIA. The Triassic formation just described as present in Maryland also traverses the south-eastern area of this State.

The chief quarries are situated on the southern flanks of Hummelston Hill, in Dauphin County. The specimen shewn comes from one of this system near Waltonville, and is usually known as "Hummelston Brown Stone." It will be observed on comparing the specimens that this rock is much coarser in the grain than the stone from Seneca Creek in Maryland.

SOUTH AMERICA. Near the town of Colon, in the Province of Entre Rios, in the Argentine, there is an outcrop of rock belonging to the Triassic system, which is quarried and utilized occasionally in that region for building purposes, but it is of very limited area. The specimen indicates that, in colour and texture, this rock bears a striking resemblance to the building stones of the British Isles belonging to the same formation. This stone has been extensively employed for ecclesiastical buildings which have recently been erected at Injár, near Colon. The rock is said to be soft when first quarried, but rapidly hardens when exposed to the atmosphere.

The scarcity of rock, suitable for building purposes, and the general employment of bricks formed from the alluvial deposits of the immense plains of the Argentine, account for the rare occurrence of a building constructed with stone in that region.

JURASSIC.

The LIAS group of the Jurassic system, present in the British Isles, is of little importance as a source for yielding building stones. The deposits are chiefly utilized for lime and cement making. A few specimens however will be found in the collection representing this group, which are examples of those used for building purposes in England.

The specimen from the Seaton Mandeville quarries exemplifies the dark blue-grey rock of this series, which is frequently used for pillars and monumental work, and is also employed for steps and flooring. The clustered pillars supporting the vaulting of the Lady Chapel of Bristol Cathedral are composed of Dark Lias. The tracery of the windows of the new nave of the same Cathedral, built 1867–78, is also composed of this stone[1]. The steps of the Medical School in Cambridge are of Seaton Mandeville Lias.

Somersetshire is the chief county where the Liassic deposits are quarried for building. The stone from the group of quarries near Shepton Mallet, as the specimens indicate, is a light cream colour, and even in the grain. It has been extensively employed for bridge building on the local railways. The examples from the Bath Road quarries at Shepton Mallet have a stratigraphical interest, as they represent a rock which is believed to be a transition deposit between the Upper Lias and the Lower Oolite divisions.

---

[1] *England's Chronicle of Stone*, J. F. Hunniwell, p. 130.

In Oxfordshire the Liassic rocks are exposed near Banbury, and are there quarried for building. The specimen shews this deposit has a greenish brown colour. An example of this is Hornton Church, built in the 13th century.

On the Continent of Europe the Liassic rocks, and especially those that occur in Switzerland, are of considerable importance for building. There is quite a number of quarries in the valley of the Rhone, and large quantities of the stone are used for constructive engineering work on the railways in that region.

The specimen from the quarry at Villeneuve is an example of the stone used for the widening of the Simplon Railway in 1907. The specimens indicate that some of the beds of rock produce a stone that takes a good polish, and is sometimes employed as marble for decorative work. The usual colour of these rocks is a blue-grey with occasional white veins. One quarry at Villeneuve yields a dark purple stone.

The OOLITES, which follow in ascending order, make up for the deficiency of the Lias group. These rocks have furnished valuable building material during a long period in nearly every country of the world. They are well developed in England, occupying a broad belt from the Yorkshire coast on the one side to that of Dorset on the other. These limestones are utilized for building in France, Germany, Austria, Switzerland, and Italy; indeed there is scarcely a country in Europe that is not represented in the collection. There are also specimens belonging to the Jurassic system from India, Australia, New Zealand, and America.

The Inferior Oolite of Somersetshire and Gloucestershire furnishes very many important building stones, some of which have been used for structural purposes since time immemorial. The specimen of Ham Hill

Stone from quarries near Norton in Somerset, is a good example. It is a brown limestone mainly composed of comminuted shells cemented together by ferruginous matter. The two specimens in the collection are from the same quarry; they differ slightly in colour, one having a light brown, or buff tint, and known as the "Yellow Bed"; the other being much darker and distinguished as the "Grey Bed." This stone has been worked since the time of the Roman occupation, some Roman coffins made of it having been found[1].

The specimens from the Doulting quarries, near Shepton Mallet, are other examples representing the Oolites in this county. The rock is in two beds, the "Chelynch" or "Weather Bed," and the "Fine Bed." The former, as the specimen shews, is much coarser than the latter and is more suitable for exterior construction, whilst the "Fine" is better adapted for interior work. This stone was also used in old times, Wells Cathedral erected in the 13th century, and Glastonbury Abbey founded in the 12th century, being both built of it. Doulting Stone has from time to time been employed for restoration work in many cathedrals, including Canterbury, Bristol, Exeter, Winchester, and Llandaff. It was also used in the recent extension of Pembroke College, Cambridge, the stone from the Chelynch bed being employed. The rock from the Fine bed has been used for the restoration of the interior of Canterbury Cathedral. The specimen from the Cary Hill quarries is an example of the stone used for the construction of Castle Cary Church.

In Gloucestershire these rocks yield a stone which is sometimes lighter in colour and finer in grain than that of Somerset. The specimens from the Painswick Hill quarries are examples, and may be specially mentioned as being almost identical in colour and texture with the celebrated Caen Stone, which will be described later when reference is made to the Foreign

[1] *Memoirs of Geol. Survey of U. K.* Vol. IV. 1894, p. 475.

Oolites. Painswick Stone was used for restorations in Gloucester Cathedral in 1894, and for building the interior of the new War Office in London.

The specimen of Nailsworth Stone from the Ball's Green quarries, it will be observed, is almost identical with Painswick Stone both in colour and texture. Until recently the quarries have not been fully developed and the stone has been used only locally. The Chipping Campden Stone, another example of these rocks, has been used in building the West Front and New Tower of Llandaff Cathedral.

The stone from the Guiting quarries in Gloucestershire differs, as the specimens shew, both in colour and texture from those just mentioned. It possesses a rich orange colour and is coarser in the grain than the Painswick rock. This stone was used for building Guiting Church in the 15th century, and St John's College and Christ Church at Oxford.

The specimens from Leckhampton represent the Inferior Oolite quarried near Cheltenham. They are sometimes called the "Cheltenham Beds." One specimen is an example of the "Weather Bed," a tough-grained ferruginous Oolite, while the other resembles the fine grained Painswick Stone. Both these rocks are worked from the same quarry and are largely used for building in Cheltenham and Gloucester.

The Inferior Oolite of Northamptonshire, Rutland, and Lincolnshire, which belong to the Lincolnshire Limestone stage, yields stone which has been famous for building from very ancient times.

The Barnack Stone, which occurs at the base of these limestones[1], in Northamptonshire, claims precedence, as regards antiquity, for building purposes. No doubt quarries in this formation were among the earliest worked. It is interesting to read of the freestone of "Barnack, Barneck, or Barnock, from whence King Wolfere, A.D. 664, built Peterborough (Cathedral).

---

[1] *Memoirs of Geol. Survey of U. K.* Vol. IV. 1894, p. 477.

Here eight pair of oxen were required to move one block[1]." The quarry at one time must have belonged to the See of Peterborough, for it is recorded that William the Conqueror issued his precept to the Abbot of Peterborough, commanding that the Abbot and Convent of St Edmund in Suffolk should be permitted to take sufficient stone for the erection of their church from the quarries of Barnack in Northamptonshire, granting at the same time an exemption from the usual tolls chargeable on its carriage from that place to Bury St Edmunds[2]. Barnack Stone was held in great reputation as a building material between the 11th and 12th centuries, being then known as the Freestone from the "Hills and Holes of Barnack[3]." It would appear that these quarries became exhausted before the 15th century, since in Barnack Church itself the alterations of that period are in a different stone, and not in the old Barnack Stone of which the church is built[4]. The inclusion of Barnack Stone specimens is a deviation from the usual procedure observed in regard to this collection. This was that all specimens must represent building stones which can still be obtained and are thus related to the economics of the present day. But the fact that Barnack Stone was the chief material employed in the erection of some of our noblest abbeys and cathedrals in East Anglia, and is said to have been largely employed in the erection of the early colleges of Cambridge, gives it a special interest, and justifies a breach of the rule. Although no documentary evidence is available, tradition has it, that Barnwell Abbey in Cambridge, Thorney Abbey, also Ramsey Abbey in Huntingdonshire, were all built of Barnack Stone. So too is Ely Cathedral (1174–89). Barnack Stone also was undoubtedly extensively used in building some of the

[1] *Character of Moses*, Townsend, p. 151.
[2] *Architecture of England*, T. Rickman.
[3] *Handbook to Eastern Cathedrals*, J. King, p. 56.
[4] *Ibid.*

oldest colleges in Cambridge, but the absence of documentary evidence, combined with the difficulty of
identifying the stone, renders it impossible to declare,
with any certainty, where it was employed.   The south
wall of the church of St Benedict is said to be built of
large blocks of Barnack Stone.   It is recorded that in
the year 1579, 182 loads of stone were sent from Barnwell Abbey by Mr Wendy, the lay-improprietor of the
Priory, also 146 tons from Thorney Abbey, by the Earl
of Bedford, for building Corpus Christi College[1], and a
considerable sum of money was paid by Dr Caius for
stone from Ramsey Abbey, for erecting buildings at
Gonville and Caius College in 1573[2].   The three shafts
of the Piscina in the south wall of the chapel of
St John's College, which were discovered in an ancient
building to the north of the old chapel, are, it is stated,
composed of Barnack Stone[3].   There are specimen
cubes of stone in the collection, from Ely Cathedral,
Bury St Edmunds, Ramsey, Thorney, and Barnwell
Abbeys, all believed to be from the original quarry at
Barnack, and although it will be seen that there is a
similarity in their general appearance, they are not
identical either in colour or texture.   But the influence of the weather during so long a period, under
conditions which may not have been equal, can account
for the variation in colour, and the dissimilarity in
texture is not to be wondered at, seeing we have
evidence of inequalities of degrees of fineness of stone
constantly occurring in the same quarry at the present
day.

Weldon Stone, quarried in Northamptonshire, also
belongs to the Lincolnshire Limestone group. It occurs
at a somewhat lower horizon than the well-known

---

[1] *Architectural History of Cambridge*, Willis and Clark, Vol. I.
p. 290.
[2] *Ibid.* p. 174.
[3] "Geology of St John's College Chapel," T. G. Bonney, *Eagle*,
Vol. XXVIII. p. 175.

Ketton Stone of Rutland[1], and has long been employed
for building operations. There is good reason to believe
that Weldon Stone was used to build old St Paul's
Cathedral in London, which was destroyed by fire in
1666[2]. As previously mentioned, when describing the
Magnesian Limestones of the Permian system, Weldon
Stone was employed for building the upper portion of
King's College Chapel in Cambridge, and as the stone
work here is said to have been completed in 1515, it
must have been previous to this date. It is believed
that the reason Weldon was substituted for the Magnesian Limestone of Huddlestone was because King
Henry VII presented a quarry, yielding Weldon Stone, to
King's College. There are many examples of this stone
in Cambridge. It was employed for building Trinity
College Chapel (1560–61); Caius College (1564–73); an
extension of Jesus College (1638) and Clare Hall (1641)[3].
It was also used for the recent extensions at Sidney
Sussex, and Gonville and Caius Colleges. It may be seen
from the specimen that this stone is more shelly and
open in its texture than most Oolites. The latter
characteristic is said to enhance the value of this stone,
rendering it better able to resist the destroying
influence of frost than the close-grained rocks[4].

The stone from the Wansford quarries, Northamptonshire, as the specimen shews, differs considerably in colour and texture from the ordinary
Lincolnshire Limestones of this county, being a brown
arenaceous stone, darker in colour and more uneven
in the grain than most of this group. It was used for
restoring Thorney Abbey in 1850.

The Glendon Stone, quarried a little north of Kettering, is somewhat similar in colour to the stone just

---

[1] *Memoirs of Geol. Survey of U. K.* Vol. IV. 1894, p. 477.
[2] Bridge's *History of Northamptonshire*, published in 1791,
p. 354.
[3] *Architectural History of Cambridge*, Willis and Clark, Vol. I.
p. 94.
[4] See p. 6.

mentioned, but as the specimen indicates, it is much finer in grain. This stone is much used in Kettering for building. All Saints' and St Mary's Churches are both constructed of it.

The specimen of King's Cliffe Stone, a light yellow shelly limestone of this group, would not have found a place in this collection, as the quarry has long since ceased to exist, but from the fact that a local interest is attached to it from its use for building the Gate of Honour at Gonville and Caius College, Cambridge, in 1573. The porch of Corpus Christi College Chapel, built in 1583 (pulled down 1823), was largely composed of King's Cliffe Stone, also the fountain in the Great Court of Trinity College and the building forming the kitchen, erected in 1605. The specimen in the collection was procured from an old building in the village of King's Cliffe.

Some of the Lincolnshire Limestones quarried in the county of Rutland, which occur at or near the top of the group, belong to what is usually known as the Ketton Stone. This is a trade name, recognised by architects and others, notwithstanding that the original Ketton Stone was exhausted several years ago. There are many examples of the original Ketton in Cambridge. The New Hall of Clare College (1638); the New Chapel of Emmanuel (1667–8); extensions of Peterhouse (1736); and the Library of Trinity (1776–88) are all built of it. The clunch walls of Christ's College were cased with Ketton Stone in 1714–40[1]. Downing College, built 1818–20, is believed to be the latest example of the use of Ketton Stone in Cambridge. Two specimens labelled "Ketton Stone" are in the collection. One is an example of the original Ketton, which was procured from the walls of a chapel in Cambridge, recently demolished to make room for a more modern edifice. The other represents the New Ketton Stone, at present worked close to the site of the old quarries. This stone was used for the

[1] *Architectural History of Cambridge*, Willis and Clark, Vol. II. p. 224.

new Wing of Pembroke College, Cambridge, 1907. It is also being used for the repairs of York Minster. Exeter Cathedral has been restored with it. The two specimens are very similar in appearance, although the New Ketton Stone is not quite so fine-grained as the original, neither has it so warm a pink tint as the old Ketton possesses, but this difference in colour may have arisen from weathering.

The specimens of Edithweston Stone and Casterton Stone are also examples of this group of rocks. The Edithweston specimen indicates that the stone bears a striking resemblance to the original Ketton as regards texture, although it lacks the pink tint of the older stone. The quarry is about 500 yards from the site of the original Ketton quarry. Edithweston Stone was selected for constructing the Prince Christian Memorial at Windsor. The quarry yielding Casterton Stone is about one mile to the east of the site of the original Ketton quarry. The specimen shews that in its physical character it resembles the rest of this group, except that it is slightly coarser in the grain. This stone has been employed in the building of Pembroke College, Cambridge.

Clipsham Stone, as the specimen indicates, is less even in the grain than most of the Ketton group, and has a few shells scattered through the mass. This stone has been employed for building in many of the colleges in Cambridge. It was used together with Weldon Stone for the upper part of King's College Chapel[1], and some of that stone worked into the fountain in the Great Court of Trinity College, Cambridge, was brought from Clipsham[2]. The plinths and cornices of the Sedgwick museum, and those of the new buildings of Pembroke College, erected in 1870-5, are made of it. Clipsham Stone has been employed for restoring Ely, Peterborough and Norwich Cathedrals.

[1] *Architectural History of Cambridge*, Willis and Clark, Vol. I. p. 489.
[2] *Ibid.* Vol. II. p. 629.

The Ancaster Stone also belongs to this group. It is quarried in Lincolnshire, and as the specimens shew, there is more than one variety, both as regards texture and colour. The stone from the Weather Bed is of a harder nature than the Brown Bed, being much more crystalline. For this reason the former is generally employed for that part of a building which is likely to be exposed to the abrasion caused by traffic. An example of this adaptation can be seen in the south wall (that facing Downing Street) of the new Medical Schools in Cambridge. The base courses of this wall above the flagging are built of Ancaster Weather Bed Stone, which can easily be distinguished by its mottled appearance, the colour being alternately grey and yellow. The upper portion of the building is constructed with Ancaster Brown Bed Stone, which is of a uniform brown colour. The specimens in the collection shew these variations. Ancaster Stone has also been used for building Gonville and Caius and Pembroke Colleges in Cambridge, also the interior walls of St John's College Chapel[1].

Lincoln Cathedral, built in the 12th century, is constructed of a stone belonging to the Lincolnshire Limestone group. The quarries from which the stone was procured are close to the cathedral, but they have long since been exhausted. The specimen has however been thought to have sufficient interest locally to entitle it to a place in the collection. It was procured while some recent repairs were being executed. The sculptures of the Cathedral, most of which were carved in 1209–35, are in a very good state of preservation[2].

The specimen from the Haydor quarries, near Grantham, is an example of the stone that has been extensively employed for restoring Lincoln Cathedral.

The specimen of calcareous sandstone from the Swerthow quarries, near Aislaby, in the North Riding

---

[1] "Geology of St John's College Chapel," T. G. Bonney, *Eagle*, Vol. XXVI. p. 175.

[2] *England's Chronicle of Stone*, J. F. Hunniwell, p. 163.

of Yorkshire, represents the rock from the Lower Estuarine beds of the Inferior Oolite known to geologists as the "Great Sand-rock." It is largely used as a building stone in Eskdale and other districts of Yorkshire. Whitby Abbey, erected in the 13th century, is built with it. The north, or Cockerell's wing, of the University Library in Cambridge, erected in 1837, is another example.

Passing on to the Great or Bath Oolites, we have stone quarried for building that is contemporaneous with the founding of the town from which it derives its name. The massive structures, around the hot mineral springs of Bath, erected by the Romans nearly 1800 years ago, are built of Bath Stone, so also is the stately Abbey Church. Malmesbury Abbey, built in the 8th century, was constructed with Bath Stone from the Box Ground quarry, formerly known as the Haslebury quarry. It is interesting to record that repairs and restoration are now going on at the Abbey with stone from the same quarry. This is a truly remarkable evidence of historical continuity, so far as building stone is concerned.

John Aubrey, a Wiltshire antiquary, living in the 17th century, in his "Description of Wiltshire" under the heading "Haslebury (in Box)" makes the following remarks, "Haslebury Quarrie is not to be forgott, it is the eminentest freestone quarry in the West of England, Malmesbury, and all round the country of it. The old men's story that St Aldelm riding over there, threwe downe his glove, and bade them dig, and they should find great treasure, meaning the quarry." The pious Abbot of Malmesbury, St Aldhelm, built the little church at Bradford-on-Avon, which is said to be one of the most perfect specimens of Saxon ecclesiastical architecture in England. The stone for this was also quarried at Haslebury. As will be seen from the label on the specimen, the stone is now known in the building trade as St Aldhelm Box.

The Bath Oolites are very varied in texture, as the numerous specimens in the Museum indicate. While the Box Ground Stone is coarse-grained, the Corsham Down Stone is very fine in texture.

All the rocks belonging to this group, when freshly quarried, are soft, and easy to work; they are of a warm yellow colour, but when exposed to the atmosphere, they harden and turn whiter. This hardening on exposure is common to all Oolites.

Bath stone is not wrought in open quarries; the workings are underground, and are reached either by adits from the face of the escarpment, or by inclined shafts. It has been found that larger-sized blocks can be obtained in underground workings, than from open quarries. The workings are sometimes very extensive. The Bath stone firms, who, as will be seen in the catalogue, have presented several specimens to this collection, are said to have at least sixty miles of underground excavations. The method of working is the reverse of that usually adopted in coal mining: there is no undercutting of the stone, operations being commenced by cutting away the roof. It is interesting to observe the enormous piles of blocks of all sizes, stacked on the ground adjacent to the mouth of the quarry shaft, undergoing the process of hardening.

The Roman Catholic Church in Cambridge, built 1885-90, is a striking example of the use of Bath Oolite, its delicate carvings and sculptures being indicative of the value of this stone for decorative purposes. The exterior portion is built with stone from Corsham Down, and the internal walls are constructed of material from Darleigh Down. The specimen from the Stoke Ground quarries represents the Bath Oolite stone employed in building Buckingham Palace, and the cube of Farleigh Stone represents that used in Lambeth Palace. Combe Down Stone was employed in restoring Henry VIIth Chapel Westminster Abbey (1808). Bath Stone has been exported to all parts of the world for building purposes. That for the New Town Hall in Cape Town came from the Monks Park quarries. The specimen of this variety shews it to be finer in the grain than most of the Bath Oolites.

The specimen of Taynton Stone is an example of the Great Oolites of Oxfordshire, which are quarried near Wychwood. This stone was largely employed for building the early colleges in Oxford (St John's College is an instance), most of which were erected in the 12th, 13th, and 14th centuries. These are still in fairly good preservation, the mouldings being sharper and less weathered than those of the buildings erected in the 17th and 18th centuries[1]. The remarkable false bedding of the Taynton Stone has led some to suggest that this stone when used for building is best face-bedded, contrary to the usual method of horizontal bedding. It was employed for constructing Eton College 1448-50[2]; also for the interior of St Paul's Cathedral.

The Limestone Beds of the Middle or Oxford division of the Oolitic group, suitable for building purposes, occur over a limited area only. The chief source of supply was in Oxfordshire, but this rock has almost entirely ceased to be used as a building material, although at one time it was extensively employed in that county.

By far the largest proportion of the colleges and ecclesiastical buildings in Oxford, erected during the 18th and 19th centuries, were built with stone from the Upper Corallian beds[3], which was quarried at Headington Hill near Oxford. Visitors to that city cannot fail to notice the prematurely ancient appearance that these buildings present. This unfortunate

---

[1] *Memoirs of Geol. Survey of U. K.* 1894, Vol. IV. p. 479.
[2] *Architectural History of Cambridge*, Willis and Clark, Vol. I. p. 397.
[3] *Memoirs of Geol. Survey of U. K.* 1894, Vol. IV. p. 479.

weathering of comparatively modern structures is attributable, in the opinion of some, to the stone being "face-bedded" instead of being placed with the planes of the bedding in a horizontal position, a grave error which has been already referred to[1]. There is however evidence of the stone decaying, both on the natural bed, as well as on the face; consequently the disintegration cannot be entirely owing to the cause just mentioned. There are two specimens of this rock, one a hard, semi-compact stone, which is usually known as "Headington Hard," and the other, a more cellular and lighter coloured rock, generally distinguished as "Soft Headington Stone." Both rocks are almost entirely made up of comminuted shells, sea-urchins, corals, and other marine organisms. In the buildings of Oxford, the hard stone has been chiefly employed for plinths, steps, and base structures, the softer stone having been used for the superstructures. There is not any evidence of this stone having been used for building in Cambridge. The walls of the Ante-Chapel of Eton College were built entirely of it in 1479, but the existence of Headington Stone is scarcely to be suspected as it was concealed by a casing of Bath Stone in 1876–7[2]. As previously mentioned, the quarries are now practically disused. The specimens in the collection were taken from stone which was removed from the Clarendon Press buildings in Oxford, erected in 1712, and replaced with a stone from Rutlandshire, belonging to the Inferior Oolite.

Passing upward, we come to the rocks belonging to the Upper or Portland group, which derives its name from Portland Island on the Dorsetshire coast, where are the chief quarries, all yielding a valuable building material. A large portion of the Island is crown property, and has been so for very many years.

[1] See p. 9.
[2] *Architectural History of Cambridge*, Willis and Clark, Vol. I p. 397.

The formation includes three distinct beds of Oolite. The Base, or bottom bed, the Whitbed, and the Roche, which last forms the top of the Portland limestone[1].

The stone of the Basebed is of a whitish-brown colour when first quarried, but becomes paler when exposed to the atmosphere. The Whitbed is slightly darker in colour. The Oolitic structure of both rocks is very marked, as can be seen from the specimens. As the grains are very small, both the Basebed and the Whitbed must be classed as fine-grained stones.

The Roche is a coarse tough and vesicular rock, being full of holes, sometimes rather large, from which fossil shells have been removed by the percolation of acidulated water, but the finer-grained part of the stone is very similar to that of the Whitbed. In many cases the interior casts of the shells remain. The stone from this bed is largely used for constructive engineering work, for which it is very suitable, but it is obviously not adapted for any fine decorative architectural work. An example of the employment of this rock for construction, is the waterfall from the Serpentine river in Hyde Park, London. It was also used for Portsmouth Docks.

The rocks from the Basebed and Whitbed, on the other hand, are peculiarly adapted for all sorts of structural work, both ornamental and otherwise, and enjoy a good reputation among architects and builders.

On the Island of Portland there are the remains of several buildings, constructed with material from neighbouring quarries, which date many centuries back. Henry VIII erected a castle on the Island, also one on the opposite shore near Weymouth, and the stone of which they are constructed has not undergone any decomposition worth notice. James I appointed Inigo Jones his chief architect and surveyor-general; under this appointment he had to survey the crown lands at Portland, and his discrimination very soon

---

[1] *Memoirs of Geol. Survey of U. K.* 1895, Vol. v. p. 198.

led to the introduction of Portland Stone for buildings in and about London. In 1619 he employed Portland Stone for the construction of the great Banqueting Hall in Whitehall, and a little later (1631), in the restoration of the north and south fronts of St Paul's Cathedral, and for his new grand Corinthian Portico at the west front[1].

Portland Stone was the material largely employed by Sir Christopher Wren and other architects, for rebuilding London after the great fire of 1666, which destroyed many old and interesting buildings, including St Paul's Cathedral. Wren used Portland Stone for the erection of the new Cathedral. He seems to have been influenced in his selection by the size of the blocks obtainable. In the life of Wren, his son writes, when referring to the stone used for the Cathedral, "All the most eminent masons of England were of opinion that stone of the largest scantlings were there (Portland) to be found, or no where. An inquiry was made after all the good stone that England afforded; and next to Portland, Rock Abbey stone, and some others in Yorkshire, seemed the best and most durable; but large stone for the Paul's works was not easily to be had even there[2]." In 1684 the quarries of the Isle of Portland were devoted exclusively to the rebuilding of St Paul's, and by the King's command no stone was allowed to be taken away from them without the express order of Sir Christopher Wren[3].

A document exists, which contains the contract for raising Portland Stone for St Paul's, dated 4 Feb. 1675, and shews how Wren attended to quality. It sets forth that "Thomas Wright citizen and Freemason of London, is at his owne charge, within eleven months after date to deliver on board vessels provided by the Commissioners three thousand tons of well conditioned Portland

[1] "Stone for Building," C. H. Smith, *Royal Institute of British Architects*, Vol. I. Pt. 2.

[2] Wren's *Parentalia*, Pt. 2, Sec. VI, p. 288.

[3] *St Paul's Cathedral*, W. Longman, p. 5.

stone, part in ordinary blocks and part in scantling stones, to be by him well scappell'd. [To scapple a stone is to reduce it to a straight surface without working it smooth, usually done by chopping immediately it is dug in the quarry.] No one stone is to exceed three tons, all is to be of the best, no flinty or rag beds to be seen on the face of any scantling or ordinary block, and none is to be sent that is tainted with any salt bed, and if any such happen to be sent, that it be utterly rejected[1]."

The Parliamentary Report on the stone suitable for the Houses of Parliament, mentioned when referring to the Magnesian Limestone as a building material, states that several frustra of columns, and other blocks, which were prepared at the Portland quarries, at the time of the erection of St Paul's Cathedral, and were then lying near the old quarries, were seen to have preserved the original chisel marks, in spite of their exposure to the damp atmosphere of the sea for nearly two centuries[2].

A bard of that period has sought to immortalize the facts just recorded, by leaving for posterity a quaint little book of poetry entitled "Portland Stone in Pauls Churchyard. Their Birth, their Mirth, their Thankfulness, and Advertisement. written by Hen. Farley, a Freeman of London, who hath done as freely for Free Stone within these Eight yeares as most men, and knows as much of their mindes as any man. Buy or goe by. Printed by G. E. for R. M., and are to be sold at the great South Doore of Paules, London 1622." The opening verses in it may be interesting:—

"Ere since the Architect of Heauen's fair frame
Did make the World, and man to vse the same ;
In Earth's wide wombe, as in our nat'rall bed,
We have beene hid, conceal'd, and covered,
Where many thousand ships haue sailed by,
But knew vs not, and therefore let vs lye.

[1] *Church Times*, June 12, 1896.
[2] *Report of the Building Stone Commission*, 15th July, 1839.

> Till at the last, and very lately too
> (Some Builders hauing building to doe,
> And time being come we could no longer tarry,
> But must be borne from out our earthly quarry),
> We were discouer'd, and to London sent,
> And by good Artists tryde incontinent;
> Who (finding vs in all things firme and sound,
> Fairer and greater than elsewhere are found;
> Fitter for carriage, and more sure for weather,
> Than Oxford, Ancaster, or Beer-stone eyther),
> Did well approue our worth aboue them All
> Vnto the King for service at Whitehall."

Many other buildings of importance in London are built of Portland Stone, including the British Museum (1753), General Post Office (1829), Somerset House (1776—92)[1]. Smeaton employed Portland Stone for building Eddystone Lighthouse[2]. It was used by Labelye in 1760 for the construction of Westminster Bridge.

There are several examples of Portland Stone in Cambridge. The Fitzwilliam Museum (1837)[3], the Senate House (1730), the façade of the University Library (1755), and the Fellows' Buildings of King's College (1724) are all built of it. The handsome façade of the Offices which the Cambridge Gas Co. has recently erected in Sidney Street is of Portland Stone from the "Whitbed." Although the building material produced by the Upper, or Portland Oolites is usually associated exclusively with the stone quarried in the Island of Portland, and its immediate neighbourhood, specimens in the collection representing that formation demonstrate that in other districts it is capable of furnishing a useful building stone.

Oolitic limestones, belonging to the Portland Beds also occur in the vale of Wardour, Chilmark Stone

---

[1] *Memoirs of Geol. Survey of U. K.* 1895, Vol. v. p. 311.
[2] See p. 18.
[3] The Foundation Stone of Fitzwilliam Museum laid on 2nd Nov. 1837 was a block of Portland stone weighing nearly 5 tons.—Extract from *John Bull*, Nov. 6, 1837.

being a notable example[1]. This rock is well known as
a building material, the use of which dates back to
a very early period. Chilmark Stone was employed for
building Salisbury Cathedral in the 13th century[2].
Rochester Cathedral is also partly constructed of it,
and it was used for restoring Westminster Abbey
1873—75.

Brief reference will now be made to the specimens
representing the estuarine Purbeck Beds, which are
generally found above the Portland group. This
formation has two subdivisions, both of which yield
a stone useful for building purposes. That from the
upper is known as "Purbeck Freestone," and the lower
is often distinguished as "Purbeck-Portland Stone."
Both these rocks are quarried in Dorsetshire, in the
neck of land east of the Island of Portland, sometimes
known as the Isle of Purbeck. The Purbeck Freestone
quarries are practically mines, worked by means of an
inclined shaft, the stone being drawn up by a horse
windlass. As will be seen in the specimen, this rock
is composed almost entirely of shelly material in
long irregular layers. Both these rocks have been
utilized for building construction for many years.
The Seacombe quarries, from which a specimen is
exhibited, have been working for upwards of a century.
The stone is extensively employed for kerbs and steps.
The staircase leading to the upper stories of the
Sedgwick Museum, Cambridge, is Purbeck-Portland
Stone. The lighthouse at Margate is built of it, also
part of Dover Pier.

There is little doubt that the restoration of London,
after the great fire of 1666, inaugurated a period of
great demand for Portland, and other Oolitic rocks,
and the reputation then gained, continued, so that the
stone was largely used by successive architects during
the 18th and 19th centuries, although the decision

[1] *Memoirs of Geol. Survey of U. K.* 1895, Vol. v. p. 312.
[2] *England's Chronicle in Stone*, J. F. Hunniwell, p. 126.

of the Royal Commission[1] called attention to stone belonging to the Permian system. This deviation of opinion appeared however to be only temporary; for we gather from the report of the second Royal Commission, appointed in 1861, to investigate the reason of the decay in the stone of the Houses of Parliament, that, after careful consideration, they thought Portland Stone more suitable than Magnesian Limestone, to withstand the smoky atmosphere of London, and other large cities[2].

Several specimens of Jurassic rocks from quarries on the Jura Mountains will be found in the collection. They are chiefly hard compact limestones, capable in many instances of taking a good polish, and will be described more in detail when reference is made to Switzerland.

AUSTRIA. The specimen of Oolitic Limestone from the quarries near Budapest, is a typical example of the stone from the Jurassic system, which outcrops through the Tertiary deposits on the plains of Hungary. As the specimen shews, this division of rocks somewhat resembles the Bath Oolites of England both in colour and texture. It is employed for all sorts of structural work in Budapest and its neighbourhood.

FRANCE. The celebrated Caen Stone of Normandy belongs to the Bath Oolites. The specimens shewn indicate that this rock is more closely allied to the English Oolites than to the general type of Jurassic rocks in France. Like the Bath Stone, it is of a yellowish white colour, although finer in texture and softer when first quarried, but hardening rapidly in the air, and turning white on exposure. The quarries are of great antiquity, and large quantities of stone were imported into England at a very early date. Instances of its use are known during the Norman period. It

[1] See p. 145. [2] *Parliamentary Reports*, 1861, Vol. xxxv.

was employed in the building of St Albans Cathedral in the 11th century. William the Conqueror made Paul of Caen, Abbot of St Albans in 1077, and it was he who built the Cathedral Church of Caen, so that doubtless it was by his command that Caen Stone was brought over to England for St Albans, thus inaugurating its use in England. The old Cathedral of St Paul's in London, rebuilt in 1087—8, was partly constructed of Caen Stone brought from Normandy[1]. It is present in Canterbury Cathedral; the buildings erected after the fire of 1175, are believed to have been constructed entirely of Caen Stone, imported for that purpose by the French architect, William of Sens, to whom the Monks of Canterbury entrusted the restoration of the Cathedral. It was also used in Westminster Abbey which dates from the 13th century. It is recorded that 407 tons of Caen Stone were employed for the building of Eton College, in the year 1443[2]. Buckingham Palace is also partly built with this stone. In England it has been chiefly employed for internal decorative work, the close and soft nature of the stone making it specially suitable for such purpose. Of the numberless instances of its application for delicately carved ecclesiastical ornamentation it may be enough to cite the elaborate altar screen in Durham Cathedral, erected in 1380[3], and the pulpit in the nave of Chichester Cathedral, raised to the memory of Dean Hook.

Beds of the Middle division of the Jurassic system occur in the Department of Meuse, yielding a singularly attractive fine-grained, white stone. The specimen from the Euville quarries is a good example. The stone was used for building the Hôtel de Ville, also the Austerlitz Bridge, in Paris.

The formation extends in a southerly direction, and is exposed in the Departments of Yonne and Côte d'Or,

[1] *St Paul's Cathedral*, W. Longman, p. 5.
[2] *Architectural History of Cambridge*, Willis and Clark, Vol. I. p. 386.
[3] *England's Chronicle in Stone*, J. F. Hunniwell, pp. 146, 225.

where there are numerous quarries yielding valuable building material. The specimens from this group indicate that the rock has a delicate colour, and an even texture, rendering it suitable for highly decorative work. Many of the most important buildings in Paris are constructed of it, such as the Railway Stations of St Lazare and Lyons, and the handsome offices of the Credit Lyonnaise in Paris. The material came from the quarries at Chassignelles, and Massangis, in Yonne. The new Museum at Geneva in Switzerland is also built of "Massangis White Oolite." The specimen from the Villars quarries represents the stone quarried in Côte d'Or, which was used for part of the Lyons Railway Station in Paris. It indicates that this rock is very fine, and even in grain, admitting of a good polish.

Extending westward the same formation appears in the Department of Vienne, and several quarries exist in the neighbourhood of Poitiers, which yield a useful building stone. Both the Middle and Lower divisions are developed in this Department, the former producing a fine even grained white stone, which is in demand for building in Paris, as well as other districts. The Post Office, Angoulême, and the Palais de Justice, Brussels, are representative examples; several important buildings in Bordeaux and Poitiers also are constructed with it. The rock from the Lower division of the system, as the specimens indicate, is not so white as that from the Middle, being a light grey shade. The cube from the Chauvigny quarries is an example. This stone has been employed for many important buildings and engineering undertakings in Paris, including Alexander III Bridge, the New Railway Station, Quay d'Orsay, and the Museum of Natural History in the Jardin des Plantes.

In the adjoining Department of Charente, there is an outcrop of the Middle division of the system which yields a white, tough, rather shelly rock. The specimen from the Vilhonneur Raillats quarries is an example.

This stone has earned a reputation for ecclesiastical architectural work, for instance, the Cathedrals of Angoulême, Bayonne, and Pau, are built with it. Passing eastward the same division outcrops in the Department of Drôme. The cream-coloured compact specimen from quarries at Lens is an example.

Nearing the line of the Mediterranean coast, extensive beds, also of the Middle division, are exposed in the Department of Alpes-Maritimes, yielding compact rocks, varying in colour from a light cream to a grey and dark buff. The specimens from quarries on Napoleon's Plateau, a little to the north of Grasse, represent the cream-coloured and the grey varieties. The new premises of the Bank of France in Grasse is an example of both.

The specimens from the quarries of La Turbie, near Nice, illustrate the dark buff-coloured rock of this series, which it will be observed takes a good polish. It was used for the Museum of Oceanography at Monaco, recently erected by the Prince of Monaco.

In Hte. Savoie, on the slopes of the Salève Hills, near Geneva, there is a group of quarries which yield a compact rock belonging to the Upper division of this system, varying in colour from a light grey to a dark blue-grey. Large quantities of this stone are brought into Geneva for building. Owing to the compact and hard nature of the rock it is very suitable for foundations and the lower parts of superstructures, where abrasion, caused by traffic, is liable to occur; the upper parts of buildings in the district being usually constructed of the softer and more easily worked rocks, belonging to Tertiary formations, which will be mentioned later[1].

GERMANY. The slopes of the Swabian Alps in Würtemberg are composed of rocks belonging to the Upper division of that part of the Jurassic system, which is known to German geologists as "Franko-

[1] See p. 230.

Swabian[1]." This formation yields a hard sub-crystalline limestone, which is quarried and utilized for building in Stuttgart, and other towns in Würtemberg. The specimen from quarries near Schneistheim is an example.

ITALY. Near Brescia, at the foot of the southern slopes of the Alps the Jurassic Limestone yields a stone known as "Boticino," which is highly esteemed by the Italians for decorative building. The compactness of its texture, and its delicate dove-coloured tint, commend it to the architect and sculptor. Large quantities of this stone are at present (1910) being employed in Rome for the construction of the monument to the memory of Emmanuel II.

There are other specimens in the collection from the same region, which are usually classed as Marbles, although they are employed locally as ordinary building material. Examples of the use of these rocks as decorative stone can be seen in the Entrance Hall of the New Government Offices at Westminster.

PORTUGAL. Except the granitic rocks forming the lofty Serra da Estrella, in the Province of Beira, the building stones of Portugal are practically limited to limestones belonging to the Upper Jurassic and Lower Cretaceous systems. These occur chiefly within a short distance of Lisbon. Numerous quarries are worked on the slopes of the hills south of Cintra, the pretty summer resort of the upper classes of the Portuguese metropolis. Most of these limestones, although geologically speaking they must be considered as marbles, are largely employed for ordinary building in and about Lisbon, and therefore claim a position in this collection. The specimens include several examples of variegated marble-like rocks, among which the cube labelled "Ecarnado" (Emperor's Red) is worthy of special notice. This beautiful decorative rock has

[1] *Text Book of Comparative Geology*, Keyser and Lake, p. 239.

a world-wide reputation for its delicate pink tint, which is almost unique.

The Cambridge student may be interested to learn that it was intended by Sir Gilbert Scott, the architect of St John's College Chapel, that the altar slab should be made of Emperor's Red, and Dr Bonney, in his description of the "Geology of St John's College Chapel," explains why this desire was not carried out[1].

SWITZERLAND. The Swiss flanks of the Jura range yield material which is extensively employed for building purposes in that country, and is also exported to Germany and Italy.

Geneva draws part of its supply of Jurassic rocks from numerous quarries in Canton Vaud. The specimen from the Trofon quarries illustrates the stone from this group. On the slopes above Lake Neuchâtel there are numerous quarries which yield a light drab compact limestone from the Upper division of the system. This is employed for building in the town of Neuchâtel as well as other parts of the Canton. The stone is chiefly used for foundations, and for the lower parts of structural work, where there is a liability to wear and tear through traffic; while the softer Molasse of the Miocene system is employed for the upper parts of the buildings, as it is easier to dress, and the quarries are more accessible[2], most of those in the limestone being at a considerable height above the level of the lake. The specimen from the Chaumont quarries is a good example. These quarries are situated about half-way up the Chaumont Hill which forms a well-known point of attraction in that region. The lower structure of the Post Office, Neuchâtel, is of this stone.

The specimens from the group of quarries near Soleure (Solothurn), in the valley of the Aare, deserve special notice. It will be observed that these compact

[1] "Geology of St John's College Chapel," T. G. Bonney, *Eagle*, March, 1907, p. 183.
[2] See p. 230.

Jurassic Limestones from the Upper division of the
series, admit in some cases of a fine polish, and display
a great variety of colour. The rock can therefore be
employed as a marble for decorative architectural work,
as well as for ordinary building. The Federal Palace
in Berne is a good illustration of both. The General
Post Office of Thun, and that of Interlaken, are built of
the stone from Soleure. The quarries are distributed
along the lower flanks of the picturesque Weissen-
stein, the summit of which commands an exceptionally
fine view. On the lower slopes erratic blocks of Protogine granite from the Alps are abundant.

Travelling further to the eastern extremity of the
Jura range, towards Basle, the rock is again exposed,
and a compact limestone for building is worked in
numerous quarries. The specimen from the Salz
quarries at Muttenz is typical of this group. The
cube from the Homburg quarries near Arlesheim is
another example. Immense quantities of this useful
building material are at present being used in Basle
for the extension of the Railway Station, and for the
enlargement of many of the hotels in that centre
of railway traffic. The High School for Girls at Basle
is built with stone from the Salz quarries.

The specimen from the Merligen quarries on Lake
Thun is a typical example of the Jurassic Limestone
worked on its shores. The outcrop lies close to the
water, against the cliff on the north shore, facing the
pyramidal Niesen.

There are also quarries on the south shore of the
lake, from which the stone is transported by water to
Spiez, Thun, Interlaken, and other familiar tourist
centres in the Bernese Oberland.

AFRICA. Morocco. The western portion of the
Atlas Mountains, which form the backbone of the
country, is largely composed of a broad belt of Jurassic
Rocks.

An enormous deposit of boulders occurs along the

steep slopes of the mountains and in the lateral valleys, which is supposed to be the product of former glacial action. The blocks are scattered over the lowlands of North Morocco, and extend west as far as the coast, being found near Old Tangier, and in the bay, two miles south-west of the town of Tangier. They are broken up by the Moors and conveyed to the towns, where the stone is utilized for building purposes. Nearly all the buildings in the country are constructed of this material. Although the specimens seem to indicate that this limestone is a durable material, and likely to resist the influence of weather, the outer walls built of it are invariably coated with cement or some kind of mortar.

AUSTRALIA. VICTORIA. The coastal plains of this State, forming the level country between Geelong and the southern coast, are interrupted by a fragment of the Otway Ranges, known as the Barrabool Hills, where Jurassic rocks are exposed[1]. This outcrop yields a useful building stone which is extensively employed in Melbourne and its surroundings for structural purposes. As the specimen shews, this stone is darker in colour than most of the rocks from the Jurassic series. The Ormond College in Melbourne is an example of a building constructed of this stone.

WESTERN AUSTRALIA. The specimens from the Donnybrook quarries are representative of the sandstones belonging to the Mesozoic rocks of this State, and usually known as Donnybrook Freestone. They occur on the Bridgetown Railway, 132 miles south of the capital, at an altitude of 208 feet above sea level. They are of various degrees of texture, as the numerous specimens indicate, but chiefly fine-grained. They also vary in colour from a nearly pure white to a yellow with occasional brown bands. Donnybrook Freestone is coming into much favour for building purposes. It was used for the construction of the Houses of Parliament, the Law Courts, the Museum, and for several private offices in the city of Perth[1].

UNITED STATES. The Oolitic Limestones of America suitable for building purposes are chiefly to be found in the States of Iowa, Indiana and Kentucky[2].

KENTUCKY. The rocks from the Oolitic Beds in this State are said to be equal, if not superior, to any other stone of that system. The specimen in the collection, judging from appearance only, seems to bear out this statement. In colour the stone is pleasing to the eye, and the fineness and evenness of its grain commend it as being very suitable for architectural decorative work. Up to the present however it has enjoyed only a local reputation, although the supply is said to be inexhaustible. It is locally known as "White Stone." The Custom House at Nashville in the adjoining State of Tennessee, is built of it; also the Cotton Exchange, New York.

[1] Information given by the Government geologist, A. Gibb Maitland, Esq.

[2] "Building and Ornamental Stones," G. P. Merrill, *Annual Report of U.S. National Museum*, 1886, p. 372.

CRETACEOUS.

Most of the rocks belonging to the Upper Cretaceous system in the BRITISH ISLES are chalk or marl, and as these are soft and friable, they are not generally suitable for building. Those belonging to the Lower division of this system however as the specimens indicate, yield useful building stones.

The specimen of Wealden Sandstone from East Grinstead, in the county of Sussex, is a good example of the ferruginous sandstone belonging to the Lower Greensand, and is generally known as "Tunbridge Wells Sands." There are several examples of this stone in buildings dating back to remote periods. An excellent illustration may be seen in the old Saxon Church of Worth, where external upright bands or pilasters of masonry are seen supporting another band which runs as a string-course round the building, the whole forming an example of the "long and short work" so characteristic of the churches of that period. The stone from this formation, like the Bath Stone of the Oolites, is soft when first quarried. The blocks are sawn from the solid rock and detached from their bed by means of wedges; when exposed to the atmosphere the "quarry water" present in the stone evaporates, and it becomes fairly hard and serviceable building material[1]. The specimen from the North Heath quarries, Midhurst, is an example of the stone used for restoring Arundel

[1] *The Quarry*, May, 1901.

Castle; and that from the Harrow quarries, near Hastings, for restoring Battle Abbey.

The specimen of variegated greenish grey, highly fossiliferous rock, from the Goringler quarries, near Grinstead, represents a stone which is used locally for general building, and is also employed in other districts as a decorative stone, being known commercially as "Sussex Marble." The example indicates that it admits of a high polish, and its appearance shews that it is a material very suitable for ornamental architectural construction, like the "Purbeck Marble" at the top of the Jurassic system.

Kentish Rag, belonging to the Lower Greensand, deserves special notice. It is a rock which from time immemorial has been used as a building stone in the south of England. In the life of Sir Christopher Wren, written by his son, reference is made to the restoration of St Paul's Cathedral in London, after the great fire of 1666. Wren confidently asserts that the Cathedral then destroyed by the conflagration was the fourth building which had occupied that site, and that each of them was erected on the old foundations, which were of Kentish Rag[1]. Rochester Castle (1078) was built chiefly of this stone[2]. Quite a large number of churches were built in London during the early part of the 19th century of a combination of Kentish Rag and Bath or other Oolite. The Kentish Rag is hard and tough, it is therefore difficult to dress, and consequently expensive to use; it is principally employed for rubble masonry, while the softer Oolite is used for the quoins, jambs and other facings.

Carstone, an indurated ferruginous sandstone[3], high up in the Lower Greensand, occurs in Bedfordshire and some parts of Cambridgeshire, as well as in Norfolk, in which county it is used for building. Being of a somewhat soft and friable nature, as indicated by

---

[1] Wren's *Parentalia*, Pt. 2, Sec. 3, p. 272.
[2] *England's Chronicle of Stone*, J. F. Hunniwell, p. 117.
[3] *Geology of Cambridgeshire*, F. R. C. Reed, p. 50.

the specimens in the collection, most of the buildings where it is employed are constructed with quoins and facings composed of some harder and more durable stone. Of the two specimens in the Museum one is considerably harder than the other, though both are from the same quarry. The harder rock is generally found in thin beds above the softer stone, which is the more massive. The chief quarries in Norfolk are within easy distance of the royal estate of Sandringham, and pleasing examples of the stone can be seen in the many picturesque buildings erected there.

The specimen of St Boniface Stone from the Upper Greensand of the Isle of Wight, represents practically the only building stone at present quarried on that island. This stone, besides having been extensively employed there for structural purposes, has occasionally been used elsewhere. Winchester Cathedral, for example, was restored with it (1825).

The Upper Greensand series which occurs in Buckinghamshire produces a building stone known as the Denner Hill Stone. This rock is found in detached isolated concretionary masses in clay or sand. These vary in size, the largest approaching 300 tons in weight. As the specimen indicates, it is a tough, hard stone, somewhat of the nature of quartzite.

Merstham Stone, from the Upper Greensand in Surrey, which is better known as " Firestone," has for many centuries been employed for building construction. It is a fine-grained sandstone containing a large amount of colloid silica[1], very porous, and remarkable for its fire-resisting qualities. So much importance was attached to this that the quarries were at one time the exclusive property of the Crown, and the stone was employed only in the erection of royal and ecclesiastical buildings. Old London Bridge was constructed with this stone (1176), it was also used for Hampton Court (1520), and Windsor Castle,

---

[1] *Memoirs of Geol. Survey of U. K.* 1900, Vol. I. p. 419.

also the chapel of Henry VII in Westminster Abbey. It is recorded that 681 tons of Merstham Firestone were employed during the first year of the building of Eton College in 1443[1]. The specimen from quarries near Reigate is another example of the Merstham Firestone in the same county. The specimen from Milford quarries, near Witley, is a third example of building stone from the Upper Greensand in Surrey. It is known there as "Bargate Stone," and is extensively used in Godalming and the surrounding district.

The example from the Windrush quarries in Hampshire, is representative of the "Firestone" of that county, and better known there as "Malmstone." There are two beds, both of which yield good building material. The upper, or White Bed, furnishes a light creamcoloured siliceous rock, which is used for all kinds of structural work, and is specially suitable for decorative building. Underlying this is the Blue Bed, which yields a darker calcareous stone less suitable where carved work is required, but useful for walling and other ordinary constructive purposes.

Beneath the Chalk Marl, and overlying the Lower Greensand in the county of Norfolk, is a red limestone, better known as the "Red Chalk of Hunstanton." The bed is three feet thick, and is divisible into three more or less distinct layers: (1) Hard lumpy reddish chalk or mottled red and white; (2) Rough nodular red limestone passing down into (3) Deep-red gritty rock, softer at the base[2]. The specimen in the Museum is an example of the stone from the middle layer, which is that usually selected for building purposes. The chemical analysis of the Red Chalk discloses that it is a marly limestone with a high percentage of peroxide of iron which forms the colouring matter. No quarry exists from which this stone can be worked, but the band of rock forms part of the cliffs at Hunstanton, on

---

[1] *Architectural History of Cambridge*, Willis and Clark, Vol. I. p. 386.

[2] *Geology of Cambridgeshire*, F. R. C. Reed, p. 84.

the north coast of Norfolk. There is a gradual, but constant disintegration going on, through atmospheric influences, which causes the Red Chalk to be loosened and precipitated on to the shore, whence it is removed for building purposes. The supply is very limited; indeed it has almost ceased to be used, and had it not been that the Red Rock of Hunstanton is of much geological interest it would hardly have been worth while to place a specimen in the collection.

The rocks belonging to the Lower Chalk yield a building material which is popularly known as "Clunch." In geological nomenclature this appellation refers only to the harder bands of the Chalk Marl series of Cambridgeshire[1], but among builders and masons the name "Clunch" is applied, as a generic term, to all chalk rocks that can be utilized for building purposes.

There are proofs that chalk was used as building stone in Cambridgeshire from very early times. Quarries near Reach were worked for clunch during the Roman occupation, and used in the basement of a Roman villa near where the Mildenhall Railway crosses the Devil's Dyke[2].

As the stone is very accessible, and is easy to quarry and carve, it was used in considerable quantities for the erection of several of the early colleges of Cambridge University. Documents preserved in the colleges record that clunch was used for the building of Peterhouse (1307), Trinity Hall (1350), Christ's (1505), and Caius (1557), the last mentioned college being supplied from quarries at Haslingfield[3]. The founders of Corpus Christi College obtained a grant to work a quarry at Cherryhinton for a supply of stone to build that College[4]. Though the name of the stone is not mentioned in the historical

---

[1] *Geology of Cambridgeshire*, F. R. C. Reed, p. 122.

[2] *Cambridgeshire*, T. McK. and M. C. Hughes, p. 113.

[3] *Architectural History of Cambridge*, Willis and Clark, Vol. I. p. 174.

[4] *History of Corpus Christi College*, H. P. Stokes, p. 23.

record, it must be clunch, for none other is found in that neighbourhood. It is also recorded that a licence was granted by Henry VI to the wardens and scholars of King's Hall, "to dig in an acre of land in the 'Downe-field' of Hynton (Cherryhinton), Co. Cambridge, and extract stones therefrom and carry the same away for the building of their Hall[1]." The specimen in the Museum is from a quarry at Cherryhinton. Most of the College buildings where clunch was used have been rebuilt, or have had the exterior walls faced with a more enduring material, therefore there are very few instances to be seen of its existence as a building material. The part of Christ's College which was begun early in the 16th century, was originally constructed of blocks of clunch in courses alternating with red bricks. In less than two hundred years, owing to the decay of the clunch, it became necessary to rebuild, or to repair the walls. A facing of Freestone was adopted which was begun in 1714, and this process of ashlaring was continued with the result that except in a few isolated places, the clunch is entirely hidden from view. This work of restoration was evidently a necessity, because it is recorded that the College presented so ruinous and repulsive an appearance that persons were deterred from entering students therein[2]. The collection contains a specimen of clunch taken from one of the walls of Christ's during a recent alteration. The north wall of the Old Court of Peterhouse, built in 1473 still remains, and the white clunch composing it can easily be seen from the churchyard of Little St Mary's. The north wall of Sidney Sussex College grounds, facing Jesus Lane, is built of large blocks of clunch. The wall is fully three feet thick, and we may observe that the lower portion has been cased with bricks to protect it from the risk of abrasion caused by the traffic of the lane. Clunch was used for building the

[1] *Calendar of Patent Rolls*, 27 Henry VI. (1449), p. 257.
[2] *Architectural History of Cambridge*, Willis and Clark, Vol. II. p. 224.

new kitchen of King's Hall, Trinity College, in 1412[1]. It was extensively employed in the construction of Barnwell Priory in the early part of the 13th century[2]. Madingley Hall, erected in 1548, the residence of the late King when a student at Cambridge, was built of clunch, but very little of the stone is now discernible, as it also has been cased with a more durable stone and bricks to prevent decay. There are several instances in Cambridge of clunch being used for internal decorative work, for instance in the vaulted roof of the cloisters of the new court of St John's College, and in the interior of the new Chapel of Queens' College.

Some of the interior mouldings and ornaments of the superb Lady Chapel in Ely Cathedral, built in 1321 —49, with its elaborate sculptures, one of the finest specimens of late decorated architecture in the kingdom, are worked in clunch, as well as the groins of the West Porch. We may note that this stone, when mentioned by early historians, is described as "Whyte Stone," under which name it is frequently mentioned in the early annals of the colleges of Cambridge[3].

In Bedfordshire, an arenaceous chalk, known as Totternhoe Stone, overlies the Chalk Marl. In Cambridgeshire it is often called the Burwell Rock[4]. The hard beds of the formation are used to a considerable extent as a building stone, while the softer variety is employed for making lime. The chief quarries are situated on Totternhoe Hill, whence the specimen in the collection comes. The numerous and extensive excavations on the Hill shew that this stone has been quarried for a very long period. This was formerly done by means of tunnels, which commenced at the outcrop of the stone and were carried a considerable

[1] *Architectural History of Cambridge*, Willis and Clark, Vol. II. p. 440.

[2] *Cambridge Antiquarian Society*, J. W. Clark, Vol. I. N. S. p. 257.

[3] *Annals of Gonville and Caius College*, Dr Venn, p. 186.

[4] "Geology of Bedfordshire," R. H. Rastall, *Geological Assoc.* 1909, p. 159.

distance into the hillside; but lately the system of open quarrying has been substituted, the whole of the overlying bed being removed and used for making lime. Totternhoe Stone was formerly extensively employed in local architecture, numerous churches in the county affording examples of its use: Dunstable Priory, also Woburn Abbey, both built in the 12th century, are illustrations. Totternhoe Stone however is not in such demand as it was formerly; for when exposed to the action of the climate it has been found to suffer, as for instance in the west front of Dunstable Priory Church. For internal decorative work it is still in repute.

The specimen of Beer Stone which forms the basement bed of the Middle Chalk group in Devonshire[1], resembles in appearance the clunch of East Anglia, being like it, nearly white, and composed of carbonate of lime, with the addition of some argillaceous and siliceous matter, and a few scattered particles of green silicate of iron. When first quarried this stone is somewhat soft and easily worked, owing to the presence of disseminated water, but becomes hard when this water has been evaporated by exposure. The Beer Stone has been quarried for ages, and some of it is stated to have been employed in the old work in the interior of Exeter Cathedral[2] (14th century). An example of this stone can be seen at Emmanuel College, Cambridge, where it forms the string-coursing of the " Brick Building " at that college, erected in 1668. It can also be seen in St Stephen's Crypt, Houses of Parliament, London. St Andrew's Church, Charmouth, (1836) is built of it.

The great Chalk Upland which occupies about two-thirds of the surface of Wiltshire, including the undulating elevation of Salisbury Plain, yields a soft Cretaceous rock which is occasionally utilized for build-

---

[1] *Stratigraphical Geography*, A. J. Jukes-Browne, p. 455.
[2] *Report on the Geology of Cornwall, Devon, &c.*, H. T. de la Beche, p. 488.

ing purposes, and is known as "Chalk Stone." The specimen from quarries near Devizes is an example.

The siliceous nodules present in the Upper Chalk have been utilized for building purposes from a very early date. Many churches, dotted about Kent and East Anglia, some of great antiquity, are built with Flints from the Chalk quarries in these regions. Frequent examples of a similar use of flint are found in the remains of the numerous structures on the south coast of England, erected during the Roman occupation, and generally known as the "Roman Ports." *Rutupiae* (Richborough), about two miles from Sandwich, on the Kentish coast, built about A.D. 205, and believed to have been the chief sea station of the Romans in Britain, was largely formed of this material. In 1849 extensive excavations at this station brought to light the outer wall of an amphitheatre 3½ feet thick, built of flints, with a facing of chalk and tiles at the angles of the doorways. The Castrum of *Regulbium*, now called Reculver, at the northern end of the old channel between the Isle of Thanet and the mainland (A.D. 194 —211), was enclosed with walls 12 feet thick, the remaining fragments of which shew that they were built with flints and pebbles[1].

Passing on to a later period, we have an example of flint having been used for the building of St Albans Cathedral, founded in A.D. 793. The great five-sided area on which is built the Tower of London is enclosed by walls composed chiefly of flints[2]. The same material was used in the building of Eton College in 1444, which was procured from a quarry near Windsor Castle[3].

Flint, together with Chalk, has been extensively used in Kent for building. St John's Church at Burham in Kent is an example. The flint nodules are usually

---

[1] *England's Chronicle in Stone*, J. F. Hunniwell, p. 26.
[2] *Ibid.* p. 80.
[3] *Architectural History of Cambridge*, Willis and Clark, Vol. I. p. 391.

subjected to the process of "Knapping" or splitting, before they are employed for building, an art which is somewhat difficult to acquire. Flint usually breaks with a conchoidal fracture; this has to be avoided in knapping, as a plain surface is what is required for building. This manipulation, as well as the use of flint in any form for building purposes, is rapidly becoming a thing of the past[1]. A good example of a knapped-flint building in Cambridge is the Lodge at the entrance gates of the Cemetery in Mill Road; also the vestry and adjoining boundary wall of St Paul's Church (1857).

Numerous churches in Norfolk and Suffolk are built entirely of flint, and in some cases it has been used for ornamenting the exterior walls of churches built of other material. The flints are knapped into shape, shewing a face of squares, hexagons, and other geometrical devices, in a pattern with other stone, thus forming attractive mosaic-like work. The squared specimen from the ruins of Covehithe Church, Suffolk, is an example of flint used for this purpose. Others of various sizes and designs, procured from the same building, will be found in the collection.

The numerous specimens representing the "Colonial and Foreign Building Stones" of the Cretaceous system are evidence that this system occupies a more important position abroad, as a producer of structural material, than it does in the British Isles.

AUSTRIA. Rocks of the Cretaceous system that are useful to the builder in Austria are chiefly found in the peninsula of Istria. The collection contains several specimens, ranging in colour and degree of fineness. These limestones, which belong to the Upper division of the system, have been quarried for building purposes since the days of Imperial Rome. The proximity of the

[1] The donor of the Knapped Flint from Kent writes that an operative capable of knapping is now difficult to find.

quarries to the port of Pola, then one of the chief arsenals of the Romans, was doubtless an inducement to export this building product to Italy, and the comparatively short voyage across the Adriatic, rendered the transport easy and inexpensive. When Honorius transferred the seat of government to Ravenna, early in the 5th century, the town increased, and the Istrian Limestone was employed to enlarge the new capital of the then languishing Roman Empire.

The specimen from the Vincurial quarries, near Pola, represents the stone which is believed to have been used by the Romans for constructing the amphitheatre at Pola.

Large quantities of Istrian Limestone are sent to Vienna to be employed for building, its delicate shades of colour causing it to find favour with the inhabitants. The light fawn-coloured stone from the San Stefano quarries, and that from the Nabresina quarries, near Trieste, both admitting of a good polish, are useful for monumental work. The façade of the Hofburg Theatre in Vienna is an example of the stone from the quarries of Melera, near Pola.

Istrian Limestone has also been extensively employed for building in Venice, and that for a long period. From the 11th century onwards it has been used for plinths, angle quoins, window tracery, and other decorative work. It is used for example in the magnificent series of arcades in the Ducal Palace[1].

DENMARK yields limestones from the Upper Chalk, which are useful as building stones, and are popularly known as "Faxe Chalk." They are extensively employed in Copenhagen for decorative work, and the specimens shew that they vary considerably in texture. One example from quarries near Faxe is a coarse-grained cellular rock, while the other from the Cliff of Stevns is closer-grained and of a more chalky nature. Both varieties are used for structural purposes

---

[1] *Encycl. Britannica*, 9th ed. Vol. XXIV. p. 149.

in Copenhagen. The buildings in the Botanic Gardens, erected in 1890, are examples of the use of the fine-grained variety.

FRANCE contributes a considerable number of specimens. In the north of that country the Upper Cretaceous beds are precisely like those of England, but on the south side of the axis between Brittany and Auvergne the soft chalk is replaced by hard limestones, containing many of the fossils called Hippurites. These rocks are valuable for building. They occur on the north flanks of the Pyrenees. The white chalky specimens from quarries near Nay in the Basses-Pyrénées are typical examples. The Lower beds of the Cretaceous system are also present in this Department, and the specimens indicate that the stone from them is compact and blue-grey in colour.

Extensive deposits, underlying the great basin of the Garonne, supply the community of Bordeaux and its surroundings with a valuable building material, which is said to resist, in a marked degree, the decaying influences of the salt-laden winds from the Bay of Biscay. The specimen from the Nersac quarries in the Department of Charente is typical of the stone used for building the Church of St Louis at Bordeaux. It is said to turn white after exposure to the air.

There are numerous quarries on either side of the valley of the Rhone. The rock extends so far south as to underly the fringe of the great alluvial delta to the west of Marseilles. The builders of that important commercial town are largely dependent on this Cretaceous bed for stone. The same rock occurs in the Department of Savoie. The specimen from quarries at Aix-les-Bains is an example. Many of the buildings in that town are constructed of it.

GERMANY obtains much building material from the Cretaceous system.

The Lower division occurs in the Province of Lorraine, and quarries are worked on the banks of the

Mosel which yield a useful building stone for structural work in Metz and other towns of the district. The specimen from the Jaumont quarries, near Metz, is an example.

The picturesque crags of the Saxon Switzerland, forming the great gorge of the Elbe, yield masses of sandstone belonging to the Upper division of this system which are extensively employed for structural purposes. It will be observed, by the specimens, that these rocks are chiefly yellowish drab in colour. The specimen from quarries near Schöna is an example of these sandstones, which are employed for building in Dresden and Leipzig.

A useful Upper Cretaceous stone is quarried in the neighbouring Province of Silesia. This rock being very fine-grained, and having a pleasing colour, as the specimens indicate, is a valuable material where decorative architectural work is required.

GREECE. There are several compact limestones belonging to the Cretaceous system in Greece, which are not hard enough to take a polish, but make useful building material[1]. They occur chiefly in Attica, and the specimens were procured from quarries near Athens, in which city the stone is at present much used for building purposes. These rocks were extensively employed in Athens for domestic architecture during the classical period, as is proved by the remains discovered in the valley between the Acropolis and the Pnyx.

ITALY. Besides the marbles of Carrara, the geological age of which is uncertain, the western slopes of the Apennines yield a useful limestone belonging to the Cretaceous system, which is employed for ordinary building and sometimes for decorative architecture.

[1] The valuable marbles from Pentelicus and other localities used in the classic buildings and sculptures of ancient Greece are of much greater geological antiquity, but these, for reasons already given, are excluded from this collection.

The specimens from quarries in the district of Campania, and from Stabia, are good examples of this rock, which finds a ready sale in Naples and its surroundings for building purposes, being chiefly employed for internal decorative work in many of the churches.

PORTUGAL. Although Cretaceous rocks are not so much in evidence in Portugal as those of the Jurassic, the specimen from the Collares quarries, near Cintra, is proof that they yield a useful building stone, which is extensively employed by the Portuguese for constructive purposes.

RUSSIA. In the Government of Kutais, which occupies the north part of Transcaucasia, the name given to that portion of the Russian Empire which lies south of the main range of the Caucasus, limestones belonging to the upper part of the Cretaceous system are utilized for building purposes.

The quarries are situated on both banks of the Tskhal-Tziteli near the village of Tchognari, about five miles from the town of Kutais. There are several specimens in the collection from this group, and they indicate that the rock varies considerably in colour, even in the same quarry. Some are light grey and drab, whilst others are of a delicate pink tint which is sometimes intermingled with brown. They are all fine-grained, the pink rock being more arenaceous than the grey and drab; the latter are compact limestones, and most of them take a fairly good polish. The beds are found in almost horizontal layers from about four to nine feet thick. This stone, especially the fine-grained compact varieties, forms good building material. At Batoum, the Cathedral, Commercial Bank, and other buildings, are constructed of it; at Tiflis, the Cathedral; at Borjon, the Grand Duke's residence, as well as many other buildings; also most of the houses at Kutais have been built of this stone.

SWITZERLAND. The well-known limestone (Pierre Jaune) from the Neocomian group of the Cretaceous system, present in this country, is worked for building purposes on the lower slopes of the Jura. Extensive quarries exist on the north banks of the Lake of Neuchâtel, and the stone is employed for building in the town of Neuchâtel, and in other places in the Canton. The specimen indicates that the stone is somewhat soft and chalky; for this reason it is only useful for superstructures. The foundations of buildings in that region are usually constructed of the harder limestones from the Jurassic system, which are also present in the Canton[1].

AFRICA. Rocks of the Cretaceous system occur in TUNIS, and furnish a useful building stone. As the specimens indicate, they are fine-grained, compact limestones, their tints being light pink, yellow, and fawn-coloured. Being soft in texture, they yield kindly to the chisel, and contribute to the oriental beauty of the handsome palaces and mosques which with their elegant domes and gilt-tipped minarets fringe the Bay of Tunis.

EGYPT. The numerous specimens of sandstone displaying various shades of colour, from the Gebel Silsella quarries, near Aswan in UPPER EGYPT, are good examples of the important and extensively developed group of rocks belonging to the Cretaceous system, known as the "Nubian Sandstone," which rests on the eroded surface of the older rocks. This quartzitic sandstone usually appears in well-defined strata of red, brown, and grey colour, in many respects resembling the "Old Red" of Western England. The whole mass, both sand grains and cementing material, is siliceous, with the exception of some iron, a few small veins of sulphate of baryta and a little gypsum[2]. The stone

[1] See p. 189.
[2] "Gold Mining in Egypt," C. J. Alford, *Institution of Mining and Metallurgy*, Vol. x. (1901–2).

has usually a dark appearance *in situ*, in consequence
of the superficial oxidation of the iron, but it is much
lighter in colour when quarried and dressed.

The Nubian Sandstone was employed in even
the earliest structures of Egypt. All the temples of
Thebes are built of this material, though not the obelisks
and a few of the propylaea[1]. Large quantities of it
have been recently used for the superstructure of the
new Barrage on the Nile. It is said that the strength
of this stone increases rapidly with the density, hence
in selecting samples for building purposes those with
a high specific gravity should be sought[2].

The Nubian Sandstone varies considerably both in
colour and texture in different districts. There are
two specimens in the collection from quarries on the
White Nile, about 25 miles south of Khartoum. It
will be observed that one of them displays a delicate
pink colour, and the texture of both is much coarser
than that of the rock found near Aswan. This stone is
at present being employed for the construction of the
Gordon Memorial Cathedral at Khartoum (1910).

ANGOLA. A few miles inland from the low-lying
coastal strip of this West African territory, on undulating ground which gradually rises to a considerable
altitude, outcrops of limestone occur which are
quarried for constructive purposes. Very little is
known of the geology of Angola, but these limestones
are said by some observers to be possibly connected
with those along the south coast, which have been
referred to the Cretaceous system[3].

The specimen is from quarries at Cacuaco, the first
station on the Loanda-Malanga Railway. This stone is
used chiefly for building in Loanda, the capital of
Angola, sometimes known as St Paul de Loanda: for

[1] *Encycl. Britannica*, 9th ed. Vol. II. p. 384.
[2] Willcock, *Egyptian Irrigation*, 2nd ed. p. 456.
[3] "Geological Notes of W. Africa," O. Lenz, *Geological Magazine*, 1879, Vol. VI. p. 172.

instance the Municipal Buildings (1910). The rock differs a little in texture, the hard variety, known locally as "Cacuaco Stone," being chiefly used for forming foundations and railway engineering construction; the softer one, called "Loanda Stone," for superstructures and internal work.

SOUTH AFRICA yields a calcareous sandstone suitable for building, from the Uitenhage series of rocks in the CAPE OF GOOD HOPE, which by some geologists is accepted as the equivalent of the European Lower Greensand[1]. This building material is quarried near Port Elizabeth, where it is known as "Coega Stone," and is used for structural purposes in that town, as well as in other centres of the Cape. There are two qualities, distinguished as "Hard" and "Soft Coega Stone." The latter is used chiefly for internal structural work. As the specimens shew, the two differ in colour as well as in texture. The Post Office at Port Elizabeth is an example of a building constructed with Hard Coega Stone.

The Uitenhage series extends inland as far as the slopes of the Zuurberg Mountains. An outcrop occurs near Cathcart, where the rock is quarried for building purposes and is known commercially as "Cathcart Stone." The drab-coloured specimen from the Cathcart quarries is an example.

TURKEY in ASIA. CHIOS (ASIA MINOR). On the east coast of this island, about a mile north of the town which bears the same name as the Island (in Italian, Scio) are quarries yielding limestones of divers colours and textures. These belong to the Cretaceous system and have been employed for building and decorative purposes for a great number of years.

The prevailing colour of these rocks is grey, sometimes streaked with flesh-coloured veins and frequently

[1] *Geology of South Africa*, Hatch and Corstorphine, p. 238.

intercalated with very fine, almost invisible lines of quartz; and as the specimen indicates, they admit of a fairly good polish. The compact saccharoidal varieties are classed among the marbles, and known as "Porta Santa." These limestones are within two hundred yards of the Ægean sea-shore, and the ridge extends inland as far as Draase and Marmaro-trapesa, where it joins the higher Piganto range of mountains. The limestone quarries on the island were worked as early as B.C. 660, and in all probability before that date[1]. All the chief churches and structures on the Island are built of this stone, also the harbour works and quays. Numerous examples of elaborate architectural work, executed in this material, can be seen in and about the town, many in partial decay, bearing testimony to the luxurious tastes of the Genoese aristocracy who had possession of the Island in the 14th century, and held it until conquered by the Turks in 1566[2].

SYRIA. The beds representing the Middle and Upper Chalk divisions of the Cretaceous system in Palestine, which extend over the area to the west of the Dead Sea and are well exposed in the neighbourhood of Jerusalem, yield rocks exhibiting several degrees of hardness and texture, as well as a considerable variety of colour. These limestones have been utilized for building from time immemorial and are still extensively employed. They come from at least four distinct beds, the rocks of which are distinguishable by their colour, but chiefly by their varying hardness.

That forming the top bed is the hardest stone of the series and is known as "Mizzeh." This division is again subdivided and is classified by the builders in respect to colour and hardness. Mizzeh Yahudi is considered to be the hardest of all; next comes Mizzeh Azraq

---

[1] *What Rome was Built With*, M. W. Porter, p. 71.
[2] *Encycl. Britannica*, 9th Ed. Vol. XXI. p. 466.

(Blue M.), Mizzeh Asfar (Yellow M.), and Mizzeh Akhdar (Green M.), all hard compact rocks although not so tough as Mizzeh Yahudi. There are also softer varieties, Mizzeh Abyad (White M.), Mizzeh Ahmar (Red M.), and lastly Mizzeh Helu (Sweet M.) which is considered to be the softest variety of the top or hardest bed. The Greek Church, on the Mount of Olives, is a typical example of the Yahudi rock.

Underneath the Mizzeh comes the bed known as "Melekeh" (the royal stone), which is a softer stone than the rocks just referred to, and was more frequently used as a building material, being less expensive to work than the harder stone. It was much used in the erection of the ancient buildings of Jerusalem.

The bed known as "Nari" is the next, which yields a stone that is variable in quality and texture, sometimes being quite friable and useless for constructive purposes.

The bed lying below it, a soft chalky limestone known as "Kakuli," is employed only for internal construction, as it is unsuitable for external work owing to its inability to resist the influence of the weather. This soft stone was much more extensively employed for building in ancient times, the harder stones being difficult to work with the tools then available, but modern implements and improved appliances have latterly enabled the artificer to utilize the "Mizzeh" rock with greater ease and economy.

The chief quarries producing the soft or "Kakuli" rock are located on the white chalky slopes of Mount Olivet. One of the specimens in the collection was procured from the quarries near a village north of Jerusalem named A-Na-Ta, the Anathoth of Jer. i. 1. This stone was employed for building the Russian Pilgrims' Society's Hospital in Jerusalem (1860).

There is yet another stratum known as "Howwar," but the rock is so soft that it is practically useless for structural purposes.

It will be seen from the catalogue that all these

divisions of rocks except the last-named are represented in the collection and that most of the quarries are situated within an easy distance from Jerusalem. The specimens very strikingly illustrate the beautiful and delicate tints of these limestones, and there is little wonder that many of the buildings erected in and about the " Holy City " command the admiration of the European lover of architectural beauty.

There are several old and disused quarries in the neighbourhood of Jerusalem, the best known of which is that near the Damascus Gate. These are of considerable extent, and have been worked for the purpose of getting the "Melekeh" stone; in places the stones have been left half cut out, and the marks of the chisel and pick are as fresh as if the workmen had just gone away[1]. On the opposite side of the road is another quarry to which the name of Jeremiah's Grotto has been given. Sections are exposed on each side, shewing very clearly the manner in which the hard limestones (Mizzeh) overlie the softer beds (Melekeh).

The small cube of Petra sandstone is easily distinguishable among the specimens of rocks representing the Upper Cretaceous series in Palestine, by its beautiful and delicate tinted mauve colour alternating with bands of light rose-pink[2]. It is a good example of the rock of which the precipices are composed on either side of the modern Wády Músá, in the mountains forming the eastern wall of the great valley between the Dead Sea and the Gulf of Akaba. This is the site of Petra, the ancient capital city of the Nabateans and the great centre of their caravan trade, described by Strabo as "lying in a level place, well supplied with water for horticulture and other uses, but encircled by a girdle of rocks, abrupt towards the outer side[3]."

---

[1] *Ordnance Survey of Jerusalem*, Sir C. W. Wilson, p. 63.

[2] Petra Sandstone belongs to a lower subdivision of the Upper Cretaceous series than the Limestones just mentioned. It is correlated with the Upper Greensand of England.

[3] Strabo, Book XVI. p. 779.

The valley of Petra is nearly two miles across in its broadest part, whilst at places it shrinks to a narrow gorge hardly twelve feet wide. On either side are towering cliffs, some upwards of 100 feet high, all composed of this attractive rock, in which are hewn vast numbers of palatial dwellings, temples and grottos, adorned with elaborate classical façades and standing out in bold relief against the mountain walls. Chief among these interesting excavated houses are the Theatre, Ed-Deir (the monastery), and the so-called Khazna or "Treasury of Pharaoh," none of which are built, but all hewn out of the rocky cliff. Most of these magnificent cave dwellings are now in a dilapidated state, but in some cases the richly carved decorative façades are in good preservation. The plain itself is covered with ruins of temples and other buildings[1]. The pink, lilac, rose, and rich browns, which are the prevailing tints, seem to blend into each other with exquisite softness. This beauty has long made the cliffs of Petra famous.

About 16 miles north of Jerusalem, a little west of Jericho, a hard, compact, slightly bituminous, blue-black rock, said to belong to the Lower division of the Upper Cretaceous system, is quarried for building purposes, and for carved decorative work. This rock is called "Nebi Musa" (Prophet Moses) by the natives, who firmly believe that the great lawgiver of the Hebrew race was buried on the spot where the quarry now is[2].

The western slopes of the hill regions, in the south of Judæa, consist in the main of Cretaceous limestone, and it is quarried in the neighbourhood of Beersheba, the stone being used locally for building. It is also conveyed to the coast and is employed for structural

---

[1] The stone at present has ceased to be utilized, there being no inhabitants in the ruined city, but it is believed that the day is not far distant when facilities will be provided to transport this building material to Jerusalem.

[2] *Customs and Traditions of Palestine*, E. Pierotti, p. 67.

purposes in and around Gaza, where the Crusaders' Mosque is built of it. The specimen from the Chalasa quarries, near Beersheba, is a good example and indicates that these beds are much softer than the Mizzeh rock existing nearer the Dead Sea and in the neighbourhood of Jerusalem.

UNITED STATES. CALIFORNIA. In Colusa County on the right bank of the river Sacramento, are beds of sandstone belonging to the Upper Cretaceous system (known locally by American geologists as the Chico formation). These yield a useful building stone which is extensively employed for structural purposes in San Francisco and elsewhere. Of this the greenish grey specimen from the Sites quarries is a good example.

CANADA. Along the coast line of British Columbia, and on the adjacent Islands, there is an immense quantity of good building material in the form of sandstones of the Cretaceous age[1]. Numerous quarries have been started although they have not been worked to any extent, but as the erection of stone buildings in that region is rapidly increasing, the demand for this stone must necessarily follow. The blue-grey specimen cube from Saturna Island is a good example of these rocks. The southern end of Saturna Island rises abruptly from the sea and is composed of sandstone interbedded with conglomerate. The quarries are on the water edge, and in calm weather the stone can be loaded directly into scows. There is no regular working face, the stone being taken out along the bedding planes at different levels wherever it occurs of the size and tint desired. The rock is of two colours, buff and blue-grey, the former being on the top. Considerable quantities of it have been quarried for building purposes. The Carnegie Public Library, Victoria, is a recent building constructed with Saturna Island stone.

---

[1] *Report of Minister of Mines, British Columbia*, 1904, p. 250.

TRINIDAD. Compact limestones of Cretaceous age, which stretch very nearly midway across the Island from west to east[1], form its chief building stone. The dark blue-grey specimen from the Laventille quarries, near Port of Spain, is a good example, and indicates that the rock contains numerous veins of calcite. The Cathedral of Port of Spain is built of it.

[1] P. M. Duncan, *Quart. Journal Geol. Society*, Vol XXIV. p. 11.

TERTIARY AND RECENT.

BRITISH ISLES. Here the Tertiary formations do not yield any rocks which can be utilized with advantage for building, for as the strata largely consist of clays, sand, and gravel, they are quite unsuitable for the purpose.

The Binsted Limestone however of the Oligocene system which occurs near Ryde, in the Isle of Wight, was at one time in request for structural purposes. This was prior to the introduction of railways into the Island; now other building stones have been introduced, and the Binsted quarries are no longer worked.

The high-level beaches of the south-west of England, belonging to the Post Tertiary system, usually known as "raised beaches" in geological nomenclature, yield a rock that has been employed for building over a long period. The specimen from the cliffs at the south side of Saunton Down, in Devonshire, is a typical example of this indurated material, which Professor Hughes pronounces to be blown sand consolidated into a hard calcareous stone[1]. It is usually quarried at the foot of the cliffs. The rock is very soft when freshly worked; it is easily cut into blocks with a saw, and left to harden by the action of the atmosphere, which soon makes it fit to be used for structural purposes. The Church of St Ann's at Saunton Sands is built entirely of this sandstone.

[1] T. McK. Hughes, *Quart. Journal Geol. Society*, XLIII. (1887), p. 662.

CONTINENTAL EUROPE. The formations existing here which belong to the Tertiary era are represented by numerous specimens in the collection; thus indicating that they hold an important position as furnishers of building stones. Indeed, it is to Tertiary limestones that Brussels and Paris owe much of their architectural beauty; while the Nummulitic beds all along their outcrops have yielded building stone from the time of the Great Pyramid down to the present day[1].

AUSTRIA. The Tertiary beds lying to the north of the great "Pest Basin" in Hungary, yield a fine-grained calcareous stone belonging to the EOCENE system, which is highly esteemed for building in and around Budapest. The specimen from Beszterezebánya, quarried in the province of Sohl, is a good example of this useful building material. It will be observed that the colour of the rocks in Hungary belonging to this system varies considerably. The specimen just referred to has a light fawn tint, while that from the Gobanka quarries is a bright orange colour.

The specimen of coarse-grained limestone from near Eggenburg in Lower Austria, serves as an example of the rock which is quarried and employed for building purposes in and about Vienna. This stone was largely employed for constructing the Old Town of Vienna, indeed, it is believed to be the first building stone which was used in that city, but owing to its coarse nature and the introduction of the finer-grained rocks of the Miocene system, it is now not so much in request for building, especially where decorative work is required, although it is still employed for walling. The Hofburg of Old Vienna, built in the 12th century is a good example. It was constructed with stone from the Rekawinkel quarries, near Vienna. The specimen indicates that this bed is darker in colour than the rock at Eggenburg.

[1] *Economic Geology*, D. Page, p. 78.

Besides the Cretaceous deposits on the Isthmus of Istria previously mentioned[1], there is a bed of Tertiary Limestone, belonging to the Eocene series of that region, which yields a fine and compact rock that is used for ordinary building as well as for decorative work. The specimen shews that the stone admits of a good polish. This stone must in former days have found its way to Rome, as did the Cretaceous rocks from the same district.

The Eocene system is made up in FRANCE of the characteristic "Calcaire grossier," a mass of limestone, sometimes tender and crumbling, in other places so compact as to be largely quarried as a building stone. Special attention may be drawn to the specimens from the Palaiseau quarries, "Meulière Rouge" and "Meulière Blanc"; the former being easily distinguished by its brilliant orange colour. They are highly cellular rocks, therefore not suitable for decorative work, but useful for interior walling. Huge piles of this stone are generally to be seen on the quays and wharves of Paris, ready to be utilized for building purposes in the French metropolis.

The specimen from St Cherons is a good example of the rock from the thick sandstone deposits of Fontainebleau belonging to the Upper Eocene or Oligocene series, and which forms part of the Paris basin. This stone, as the specimen indicates, is somewhat soft and of a saccharoid texture, but being close-grained it is generally used by the Paris mason for decorative and carved work.

The limestone of ITALY, produced from quarries near Verona, is usually known as "Verona Marble," but since this stone was much used for ordinary building during the Roman period, and is still extensively employed for that purpose, it claims a position in the collection. As the specimens shew, there are three varieties of this stone, one of a deep orange colour, or perhaps better described as red with orange-coloured veins; another a lighter shade of orange with cream-coloured veins, usually known as "Almond Verona"; and a third, white, with sometimes a faint line of pink running through it. This stone is specially noticeable in the numerous churches and other public buildings of Venice. The columns and arches of the two colonnades on the south side of the Palace of the Doges, the portal of St Mark's, the Campanile and the Academy of Fine Arts, are a few of many examples of its use[1].

In Tuscany a rock, *Pietra Sirena*, belonging to the Eocene system, enjoys a good reputation as a useful building material, and is employed in Florence and its surroundings. It is also used in Rome. The specimen indicates that it is a fine-grained sandstone, its colour being of a greenish tint.

In the Abruzzi, on the western slopes of the Gran Sasso d'Italia, are quarries yielding a light cream-coloured limestone belonging to this series, which is used for building in Aquila and other parts of central Italy, including Rome. The specimen in the collection indicates that the stone is remarkably fine-grained, therefore well adapted for decorative architecture. The chapel erected for Prince Alfonso Doria Pamphili is built of it.

The Eocene rocks of SOUTHERN EUROPE that fringe the shores of the Mediterranean, yield building material which is utilized on the French and Italian Riviera. Extensive quarries are worked in the neighbourhood of Mentone and Bordighera, and the specimens from the latter district are representative examples of this sandstone. It will be noticed they vary considerably in colour, although from the same quarry. The Bicknell Museum and All Saints' Church, both in Bordighera, are built of this stone.

---

[1] *Limestones and Marbles*, S. M. Barnham, p. 191.

TURKEY. The Island of Marmora, in the sea of the same name, is wholly composed of a limestone of the Eocene system. Since it has a fine grain and takes a good polish, it is often used for decorative work, but it is also employed for general building. It is in much request for facades of important buildings and palaces in Constantinople and other towns in the Ottoman empire.

The ISLAND of MALTA yields a very fine-grained arenaceous limestone, usually of a light straw colour, sometimes white, and occasionally with a pinkish tint. An examination of the specimen in the collection will shew that, although this rock belongs to the Oligocene series of the Tertiary age, it bears a striking resemblance to the celebrated French Caen stone of the Great Oolite. It is employed for building in Valetta and other parts of the Island, and is also sent to more distant regions bordering on the Mediterranean. The stone is used for building in Naples.

AFRICA. Egypt. The Nummulitic limestones of Egypt suitable for building are well represented in the collection. Most of the quarries are situated in the hilly range of Jebel Mokhattam, which extends from immediately south-east of Cairo, for about 25 miles east from the Nile. These quarries furnish the building material for modern Cairo and its surroundings, but their chief interest is centred in their former use, as a source of supply for the structural undertakings of the ancient Egyptians. It will be observed on examining the specimens that there is a great variation in the texture of the rock; some examples are soft and chalky, whilst others are hard and compact, almost capable of taking a polish; of the latter, the specimens from the Tura Prisons quarries are typical illustrations. Vast quantities of both qualities were used for building the ancient Pyramids of Gizeh. The softer Nummulitic stone was employed for the general structural work, while the hard compact rock, like the specimen from the Tura Prisons quarries, was used for the casing of the Pyramids.

No definite evidence seems to be forthcoming at present to associate any particular quarry as the source of the supply for the construction of the Pyramids, but it seems probable that the group of quarries on the Mokhattam Hills yielded both qualities of stone for the erection of these huge mausoleums of the ancient kings.

Some students of Egyptology believe that only the fine stone used for the casing and passages of the Pyramids of Gizeh was brought from the Mokhattam cliffs, and that the bulk of the masonry was quarried in the neighbourhood. Recent explorations have brought to light some evidence which leads one to conjecture that such may have been the case. Half a mile south of the Great Pyramid a ridge of rock rises from the plain above the Arab cemetery where a group of trees forms a well-known object in the landscape. This ridge runs south for half a mile and is riddled with tombs, especially at its southern end. The plain between the hills, to the west of this ridge, is found to be covered with broken stone chips many feet thick. The site is thought to be too remote to have been used for the waste from the pyramid building; there is therefore a possibility that a bed of good stone, similar in quality to that forming the pyramids, existed there, and that this has all been worked out for the pyramids, and only the quarry waste left on the ground[1].

A modern historian, when referring to Khufu or Cheops, the builder of the Great Pyramid which bears his name, writes: "This wonder of the ancient world, the sepulchre of the monarch, was constructed at great expense and suffering. One hundred thousand men, changed every three months, being employed by forced labour, ten years, in constructing the causeway by which

[1] *Gizeh and Rifeh*, W. M. Flinders Petrie, 1907, p. 9.

the blocks of stone were transported from the Tura quarries in the Arabian chain, to the quay on the banks of the Nile, where they were transported by boat to the other bank[1]." The method adopted by the Egyptians to raise and place in position the huge blocks of stone which form the outer covering of these royal sepulchres, is still a matter of conjecture.

It may be interesting to add that the outer casings of the Pyramids of Gizeh were not composed entirely of limestone. In consequence of the covering of the Great Pyramid having almost completely disappeared, there is some doubt as to its composition, but there is evidence to shew that the casing of Nos. 2 and 3 Pyramids were composed of limestone, except the lower courses, which were of granite[2].

The Tura quarries, besides producing the material for the tombs of the ancient Kings of Egypt, supplied building stone for their splendid temples, eclipsing in size and decorative grandeur the sepulchral chambers of the Pyramids. Hieroglyphical tablets have been discovered in these quarries which furnish us with the necessary information. One example, as an illustration, may be interesting. It is in the shape of a propylaeon, and a long inscription round the lintels, although partially effaced, records, that Amenophis II, "under the titles of the Mighty Horus, the greatly vigilant, the King, the Sun, the chief of the World, has paid devotion to the gods and goddesses, and has opened the quarries to take away the good and white stone for the temples[3]."

These limestones are still extensively employed for structural purposes. Most of the important buildings in Cairo are made with them. The Tura Prison Buildings are constructed with the hard, slightly shelly rock from the quarries in the immediate neighbourhood, also the Lady Cromer Hospital.

[1] *Encycl. Britannica*, 9th ed. Vol. v. p. 582.
[2] *Pyramids of Gizeh*, W. M. Flinders Petrie, p. 32.
[3] *Ibid.*, Col. Vyse, p. 95.

The specimen from the Palomba quarries, at Abbas-sieh, about three miles north of Cairo, represents the stone that has recently been employed for constructing the Government Barrage.

Most of the buildings at Abbassieh are composed of the dark cream-coloured rock from the quarries at Giuschi, which are situated a little south of the Palomba quarries. The specimen indicates that it is much softer than the Palomba stone; consequently it is pre-ferred for ordinary building, being more easily worked.

Travelling further south are the well-known quarries of Atar el Nabi, producing a close-grained, cream-coloured rock, known commercially as " Batur Stone." The quarries are situated under the windmill plateau, near St George's, a suburb of Old Cairo. This limestone is said to command, as a building material, a good reputation around Cairo, and on account of its low absorption is well suited for piers and similar works of construction under water. Dr Hume, of the Egyptian Geological Survey, states it to be the most compact stone found round Cairo[1].

MOROCCO. A few miles to the east of Tangier, on the flanks of the Darka Hills, the highest summit of which is the well-known Jebel Musa, better known as Ape's Hill, there are outcrops of Nummulitic limestone, which are occasionally used for building purposes in Tangier, and other towns in North Morocco. The specimens indicate that this stone is not so light in colour as the rocks from the same series in Egypt, neither is it so durable, because in Morocco, when the stone is used for structural purposes it is invariably coated with cement or mortar to prevent decay, which is not the case in Egypt. The specimens indicate that the rock varies in texture; that from the quarry near Jews river is a sample of the soft variety, while the cube from the Three Saints quarry, about two miles south-east of Tangier, represents the harder rock.

[1] "Building Stones of Cairo," W. F. Hume, *Survey Depart. Paper*, No. XVI. p. 11.

ASIA MINOR. A coarse-grained pinkish grey rock belonging to the Eocene series, occurs on the eastern slopes of the Sea of Marmora, and is quarried for building purposes near Ismid. The specimen from the Heréké quarries is an example. The coarse texture of this stone does not admit of its being used when decorative work is required, but it is extensively employed locally for ordinary structural purposes. The walls of the General Post Office, and those of the Greek Church, at Kadekena, are constructed of it.

INDIA. In the Pegu Division of Lower Burma there is a sequence of rocks which is classed by Indian geologists as the Pegu group, part of which consists of Nummulitic limestone of the Eocene system. An outcrop of this Pegu rock is quarried a little to the west of Myanaung, and yields a building stone which is used in Rangoon and other towns in Lower Burma for structural purposes. As the specimen indicates, the stone is a medium-grained light yellow calcareous rock.

AUSTRALIA. Victoria. The brown coal of the Eocene beds of Victoria is followed by a series of marine beds—limestones, clays, and sandstones—extending along the south coast and running in a series of gulfs northward into the country[1]. Where the limestone occurs it yields a useful building material. As the specimens shew, it is a yellow crystalline stone, somewhat similar in appearance to the Magnesian Limestone of the Permian system in Europe. The stone enjoys a good reputation among architects in the State. St Paul's Cathedral in Melbourne is built of it.

South Australia. Rocks of the Tertiary era, which rest on the Cambrian formation of the Murray Basin in the south-east corner of South Australia, yield useful building sandstones. The specimen from the quarries at

[1] *Australasia*, J. W. Gregory, p. 418.

Murray Bridge is a good example of this stone, which is extensively employed in Adelaide for structural purposes. The Telegraph Offices are built of it.

WESTERN AUSTRALIA. The stone most generally used for ordinary and second-class buildings in this State, especially in the municipality of Perth, is the coastal limestone of Tertiary age obtained chiefly from Rottnest Island and Cottesloe. The stone of this series varies greatly in the quantity of lime present, being frequently more of a calcareous sandstone than a limestone[1]. The specimens from the Rottnest Island quarries are representative examples, resembling in physical character the calciferous sandstones of Scotland.

NEW ZEALAND. A white limestone, known as "Oamaru Stone," also distinguished among builders as "K" stone, occurs over a large area of the South Island, and is quarried in the Oamaru district of Otago. This belongs to the Oligocene system[2], and, as the specimen indicates, is a somewhat soft and chalky rock. It makes an excellent building stone, especially where decorative work is required, but is not very suitable for foundations or lower superstructures, where it is liable to abrasion. Most of the principal buildings in Dunedin and other towns in New Zealand are built of it. It has also been exported to many parts of Australia, as well as to South Africa. In the building of the Bank of New Zealand, Dunedin, the stone was used from the lower sill-course upwards. It was also employed for the facings of the Law Courts in that city. The First Church is another example which is built entirely of Oamaru Stone except the foundations.

The same system furnishes a building stone in the district of Ashburton, but of a much harder nature than the Oamaru Stone, as the specimen shews. It is

[1] Information given by the Government geologist, A. Gibb Maitland, Esq.

[2] *Australasia*, J. W. Gregory, p. 599.

also a little darker in colour. The principal quarries are situated near the small town of Mount Somers, about 20 miles from Ashburton. This is principally used for the lower courses of buildings, the softer Oamaru Stone forming the superstructure.

JAMAICA. Sandstones of the Eocene age, useful for building purposes, are found in abundance on the Island. The most notable example occurs at Kellett, in the parish of Clarendon, in the county of Middlesex. As the specimens shew, it is a reddish brown, fine-grained grit. The Sugar Works, Hospital, and other buildings of importance, are constructed of this stone, some of which are elaborately finished and display a considerable amount of carving.

The other specimens represent sandstones belonging to the same geological formation, which are quarried at Serge, in the parish of St Thomas in the East, County of Surrey. Although this rock is more laminated than the Kellett Stone it is a useful building material and has been employed in the construction of the Residence, Aqueduct, etc.[1]

ANTIGUA is chiefly composed of volcanic eject-ments, including basalts, dolerites, and tuffs.

In the north-eastern portion of the Island however Tertiary limestones occur, and these rocks are employed by the inhabitants for building purposes. They resemble in many respects the yellow limestones of Jamaica, but, as the specimen indicates, are slightly paler in colour. It is practically the only stone on the Island that is used for building. The Cathedral and the Court House in St John are constructed of it.

[1] " Geology of Jamaica," J. G. Sawkins, *Memoirs of the Geol. Survey*, 1869, p. 30.

MIOCENE.

AUSTRIA. The sandstones of the Eastern Alps and the Foraminiferal limestones of the Leitha-Gebirge, both belonging to the Miocene system, are quarried within a short radius of Vienna, and yield a good building material, which is largely employed by the Viennese for structural purposes.

There is evidence of these rocks having been used for building from a very early date. The sandstone claims precedence as regards antiquity. It must have been quarried as far back as the 12th century at any rate, the stone having been employed for building the oldest part of the Hofburg, or Imperial Palace, of old Vienna, which was founded early in the 13th century[1]. The same stone was used for building the Church of Maria-Stiegen in Vienna, an interesting Gothic building of the 14th century.

The Leitha-Gebirge limestones were not far behind in their introduction for economic purposes, because as new buildings were added to the Hofburg in the 15th century, this limestone was employed in their construction. The façade of the College of Industry built in the 16th century is another example of its use.

Both the sandstone and limestone quarries are still worked and buildings of importance have been erected in Vienna within the last 50 years with these rocks. The barracks and fortifications commenced in 1860 are illustrations, also the Royal Museum built in 1876.

---

[1] *Encycl. Britannica*, 9th Ed. Vol. XXIV. p. 220.

It will be observed that one of the limestones is a highly fossiliferous compact rock, and would be very suitable for decorative work.

The light brown and grey specimen of conglomerate from quarries at Baden, near Vienna, is a sample of the stone used by the Viennese engineers for constructing and maintaining the aqueducts in the city. This stone is said to have been extensively employed by the Romans for millstones.

The Leitha-Gebirge limestones extend into Hungary as far as Budapest, in the neighbourhood of which the stone is quarried for building. The specimens indicate that it is more compact than the rock of the same group in the Vienna district, and takes a good polish. Some of the quarries date back over a long period. That from the Duna Almas quarries represents the stone formerly used for erecting fortifications in and around Budapest, and is now extensively employed in that town for general building purposes, and it is also sent to Vienna. The cream-coloured specimen from the St Margit quarries, also near Budapest, is another example; it is chiefly used for building churches in Hungary.

In GREECE the Miocene beds furnish limestones of wide distribution and economic value. An excellent building stone, usually known as "Poros Stone," which name it also bore in ancient times, is used throughout the whole country. The rock is found in the Isthmus of Corinth in Megara, also in the Island of Ægina, and in the peninsula of Argolis. It likewise occurs on the western declivity of the wild mountain ranges of Taygetus, popularly known as Pentedaktylon, in Laconia, and in the valley of Alpheus, extending as far north as the borders of Elis[1].

The formation in the Island of Ægina, as the specimens indicate, yields white and pink rocks, some of them rather porous and cellular in their texture

[1] *Limestones and Marbles*, S. M. Barnham, p. 202.

but most of the examples are fairly compact and of a delicate cream colour. The Temple of Aphaia, the so-called Temple of Athena, on the Island, was built of this stone.

The cube from the Pireus quarries shews the stone used for building the British School at Athens.

SPAIN. Deposits of Tertiary age cover more than a third of this country. They are chiefly composed of the accumulations of the great lakes which in Oligocene and Miocene times spread over so large an expanse of the table-land of Spain.

These freshwater deposits occupy nearly all the central provinces of the country, including Madrid, Segovia, and Cuenca. The formation may be separated into three divisions. Limestone prevails in the Upper almost exclusively; the Middle consists chiefly of clay and gypsum; while the Lower is made up of sandstones and conglomerates.

A considerable portion of the limestone forming the Upper bed has been removed by denudation, but where it has not been so affected it yields a valuable building stone. The city of Madrid is largely built of it. As may be seen from the numerous specimens in the collection it varies in texture, but not in colour, a light pinkish cream being the prevailing tint. The best building stone is quarried in the neighbour-hood of Colmenar de Oreja, about eighteen miles north of Madrid[1]. The specimen from these quarries indicates that the rock of this outcrop is more compact than those from other parts of the province. The beds occurring in the Province of Segovia yield a softer and more chalky stone than those of Madrid. The specimen from the Otero quarries is an example.

These lacustrine deposits extend eastward to Valencia and Murcia, traversing those provinces and eventually reaching the sea a little north of Alicante. The specimens from the group of quarries in this region

[1] *Quart. Journal Geol. Survey*, Vol. XXII. p. 84.

indicate that, as it extends eastward, the rock loses its pinkish cream tint and assumes a light yellow or grey colour, and some, represented by the specimen from the Novelda quarries, are distinctly arenaceous, practically a very fine sandstone, and belong probably to the Lowest division of the series. This may be expected, as the different kinds of rock are found alternating with each other, and varying even within short distances. The rock is extensively quarried and used for building in Alicante, Valencia, Carthagena and other towns on the east coast.

SWITZERLAND. The "Molasse" of Switzerland belonging to the Miocene system extends in a S.W. to N.E. direction, from the Lake of Geneva as far as Würtemberg. It is extensively employed for building purposes in the towns on the lowlands of Switzerland, between the Alps and the Jura. This stone, like the "Pierre Jaune," is too friable to stand much rubbing, and is liable to abrasion when exposed to the wear and tear of ordinary traffic. Careful selection in quarrying is necessary, the quality being very variable. There are several examples in Geneva, also in Berne, where some buildings erected with this stone are sadly decayed. The compact limestones of the Jurassic system existing in that country are now generally employed for foundations and the lower courses of buildings, although it is more expensive to quarry and dress than the soft Molasse[1].

The "Molasse Marine," which is generally preferred for building, is made up largely of fossiliferous sandstones, while the freshwater Molasse is more calcareous. The specimen from the Savigny quarries, near Lausanne, represents the "Molasse Marine" used for restoring the Cathedral in that town (1906).

The same formation is quarried for building purposes in the neighbourhood of Lucerne, and that

[1] See p. 189.

from which the specimen comes is close to the famous Lion of Lucerne, which is hewn out of the rock itself.

The specimen of Molasse taken from the quarries at Estavayer on the shores of Lake Neuchâtel, is much harder and very different in texture to the rocks just mentioned. It is employed for steps and staircases in the large hotels and other important buildings in Switzerland.

INDIA. Spread over a considerable area of SIND is a mass of fossiliferous limestones and calcareous beds, easily distinguishable from the limestone of the older Tertiary formations by the absence of Nummulites[1]. This group of rocks is known as the Gáj beds, as they are called from the Gáj river, on the banks of which they are exposed. They form the low hills east and north-east of Karáchi, and furnish materials of which the houses of the town are mostly built. They chiefly consist of straw-coloured limestones of a fairly compact nature. The specimens from Hands Hill quarries, about eight miles east of Karáchi, are examples of this group of rocks, which is classed by geologists as belonging to the Miocene system.

The appearance and texture of the rock belonging to the Gáj group, in the neighbourhood of Chizree, is somewhat different from the Hands Hill deposits. The specimen indicates that it is lighter in colour and harder in texture. This stone is chiefly used for road-making, although it is occasionally employed for building by the native Indians.

In Upper Burma, in the district of Lower Chindwin. also in the Northern Shan States, the group of Pegu rocks which furnished the Nummulitic limestones mentioned when describing the Eocene system of India, also includes an immense thickness of shales and sandstones, believed by some geologists to belong to the

[1] *Geology of India*, R. D. Oldham, p. 311.

Miocene system [1]. The sandstones yield a rock which is utilized for building purposes. The Lower Chindwin rock is slightly ferruginous, shewing dark bands in the yellow stone, as the specimen indicates.

The rock from the Shan States is a beautiful example of stone for decorative purposes, possessing a fine even grain, and having a pleasing light terra-cotta tint. It is extensively employed for building in Rangoon. The appearance of the specimen suggests that the rock belongs to an older age than the Tertiary; but the region has not yet been surveyed, consequently it is not possible to associate it with the Trias, or Permian system, which the general appearance of the rock seems to suggest.

CUBA. The three specimens of Miocene limestone represent the class of stone that is used on this Island for building purposes. Granite is present, but it is partially decomposed, and the only use made of it by the Cuban builder is to grind it up and mix it with cement, to form a concrete for foundations. The specimens shew that the limestone varies in colour, from a pure white to a rich cream. The texture also alters considerably, namely from a fine white chalky stone to a coarse vesicular rock. The demand for this class of building material is very limited, because stone is only used for very important buildings. The ordinary Cuban is satisfied with his palm-leaf roofed hut and palm-bark walls. A large Hotel is at present (1908) in course of construction at Havana, where all three descriptions of limestone, illustrated by the specimens, are being employed.

JAMAICA yields a whitish soft saccharoidal lime-stone, belonging to the Miocene system, which is useful for building, and is employed for that purpose on the

---

[1] *A Summary of the Geology of India*, E. W. Vredenburg, p. 58.

Island. The chief deposits worked lie in the Parish of Portland, and the rock is popularly known as "Portland Chalk Stone," of which the Church of Port Antonio, considered to be the best specimen of architecture on the Island, is built[1]. Some varieties of this rock are streaked with light brown veins, and are capable of taking a fair polish. Of these the specimen in the Museum is an example.

[1] "Geology of Jamaica," J. G. Sawkins, *Memoirs of Geol. Survey*, 1869, p. 29.

PLIOCENE.

GREECE. The northern portion of the Morea, the ancient Peloponnesus, is composed of a calcareous conglomerate belonging to the Pliocene system. Similar rocks are also present on the north-eastern slope of the mountains which border the Gulf of Corinth, and stretch into Boeotia and Attica. The specimens are from quarries in the eastern or Attic portion of this province, and are examples of the many and various shades of colour exhibited by these conglomerates. Although useful for building walls and other ordinary structural work, these rocks, from their physical characteristics, are not adapted for facing or decorative work. They were extensively employed for foundations and sub-structural work of ancient buildings of note, including the Theatre of Dionysus and the Odeum of Herodes Atticus.

CYPRUS. The specimens of limestone from quarries near Larnaca represents the rock, belonging to the Pliocene system, used by the Cyprians for building. A rock which occurs in the neighbourhood of Nicosia is sometimes used for internal construction, but the stone from the Larnaca district is that usually adopted for general building.

It will be seen, by examining the specimens, that the stone varies in colour. The rock from the Helofaghton quarries is white, while that from the Voroklini quarries is a dark cream colour. The latter is usually considered to be superior to the former rock, although both are quarried in the same district.

A bed of rock near Limasol, belonging to the same geological series as that quarried near Larnaca, is also employed for structural purposes. The Cyprian builder generally prefers this stone to the white cellular rock from Helofagton, but reckons it inferior for building to the Voroklini stone.

AFRICA. ALGERIA. The coast of the Oran division of this French colony is chiefly composed of a semi-compact, light-coloured limestone of the Pliocene system, which extends inland for a considerable distance, occupying a large portion of the area which is known as the Tell Zone. This rock is quarried for building purposes a little to the west of the town of Oran. There are two varieties, one a hard compact stone, suitable for foundations and external work, the other much softer, and well adapted for internal construction. They are both of a light cream colour. The soft variety is represented by the specimen from the Raz-el-Ain quarries, the hard one by that from Karguentah, near Gambetta. The latter kind is not wrought in open quarries, but is excavated in underground workings, which are approached by inclined shafts, like those used in working the Bath Stones of England. This limestone is extensively used for building in and around the town of Oran, for instance the Municipal Offices, the Theatre, and the Police Station.

TURKEY in ASIA. ASIA MINOR. On the slopes of the hills on either side of the Mœander valley, a little south of Ephesus, are numerous outcrops of limestone, belonging to the Pliocene system, which are quarried for building purposes. The specimen from quarries at Sarakeni is an example. The stone is conveyed to Smyrna, about 100 miles distant, for use as a building stone.

PLEISTOCENE AND RECENT.

GERMANY. Bordering on the Black Forest, in the vicinity of Urach, are calcareous deposits, formed by the deposition of carbonate of lime from solution, and known in Germany as "Tuffstein," corresponding to the Travertine of Italy. It is soft when first quarried and can easily be sawn into slabs, but hardens under exposure and makes an excellent building stone. The dark cream-coloured vesicular specimen from the Seeburg quarries is an example, and another one will be found in the light brown specimen from Gröningen.

A somewhat similar deposit exists at Munster, on the Neckar, but this differs in texture from the Seeburg Stone. The specimen indicates that it is much more compact; consequently it has a higher specific gravity, and is more useful for heavy engineering work.

ITALY. Travertine, a rock extensively employed in the construction of ancient Rome, and still the chief building material used in the modern Italian metropolis, is a calcareous deposit of a creamy colour, precipitated from running water. The specimens from the Tivoli and Sabino quarries near Rome are typical examples of this celebrated building material, which the Romans knew as "Lapis Tiburtinus." This stone when first quarried is sufficiently soft to be cut by a saw, but hardens when exposed to the atmosphere. The colour of the two specimens is almost identical, but the texture differs considerably, one being much more cellular than the other.

As previously mentioned[1], Travertine was much used in Rome for building about the middle of the 1st century. The external walls of the Colosseum were built entirely of it, and what remains is a striking illustration of the durability of the material. The Courts of Justice in Rome are an example of modern construction with Travertine, and the walls of St Peter's Church are built of it. The columns which support the arched roof of the new Railway Station in New York (1910) are composed of Roman Travertine.

SWITZERLAND. Two specimens will be seen in the collection representing the Pleistocene system of this country, which bear a resemblance, both in colour and texture, to the Roman Travertine. It occurs in several districts throughout the Bernese Oberland, and is known by the name of "Tuffstein." Like the Travertine, this rock is soft when first quarried, and can easily be cut with a saw, but hardens rapidly when exposed to the air. It then becomes a useful building stone, and is employed for important structural undertakings, where strength is essential. It is at present being used in the construction and maintenance of the mountain railway which winds about amid the grand environs of Grindelwald.

AFRICA. South Nigeria. The geological formation of a considerable part of this territory, especially the western side, consists of late Tertiary gravelly sands. In some districts iron oxide has acted on these beds, forming an indurated ferruginous sandstone, of a warm red colour, which is occasionally used for building. Although most of the houses in the territory are constructed of mud, the inhabitants who aspire to a more substantial home employ the sandstone just described, using the mud as mortar, and sometimes covering the outside with cement. The specimen in

[1] See p. 82.

the collection was quarried near Onitsha, where the stone has lately been used for constructing the Roman Catholic Mission House.

CAPE OF GOOD HOPE. On the coast of this part of South Africa, there are considerable dune deposits which have been in part or entirely cemented to a fairly compact rock by the deposition of carbonate of lime, derived from the solution of shell fragments[1]. In some instances, by the long-continued deposition of the carbonate of lime, the sand dunes are converted into hard rock through a distance of many feet from the surface, and where repeatedly wetted and dried, as happens when the sea has encroached upon old dunes, the rock becomes intensely hard and weathers with a peculiarly jagged surface.

At Hoetjes Bay, an inlet of Saldanha Bay, the sandstone derived from these hardened dunes has been quarried for building[2]. The specimen in the collection is from the Hoetjes quarries. The General Post Office and the South African Museum in Cape Town, both built in 1895, are constructed of this stone. These buildings unfortunately shew serious signs of decay. Whether this is due to the stone not being a durable one in the Cape climate, or sufficient care has not been exercised in selecting a stone that has been thoroughly indurated, is a matter of conjecture at present.

ZANZIBAR. This Island is chiefly composed of a soft chalky coralline limestone, which is used by the inhabitants for building purposes. As the numerous specimens indicate, this formation yields rock of varying fineness and texture. Some are fairly compact, whilst others are coarse and vesicular. One specimen in particular displays the cellular character in a marked degree. This is chiefly caused by the presence of shells, many of which have disappeared, leaving casts

[1] *Geology of South Africa*, Hatch and Corstorphine, 1st Ed. p. 264.

[2] *Geology of Cape Colony*, A. W. Rogers, p. 373.

only. This specimen resembles somewhat that from the Roche bed of the Portland group of Britain.

In regard to the age of these rocks it should be said that they appear in the collection as belonging to the Pleistocene. Dr Werth of Berlin, who has recently studied the geology of this Island, refrains from deciding whether the age of this formation is Pliocene or Pleistocene. He also describes a Post-Tertiary deposit of sand on this Island[1]. In this deposit a consolidation has in some instances taken place, doubtless aided by carbonate of lime in solution, forming a fairly hard material. It is used for building, and the specimen of it in the collection suggests that it is better adapted for structural purposes than the friable, and, in most cases, cellular coralline rock. The High Court, the Post Office, and other Government buildings in the town of Zanzibar, are built of the sandstone, while for the native dwellings the soft limestone is chiefly used.

TURKEY in ASIA. In the south-west corner of PALESTINE, close to Gaza, deposits of light yellow, cellular limestone occur, which are quarried for building purposes. This rock, as the specimens indicate, resembles the Travertine of Italy, and has doubtless been formed in the same way.

Another specimen in the collection from this district represents a rock which is practically a consolidated mass of broken shells, some of a considerable size. This rock is excavated from the cliffs on the sea shore three miles from Gaza, and is employed occasionally in that town for rough building, but chiefly for constructing tombs in the burial grounds of the district.

INDIA. A specimen will be found of Pleistocene age from quarries near Cochin, on the Malabar coast, which has a singularly soft and friable appearance,

---

[1] E. Werth, *Zeitschrift der deutschen geologischen Gesellschaft*, Vol. LIII. 1901, p. 287.

somewhat resembling decayed brick. It is the product
of disintegrated gneiss, largely distributed in the
Southern Presidencies of India, and known as "Laterite." In its normal form, it is a porous, argillaceous
rock, mottled with various tints of brown, red and
yellow, and a considerable proportion sometimes consists of white clay.

In many forms of Laterite the rock is traversed by
small irregular tortuous tubes, from a quarter of an
inch to upwards of an inch in diameter. The tubes are
most commonly vertical, but their direction is quite
irregular, being sometimes even horizontal.

When first quarried the rock is so soft that it can
easily be cut out with a pick, sometimes even with a spade,
but it hardens greatly on exposure[1]. When the stone
is sufficiently hardened after being quarried, it is
then chipped or chiselled into shape and used for
building. It seldom attains a hardness which will enable it to resist the decaying influences of the weather,
so it is usually covered with plaster or mortar as a
protection. In the coastal districts, however, many
temples, some of considerable antiquity, are built solely
of Laterite, and appear to have stood well[2].

The rock is generally divided into two classes, high-
level and low-level Laterite. The specimen in the
museum is typical of the low-level Laterite from the
coast of Malabar. Mr P. Lake describes two distinct
varieties found in that district, distinguished as vesicular Laterite and pelletty Laterite[3]. The specimen
represents the former variety.

Overlying the limestone and Laterite beds in the
district of Surat, a ferruginous sandstone is found,
which is transported to Bombay, and there used for
building operations. Although, as the specimen indicates, this stone is somewhat coarse-grained, it is
frequently used for decorative work, being soft when

---

[1] *Geology of India*, R. D. Oldham, p. 371.
[2] *Ibid.*, V. Ball, Pt. 3, p. 550.
[3] *Memoirs of Geological Survey of India*, Vol. XXIV. p. 217.

freshly quarried, and possessing the property of hardening by exposure to the atmosphere.

Geologists are still uncertain whether these deposits belong to the Pleistocene or the Pliocene, but meanwhile the specimen has been placed among the rocks belonging to the former division.

On the coast of Káthiáwár a comparatively modern marine limestone is found, which is extensively employed for building in Bombay. It is generally known as Porebandar Stone, from the name of the port whence it is shipped, or Miliolite, a name proposed by Dr Carter. The typical miliolite, as the specimen shews, is a finely oolitic freestone, largely composed of Foraminifera, which form the nuclei of the oolitic grains; but near the sea coast the limestone is not infrequently mixed with a large proportion of sand[1]. This stone, it is stated, is admirably suited for building purposes, but is also said to be incapable of sustaining great pressure[2].

CEYLON. The collection contains specimens of Laterite from quarries near Colombo, and in the neighbourhood of Galle, in the Island of Ceylon. It will be noticed, on examining the specimens, that these examples are of a much more stable nature than that just mentioned from the Malabar coast. The Ceylon specimens also differ somewhat in texture one from another. This stone, which is also a product of disintegrated gneiss, and known in Ceylon as " Cabook," exists in vast quantities in many parts of the Island. The beds are often thirty feet or more in thickness. The rock is quarried and prepared for building in the same manner as that of the Malabar coast. Many native houses on the Island are built of this material. When protected from the weather by a coating of mortar, which is usually composed of two parts of sand

[1] *Geology of India*, R. D. Oldham, p. 395.
[2] *Ibid.*, V. Ball, p. 465.

to one part of coral lime, these buildings prove to be very serviceable.

STRAITS SETTLEMENTS. Among the Recent sedimentary rocks of the Island of Singapore, conglomerates occur which are utilized for constructive purposes. They are more often indurated than friable. Their pebbles consist chiefly of a dull whitish quartz, a greyish variety being also common. Black, red, and yellowish pebbles occur sparingly[1]. These characteristics appear in the specimen from the Mount Faber quarries, which are situated near the centre of the Island. This stone has been extensively employed for constructing the Sea Wall at Singapore.

Beds of coral are also occasionally quarried on the Island, and the material is employed for making the foundations of houses in Singapore. Of this the specimen, shewing some shells imbedded, is an example.

PERSIA. The hills on either side of the river Karan, a tributary of the Euphrates, in the province of Khazistan, are largely composed of rocks belonging to the gypsiferous series of the Upper Tertiary system. These beds consist of gravel and sandstone and the lower part of fine sandstone alone[2]. The stratification is horizontal. These rocks are used in the province for structural purposes, the town of Shushter being almost entirely built of them. The greyish-coloured specimen, from a quarry on the low hill overlooking the town, is an example of the material of which the Castle, or Kalah, still to be seen in Shushter, is composed. This Castle is said to have been built in the 3rd century by Shápúr I, King of Persia.

BARBADOES. The Pleistocene rocks of this Island yield a coral limestone which makes a useful building stone. It resembles the British Oolites in possessing

---

[1] *Journal of Geological Society*, 1851, J. R. Logan, p. 318.
[2] *Quart. Journal Geol. Society*, W. K. Loftus, Vol. XI. p. 260.

the property of hardening on exposure to the atmosphere. When freshly quarried it is soft and can easily be cut with a saw, and on that account it is locally known by the name of "Sawstone[1]." Porosity is its only fault as a building stone. Here and there however layers of hard and compact rock occur, which break with conchoidal fracture. The specimens in the collection represent both qualities. The highly porous nature of one specimen is very manifest, not only by its appearance, but also by its low specific gravity. When the Barbadian employs the softer quality of stone for building, he generally coats the outside walls with plaster or cement.

BAHAMAS. The specimen cube from the Nassau quarries in New Providence, is a good example of the only kind of building stone found in this group of Islands, which are all formed of coral and shell, hardened into limestone, and without a trace of igneous or other sedimentary rock. It has a hard surface perforated by innumerable cavities. Underneath it gradually softens, and furnishes an admirable building stone. It can be sawn into blocks of any size, which harden on exposure to the atmosphere[2].

BERMUDA ISLANDS. Nearly all the rocks of the Bermudas, above sea-level, and to a considerable depth below it, are made up of wind-drifted shell-sand, with material derived from corals and other organisms, such as Foraminifera, corallines, etc. These materials when consolidated, form a true aeolian limestone, sometimes friable, but in many places very hard and compact. The specimens exhibited illustrate this useful building material, in several degrees of hardness. Many of these limestones are so soft before being exposed

[1] *Geology of Districts*, Harrison and A. J. Jukes-Browne, 1890, p. 57.
[2] *Stones for Building and Decoration*, G. P. Merrill, p. 202.

to the atmosphere, that they are quarried by large chisels, and cut with ordinary saws, into regular blocks as easily as wood, but they become quite hard and suitable for building after exposure to the air for a few weeks. The softer and more friable variety is sometimes sawn into thin slabs and used as a roofing material for the houses of Bermuda[1].

[1] *Stones for Building and Decoration*, G. P. Merrill, p. 202.

BRITISH BUILDING STONES

IGNEOUS ROCKS (PLUTONIC).

1. ENGLISH GRANITE. *MUSCOVITE-BIOTITE-GRANITE*
Light greenish grey, porphyritic, felspar crystals, unusually large
Lamorna Quarries, Penzance, **Cornwall**
Presented by Messrs John Freeman Sons & Co., Ltd., Penryn
Chemical Composition (*Geol. Survey*, 1907):
SiO_2 70·17, TiO_2 0·41, Al_2O_3 15·07, Fe_2O_3 0·88, FeO 1·79,
MnO 0·12, MgO 1·11, CaO 1·13, K_2O 5·73, Na_2O 2·69,
H_2O 0·88, NaCl 0·06, F 0·15, P 0·34, S 0·04, =100·57
Weight per cub. ft. 168 lbs.

2. ENGLISH GRANITE. *MUSCOVITE-BIOTITE-GRANITE*
Light grey, coarse-grained, porphyritic
Penryn Quarries, Penryn, **Cornwall**
Presented by Messrs John Freeman Sons & Co., Ltd., Penryn
Chemical Composition (*Geol. Survey*):
SiO_2 72·84, Al_2O_3 16·25, Fe_2O_3 0·14, FeO 1·49, MgO 0·55,
CaO 1·10, K_2O 5·19, Na_2O 2·25, H_2O &c. 0·63, =100·44
Weight per cub. ft. 165 lbs. Spec. grav. 2·65. Absorp. 0·12 %
Crushing strain per sqr. ft. 1250·8 tons (Kirkaldy)

3. ENGLISH GRANITE. *MUSCOVITE-BIOTITE-GRANITE*
Light greenish grey, medium-grained
Carnsew Quarries, Penryn, **Cornwall**
Presented by Messrs John Freeman Sons & Co., Ltd., Penryn
Chemical Composition (*Geol. Survey*):
SiO_2 72·05, Al_2O_3 15·83, Fe_2O_3 0·39, FeO 1·50, MgO 0·51,
CaO 1·14, K_2O 4·79, Na_2O 2·65, H_2O &c. 0·64, =99·50
Weight per cub. ft. 168 lbs.
Crushing strain per sqr. ft. 1436·4 tons (Kirkaldy)

4. ENGLISH GRANITE. *MUSCOVITE-BIOTITE-GRANITE*
Light greenish grey, coarse-grained, porphyritic
Colcerrow Quarries, St Austell, **Cornwall**
Presented by Messrs John Freeman Sons & Co., Ltd., Penryn
Weight per cub. ft. 168 lbs.
Crushing strain per sqr. ft. 1336·1 tons (Kirkaldy)

5. ENGLISH GRANITE. *MUSCOVITE-BIOTITE-GRANITE*
Light greenish grey, medium-grained
    De Lank Quarries, near Bodmin, **Cornwall**
Presented by The Hard Stone Firms, Ltd., Bath
    Weight per cub. ft. 165 lbs.
    Crushing strain per sqr. ft. 1171 tons (Kirkaldy)

6. ENGLISH GRANITE. *MUSCOVITE-BIOTITE-GRANITE*
Light grey, coarse-grained, porphyritic
    Cheesewring Quarries, near Liskeard, **Cornwall**
Presented by Messrs John Freeman Sons & Co., Ltd., Penryn
    Weight per cub. ft. 168 lbs.
    Crushing strain per sqr. ft. 1440·3 tons (Kirkaldy)

7. ENGLISH GRANITE. *BIOTITE-GRANITE*
Light grey with pinkish tinge, medium-grained
    Tor Quarries, Merrivale, **Devonshire**
Presented by Messrs Duke & Sons, Princetown
    Weight per cub. ft. 165 lbs.

8. ENGLISH GRANITE. *TOURMALINE-BIOTITE-GRANITE*
Light grey with pinkish tinge, coarse-grained
    Princetown Quarries, Dartmoor, **Devonshire**
Presented by Messrs Pethick Bros., Ltd., Plymouth
    Weight per cub. ft. 163 lbs.

9. ENGLISH GRANITE. *TOURMALINE-BIOTITE-GRANITE*
Light grey with pinkish tinge, coarse-grained
    Princetown Quarries, Dartmoor, **Devonshire**
Presented by Messrs Pethick Bros., Ltd., Plymouth
    Weight per cub. ft. 163 lbs.

10. ENGLISH GRANITE. *TOURMALINE-BIOTITE-GRANITE*
Light grey with pinkish tinge, medium-grained
    Princetown Quarries, Dartmoor, **Devonshire**
Presented by Messrs Pethick Bros., Ltd., Plymouth
    Weight per cub. ft. 165 lbs.

11. ENGLISH GRANITE. *TOURMALINE-BIOTITE-GRANITE*
Light grey with pinkish tinge, medium-grained
    Princetown Quarries, Dartmoor, **Devonshire**
Presented by Messrs Pethick Bros., Ltd., Plymouth
    Weight per cub. ft. 165 lbs.

12. ENGLISH GRANITE. *Muscovite-Biotite-Granite*
Light grey, fine-grained, with white orthoclase crystals, porphyritic
 Gunnislake Quarries, near Tavistock, **Devonshire**
Presented by Lord Cowdray, London
 Chemical Composition :
SiO_2 70·95, Al_2O_3 17·04, Fe_2O_3 1·36, MgO 0·11, CaO 0·76,
K_2O 4·36, Na_2O 3·79, H_2O 0·67, =99·04
 Weight per cub. ft. 175 lbs.
 Crushing strain per sqr. ft. 900 tons (Kirkaldy)

13. ENGLISH GRANITE. *Muscovite-Biotite-Granite*
Greenish grey, with long porphyritic orthoclase crystals
 Blackenstone Quarries, Moreton Hampstead, **Devonshire**
Presented by Messrs J. Easton & Son, Exeter
 Weight per cub. ft. 163 lbs.

14. ENGLISH GRANITE. *Biotite-Granite*
Pink and brown, coarse-grained, large felspar crystals, porphyritic,
 known as "Dark Shap"
 Wasdale Fell Quarries, Shap, **Westmorland**
Presented by The Shap Granite Co., Manchester
 Chemical Composition (J. B. Cohen) :
SiO_2 68·55, Al_2O_3 16·21, Fe_2O_3 2·26, MnO 0·45, MgO 1·04,
CaO 2·40, K_2O 4·14, Na_2O 4·08, =99·13
 Weight per cub. ft. 160 lbs.
 Crushing strain per sqr. ft. 1200 tons (Middleton)

15. ENGLISH GRANITE. *Biotite-Granite*
Light pink, coarse-grained, large felspar crystals, porphyritic,
 known as "Light Shap"
 Wasdale Fell Quarries, Shap, **Westmorland**
Presented by The Shap Granite Co., Manchester
 Weight per cub. ft. 166 lbs. Spec. grav. 2·68

16. ENGLISH GRANITE. *Biotite-Granite*
Grey, medium-grained
 Eskdale Quarries, **Cumberland**
Presented by The Shap Granite Co., Manchester
 Chemical Composition (J. Hughes) :
SiO_2 73·573, Al_2O_3 13·750, Fe_2O_3 0·615, FeO 2·103, MgO 0·396,
CaO 1·064, K_2O 3·512, Na_2O 4·315, H_2O &c. 0·660, P_2O_5 0·012,
=100·00
 Weight per cub. ft. 162·5 lbs.

17. ENGLISH GRANITE. *MICROGRANITE*
Dark greenish grey, medium-grained
  Threlkeld Quarries, near Keswick, **Cumberland**
Presented by The Threlkeld Granite Co., Ltd., Keswick
    Chemical Composition (*Quarry*):
SiO_2 67·180, Al_2O_3 16·650, Fe_2O_3 0·559,   FeO 2·151,   MgO 1·549,
CaO 2·352, K_2O 2·914, Na_2O 4·032, H_2O &c. 1·549, CO_2 0·885,
P_2O_5 0·179,   = 100·000
  Weight per cub. ft. 164·37 lbs.   Spec. grav. 2·63
  Crushing strain per sqr. ft. 2052 tons (*Quarry*)

18. ENGLISH GRANITE. *HORNBLENDE-BIOTITE-GRANITE*
Dark red, medium-grained
  Mountsorrel Quarries, near Sileby, **Leicestershire**
Presented by The Mountsorrel Granite Co., Loughborough
    Chemical Composition (C. K. Baker):
SiO_2 67·16, Al_2O_3 16·19, Fe_2O_3 3·82,   MgO 1·58,   CaO 2·59,
K_2O 5·38, Na_2O 2·43, H_2O 1·02,   = 100·17
  Weight per cub. ft. 166 lbs.   Spec. grav. 2·661
  Crushing strain per sqr. ft. 2087·1 tons (Kirkaldy)

19. ENGLISH GRANITE. *HORNBLENDE-BIOTITE-GRANITE*
Dark grey with pinkish tinge, medium-grained
  Mountsorrel Quarries, near Sileby, **Leicestershire**
Presented by The Mountsorrel Granite Co., Loughborough
  Weight per cub. ft. 163 lbs.

20. ENGLISH GRANITE. *GRANOPHYRE*
Dark brownish red, fine-grained
  Stoney Stanton Quarries, near Narborough, **Leicestershire**
Presented by The Mountsorrel Granite Co., Loughborough
    Chemical Composition (C. K. Baker):
SiO_2 61·00, Al_2O_3 18·29,   FeO 5·02,   MgO 3·05,   CaO 4·00,
K_2O 5·28, Na_2O 2·00, H_2O 1·84,   = 100·48
  Weight per cub. ft. 170·5 lbs.

21. ENGLISH GRANITE. *GRANOPHYRE*
Dark red, fine-grained
  Croft Quarries, Croft, **Leicestershire**
Presented by The Croft Granite Co., Croft
    Chemical Composition (B. Blount):
SiO_2 62·04, TiO_2 0·43, Al_2O_3 17·38, Fe_2O_3 1·75,   FeO 2·65,
MgO 2·85,   CaO 3·54, K_2O 2·65, Na_2O 4·69, H_2O &c. 2·29,
= 100·27
  Weight per cub. ft. 171 lbs.   Spec. grav. 2·675
  Crushing strain per sqr. ft. 2359 tons (Kirkaldy)

22. ENGLISH GRANITE. *GRANOPHYRE*
Dark brownish red, medium-grained
Croft Quarries, Croft, **Leicestershire**
Presented by The Croft Granite Co., Croft
Weight per cub. ft. 157 lbs.

23. WELSH GRANITE. *GRANITE-PORPHYRY*
Light yellow and black mottled, medium-grained
Llanbedrog Quarries, **Carnarvonshire**
Presented by The Clee Hill Granite Co., Ludlow
Weight per cub. ft. 171 lbs.
Crushing strain per sqr. ft. 1331·8 tons (Kirkaldy)

24. WELSH GRANITE. *GRANITE-PORPHYRY*
Fawn-coloured, medium-grained, porphyritic
Llanbedrog Quarries, **Carnarvonshire**
Presented by The Clee Hill Granite Co., Ludlow
Weight per cub. ft. 171 lbs.

25. MANX GRANITE. *MUSCOVITE-GRANITE*
Light yellowish grey, medium-grained
Foxdale Quarries, near St John's, **Isle of Man**
Weight per cub. ft. 163 lbs.

26. MANX GRANITE. *MUSCOVITE-GRANITE*
Light yellowish grey, medium-grained
Foxdale Quarries, near St John's, **Isle of Man**
Weight per cub. ft. 171 lbs.

27. MANX GRANITE. *BIOTITE-GRANITE*
Grey, medium-grained
Dhoon Quarries, **Isle of Man**
Weight per cub. ft. 163 lbs.

28. MANX GRANITE. *GABBRO*
Dark blue-grey nearly black, medium-grained
Poortown Quarries, **Isle of Man**
Chemical Composition (Dickson & Holland):
SiO_2 47·13, TiO_2 0·58, Al_2O_3 8·48, Fe_2O_3 6·15, FeO 5·54,
MnO 0·64, MgO 13·61, CaO 11·34, K_2O 0·22, Na_2O 1·28,
H_2O 3·90, CO_2 0·47, P_2O_5 0·32, = 99·66
Weight per cub. ft. 190 lbs.

29. MANX GRANITE. *Aplite*
Fawn-coloured, fine-grained
　　　Crosby Quarries, **Isle of Man**
　　　　　Chemical Composition (Dickson & Holland):
SiO_2 74·39,　Al_2O_3 15·55,　Fe_2O_3 1·35,　MnO 0·22,　MgO 0·33,
CaO 0·48,　K_2O 2·14,　Na_2O 3·79,　H_2O 1·18,　=99·43
　　　Weight per cub. ft. 171 lbs.

30. LUNDY ISLAND GRANITE. *Muscovite-Biotite-*
　　　　　Granite
Light grey, coarse-grained, porphyritic
　　　Granite Quarries, **Lundy Island**
Presented by The Rev. H. G. Heaven, Lundy Island
　　　Weight per cub. ft. 171 lbs.

31. LUNDY ISLAND GRANITE. *Muscovite-Biotite-*
　　　　　Granite
Light grey, medium-grained
　　　Granite Quarries, **Lundy Island**
Presented by The Rev. H. G. Heaven, Lundy Island
　　　Weight per cub. ft. 171 lbs.

32. SCOTCH GRANITE. *Biotite-Granite*
Grey, medium-grained
　　　Dyce Quarries, Dyce, **Aberdeenshire**
Weight per cub. ft. 165·4 lbs.　Spec. grav. 2·65.　Absorp. 0·19%
　　　Crushing strain per sqr. ft. 1105·8 tons (Beare)

33. SCOTCH GRANITE. *Muscovite-Biotite-Granite*
Light silvery grey with mica specks, medium-grained
　　　Kemnay Quarries, near Aberdeen, **Aberdeenshire**
Weight per cub. ft. 161 lbs.　Spec. grav. 2·58.　Absorp. 0·21%
　　　Crushing strain per sqr. ft. 1211·1 tons (Beare)

34. SCOTCH GRANITE. *Muscovite-Biotite-Granite*
Bluish grey, fine-grained
　　　Rubislaw Quarries, Aberdeen, **Aberdeenshire**
　　　　　Chemical Composition (W. Mackie):
SiO_2 69·01,　Al_2O_3 17·74,　Fe_2O_3 0·97,　FeO 2·05,　MgO 0·48,
CaO 1·95,　K_2O 3·94,　Na_2O 2·73,　H_2O &c., 1·18,　=100·05
　　　Weight per cub. ft. 163 lbs.　Spec. grav. 2·61
　　　Crushing strain per sqr. ft. 1289 tons (Beare)

35. SCOTCH GRANITE. *MUSCOVITE-BIOTITE-GRANITE*
Light blue-grey, close-grained
  Sclattie Quarries, near Aberdeen, **Aberdeenshire**
Weight per cub. ft. 161 lbs. Spec. grav. 2·58. Absorp. 0·10%
  Crushing strain per sqr. ft. 850 tons (Beare)

36. SCOTCH GRANITE. *MUSCOVITE-BIOTITE-GRANITE*
Light blue-grey, medium-grained
  Dancing Cairns Quarries, near Aberdeen, **Aberdeenshire**
Presented by Messrs A. and F. Manuelle, Aberdeen
  Weight per cub. ft. 171 lbs.

37. SCOTCH GRANITE. *MUSCOVITE-BIOTITE-GRANITE*
Light blue-grey, fine-grained, with very minute quartz crystals
  Persley Quarries, near Aberdeen, **Aberdeenshire**
Weight per cub. ft. 162·3 lbs. Spec. grav. 2·60. Absorp. 0·19 %
  Crushing strain per sqr. ft. 942·8 tons (Beare)

38. SCOTCH GRANITE. *BIOTITE-GRANITE*
Bright salmon-coloured, medium-grained
  Correnie Quarries, near Alford, **Aberdeenshire**
Weight per cub. ft. 162 lbs. Spec. grav. 2·61. Absorp. 0·38 %
  Crushing strain per sqr. ft. 1284 tons

39. SCOTCH GRANITE. *BIOTITE-GRANITE*
Bright pink, medium-grained
  Correnie Quarries, near Alford, **Aberdeenshire**
Presented by Sir John Jackson, LL.D., London
Weight per cub. ft. 159 lbs. Spec. grav. 2·55. Absorp. 0·42 %
  Crushing strain per sqr. ft. 1318·3 tons (Beare)

40. SCOTCH GRANITE. *BIOTITE-GRANITE (some Muscovite)*
Bluish grey, medium-grained, slightly lined
  Tillyfourie Quarries, near Alford, **Aberdeenshire**
  Weight per cub. ft. 171 lbs.

41. SCOTCH GRANITE. *BIOTITE-GRANITE*
Pinkish, medium-grained
  Crathie Quarries, **Aberdeenshire**
  Weight per cub. ft. 176 lbs.

42. SCOTCH GRANITE. *BIOTITE-GRANITE*
Bright pink, fine-grained
  Tyrebeggar Quarries, **Aberdeenshire**
  Weight per cub. ft. 171 lbs.

43. SCOTCH GRANITE. *Biotite-Granite*
Brilliant red, coarse-grained
      Red Granite Quarries, Peterhead, **Aberdeenshire**
Presented by Messrs D. H. and J. Newall, Dalbeattie
      Chemical Composition (J. A. Phillips) :
SiO_2 73·70, TiO_2 trace, Al_2O_3 14·44 trace, Fe_2O_3 0·43,
FeO 1·49, MgO trace, CaO 1·08, K_2O 4·43, Na_2O 4·21,
H_2O 0·61, P_2O_5, = 100·39
      Weight per cub. ft. 158·5 lbs. Spec. grav. 2·54 (Beare)
Absorp. 0·29%. Crushing strain per sqr. ft. 1470 tons (Kirkaldy)

44. SCOTCH GRANITE. *Biotite-Granite (Adamellite)*
Dark blue-grey, fine-grained
      Rora Quarries, near Peterhead, **Aberdeenshire**
Presented by Messrs Heslop Wilson & Co., Boddam
      Weight per cub. ft. 203·7 lbs.
      Crushing strain per sqr. ft. 1258 tons (Kirkaldy)

45. SCOTCH GRANITE. *Muscovite-Biotite-Granite*
Blue-grey, medium-grained
      Cairngall Quarries, near Peterhead, **Aberdeenshire**
      Weight per cub. ft. 171 lbs.

46. SCOTCH GRANITE. *Biotite-Granite*
Pink, rather fine-grained
      Hill o' Fare Quarries, nr Banchory, **Kincardineshire**
Presented by H. Hutcheon, Esq., Aberdeen
Weight per cub. ft. 160·2 lbs. Spec. grav. 2·58. Absorp. 0·10%,
      Crushing strain per sqr. ft. 1360·3 tons (Beare)

47. SCOTCH GRANITE. *Biotite-Granite*
Brilliant red, fine-grained
      Abriachan Quarries, **Inverness-shire**
      Chemical Composition (W. Mackie):
SiO_2 71·25, Al_2O_3 18·03, Fe_2O_3 1·29, FeO 0·34, MgO 0·38,
CaO 2·61, K_2O 3·09, Na_2O 2·25, H_2O 0·82, P_2O_5 0·13,
= 100·19
      Weight per cub. ft. 176 lbs.

48. SCOTCH GRANITE. *Kentallenite (Olivine-Monzonite)*
Dark blue-grey, medium-grained
      Kentallen Quarries, **Argyllshire**
      Chemical Composition (J. J. H. Teall) :
SiO_2 48·00, TiO_2 0·22, Al_2O_3 12·52, Fe_2O_3 8·74, MgO 15·26,
CaO 7·94, K_2O 2·68, Na_2O 3·11, H_2O &c. 1·36, = 100·83
      Weight per cub. ft. 185 lbs. Spec. grav. 2·95

49. SCOTCH GRANITE. *Biotite-Granite*
Warm red, coarse-grained, with ragged felspars
        Granite Quarries, Ross of Mull, **Argyllshire**
            Chemical Composition (Haughton) :
SiO_2 74·48, Al_2O_3 16·20, Fe_2O_3 0·20,   MgO 0·27,   CaO 0·13,
K_2O 4·56, Na_2O 3·78, H_2O 0·60,   =100·22
        Weight per cub. ft. 176 lbs.

50. SCOTCH GRANITE. *Quartz-Diorite*
Dark grey, medium-grained
        Ben Cruachan Quarries, Loch Awe, **Argyllshire**
        Weight per cub. ft. 171 lbs.
        Crushing strain per sqr. ft. 877 tons (Rivington)

51. SCOTCH GRANITE. *Tonalite*
Grey and buff, the felspar having a brown tint, medium-grained
        Granite Quarries, Dalbeattie, **Kirkcudbrightshire**
Presented by Messrs D. H. and J. Newall, Dalbeattie
            Chemical Composition :
SiO_2 78·85, Al_2O_3 4·64, Fe_2O_3 4·56,   MgO 1·15,   CaO 1·51,
K_2O, Na_2O &c. 8·29,   =99·00
        Weight per cub. ft. 180 lbs.
        Crushing strain per sqr. ft. 904·8 tons (Kirkaldy)

52. SCOTCH GRANITE. *Biotite-Granite*
Light grey, medium-grained
        Kirkmabrede Quarries, Creetown, **Kirkcudbrightshire**
Presented by The Scottish Granite Co., Creetown
        Weight per cub. ft. 170 lbs.
        Crushing strain per sqr. ft. 1633 tons

53. IRISH GRANITE. *Muscovite-Biotite-Granite*
Light grey, medium-grained
        Dalkey Quarries, **Co. Dublin**
            Chemical Composition (Haughton) :
SiO_2 70·38, Al_2O_3 12·64, Fe_2O_3 3·16,   MgO 0·53,   CaO 2·84,
K_2O 5·90, Na_2O 3·13, H_2O 1·16,   =99·74
        Weight per cub. ft. 165 lbs. Spec. grav. 2·647

54. IRISH GRANITE. *Muscovite-Biotite-Granite*
Pinkish grey, medium-grained
        Glencullen Quarries, **Co. Dublin**
Presented by E. S. Glanville, Esq., Dublin
        Weight per cub. ft. 166 lbs.

55. IRISH GRANITE. *MUSCOVITE-BIOTITE-GRANITE*
Light grey, medium-grained
      Ballyknockan Quarries, **Co. Wicklow**
Presented by E. S. Glanville, Esq., Dublin
      Chemical Composition (Haughton):
SiO_2 70·82, Al_2O_3 14·08, Fe_2O_3 3·47, MgO 0·31, CaO 2·65,
K_2O 4·62, Na_2O 2·31, H_2O & loss 1·39, =99·65
      Weight per cub. ft. 166 lbs. Spec. grav. 2·636

56. IRISH GRANITE. *MUSCOVITE-BIOTITE-GRANITE*
Light grey, medium-grained
      Deer Park Quarries, Newtownbarry, **Co. Wexford**
Presented by R. Hall Dare, Esq., Newtownbarry
      Weight per cub. ft. 160 lbs.

57. IRISH GRANITE. *MUSCOVITE-BIOTITE-GRANITE*
Light yellowish grey, medium-grained
      Scrubb Quarries, Newtownbarry, **Co. Wexford**
Presented by R. Hall Dare, Esq., Newtownbarry
      Weight per cub. ft. 165 lbs.

58. IRISH GRANITE. *HORNBLENDE-BIOTITE-GRANITE*
Grey with pink tinge, fine-grained
      Glenville Quarries, **Co. Down**
Presented by Messrs E. Ewan & Sons, Newry
      Chemical Composition:
SiO_2 75·00, Al_2O_3 13·24, Fe_2O_3 2·52, CaO 0·69, K_2O 4·33,
Na_2O 3·07, H_2O 0·80, =99·65
      Weight per cub. ft. 170 lbs.

59. IRISH GRANITE. *QUARTZ-MONZONITE*
Light grey, medium-grained
      Moor Quarries, Newry, **Co. Down**
Presented by Messrs H. Campbell & Son, Newry
      Chemical Composition (Haughton):
SiO_2 64·60, Al_2O_3 14·64, Fe_2O_3 6·04, MgO 2·80, CaO 3·16,
K_2O 3·15, Na_2O 4·02, H_2O 1·13, =99·54
      Weight per cub. ft. 168 lbs. Spec. grav. 2·695

60. IRISH GRANITE. *BIOTITE-GRANITE*
Pink, fine-grained
      Shantalla Quarries, **Co. Galway**
Presented by The Galway Granite Quarries Co., Galway
      Weight per cub. ft. 171 lbs.

61. IRISH GRANITE. *BIOTITE-GRANITE*
Pink, coarse-grained
Ballagh Quarries, **Co. Galway**
Presented by The Galway Granite Quarries Co., Galway
Weight per cub. ft. 171 lbs.

62. IRISH GRANITE. *BIOTITE-GRANITE*
Dark grey with pinkish tinge, medium-grained
Barna Quarries, **Co. Galway**
Presented by The Galway Granite Quarries Co., Galway
Weight per cub. ft. 169 lbs.

63. IRISH GRANITE. *BIOTITE-GRANITE*
Pinkish grey, coarse-grained
Burtonport Quarries, **Co. Donegal**
Presented by Messrs Kirkpatrick Bros., Manchester
Weight per cub. ft. 163 lbs.

64. GUERNSEY GRANITE. *HORNBLENDE-GRANITE*
Light buff, tough, fine-grained
Cobo Quarries, **Guernsey**
Presented by Messrs Wm. Griffiths, Ltd., St Sampson
Weight per cub. ft. 190 lbs.

65. GUERNSEY GRANITE. *QUARTZ-DIORITE (TONALITE)*
Grey, medium-grained
Mont Cuet Quarries, **Guernsey**
Presented by Messrs Wm. Griffiths, Ltd., St Sampson
Weight per cub. ft. 176 lbs.

66. GUERNSEY GRANITE. *QUARTZ-DIORITE*
Dark blue-grey, tough, fine-grained
Mont Cuet Quarries, **Guernsey**
Presented by Messrs Wm. Griffiths, Ltd., St Sampson
Weight per cub. ft. 190 lbs.

67. GUERNSEY GRANITE. *HORNBLENDE-BIOTITE-GRANITE*
Drab with pinkish tinge, medium-grained
Grand Rocque Quarries, **Guernsey**
Presented by Messrs Wm. Griffiths, Ltd., St Sampson
Weight per cub. ft. 176 lbs.

68. GUERNSEY GRANITE. *DIORITE*
Dark grey, medium-grained
Bordeaux Quarries, **Guernsey**
Presented by Messrs Wm. Griffiths, Ltd., St Sampson
Weight per cub. ft. 180·5 lbs.

69. JERSEY GRANITE. *HORNBLENDE-GRANITE*
Pinkish grey, fine-grained
    Mont Mado Quarries, near St Martin, **Jersey**
Presented by A. Gulliver, Esq., St Helier
    Weight per cub. ft. 171 lbs.

70. JERSEY GRANITE. *HORNBLENDE-GRANITE*
Light pink, medium-grained
    Mont Mado, near St Martin, **Jersey**
Presented by A. Gulliver, Esq., St Helier
    Weight per cub. ft. 163 lbs.

71. JERSEY GRANITE. *HORNBLENDE-GRANITE*
Light brown, medium-grained
    La Moie Quarries, near St Helier, **Jersey**
Presented by A. Gulliver, Esq., St Helier
    Weight per cub. ft. 166 lbs.

IGNEOUS ROCKS (VOLCANIC).

72. BORROWDALE GREEN SLATE STONE. *VOL-CANIC ASH*
Blue-green mottled, fine-grained
    Borrowdale Quarries, **Cumberland**
Presented by Wm. Bromley, Esq., Keswick
    Weight per cub. ft. 172 lbs.

73. MELROSE AGGLOMERATE. *TRACHYTE*
Brown and white mottled agglomerate
    Melrose Quarries, **Roxburghshire**
Presented by the Duke of Buccleuch
    Weight per cub. ft. 114 lbs.

74. SCOTCH DOLERITE. *DOLERITE*
Dark grey with greenish tinge, fine-grained
    Devonshaw Quarries, Rumbling Bridge, nr. Alloa, **Kinross**
Presented by Messrs G. and R. Cousins, Edinburgh
    Weight per cub. ft. 178·5 lbs.

75. SCOTCH DOLERITE. *DOLERITE*
Dark grey, nearly black, medium-grained
    Dolerite Quarries, Alloa, **Clackmannanshire**
    Weight per cub. ft. 175 lbs.

76. TARDREE STONE. *QUARTZ-PORPHYRY*
Light straw-coloured with smoky quartz
    Tardree Mountain Quarries, **Co. Antrim, Ireland**
Presented by B. Meenan, Esq., Muckamore
    Chemical Composition (Player):
SiO_2 76·4, Al_2O_3 14·2, Fe_2O_3 1·6,   CaO 0·6, K_2O 4·2,
Na_2O 1·8, H_2O 1·5,   = 100·3
    Weight per cub. ft. 147 lbs.   Spec. grav. 2·43 to 2·46 (Cole)

METAMORPHIC ROCKS.

77. POLYFANT STONE. *Serpentinous Diabase*
Dark greenish grey with brown ferruginous spots, compact
Polyfant Quarries, near Launceston, **Cornwall**
Presented by F. Nichols, Esq., Polyfant
Chemical Composition (Collins):
SiO_2 36·90, Al_2O_3 11·80, Fe_2O_3 12·00, FeO 3·56, MgO 15·03, CaO 2·80, K_2O 3·64, Na_2O trace, H_2O 13·16, = 98·89
Weight per cub. ft. 171 lbs.

78. SALCOMBE STONE. *Diabase Schist*
Light greenish grey, argillaceous
Salcombe Quarries, near Salcombe, **Devonshire**
Weight per cub. ft. 166 lbs.

79. ST CATHERINE'S STONE.
Blue-grey, fine-grained
St Catherine's Quarry, Loch Fyne, **Argyllshire**
Presented by the Duke of Argyll, Inverary Castle
Weight per cub. ft. 178 lbs.

SEDIMENTARY ROCKS.

CAMBRIAN AND SILURIAN.

80. HARTSHILL FREESTONE. *Lower Cambrian*
Dark purple, close-grained quartzite
Hartshill Quarries, Atherstone, **Warwickshire**
Presented by The Hartshill Quarry Co., Atherstone
Chemical Composition :
SiO_2 81·90, Al_2O_3 7·60, Fe_2O_3 3·14, MgO 0·55, CaO 3·45,
K_2O & Na_2O 1·25, H_2O &c. 2·20, =100·09
Weight per cub. ft. 174 lbs. Spec. grav. 2·80
Crushing strain per sqr. ft. 2066·6 tons

81. HARTSHILL FREESTONE. *Lower Cambrian*
Light pinkish purple, close-grained quartzite
Hartshill Quarries, Atherstone, **Warwickshire**
Presented by The Hartshill Quarry Co., Atherstone
Chemical Composition :
SiO_2 81·90, Al_2O_3 7·60, Fe_2O_3 3·14, MgO 0·55, CaO 3·45,
K_2O & Na_2O 1·25, H_2O &c. 2·20, =100·09
Weight per cub. ft. 174 lbs. Spec. grav. 2·80
Crushing strain per sqr. ft. 2066·6 tons

82. SOUDLEY SANDSTONE (Lower Bed). *Lower Silurian*
Greenish yellow, medium-grained
Soudley Quarries, Hope Bowdler, **Shropshire**
Weight per cub. ft. 152 lbs.

83. SOUDLEY SANDSTONE (Upper Bed). *Lower Silurian*
Greenish yellow with dark purple bands, medium-grained
Soudley Quarries, Hope Bowdler, **Shropshire**
Weight per cub. ft. 157 lbs.

84. SOUDLEY SANDSTONE (between Lower and Upper Bed). *LOWER SILURIAN*
Greenish yellow, slightly banded, medium-grained
    Soudley Quarries, Hope Bowdler, **Shropshire**
    Weight per cub. ft. 152 lbs.

85. HOAR EDGE GRIT. *LOWER SILURIAN*
Orange-yellow, coarse-grained
    Hoar Edge Quarries, near Leebotwood, **Shropshire**
    Weight per cub. ft. 142 lbs.

86. HORDERLEY STONE. *LOWER SILURIAN*
Dark purple, fine-grained
    Long Lane Quarries, Horderley, **Shropshire**
    Weight per cub. ft. 144 lbs.

87. HORDERLEY STONE. *LOWER SILURIAN*
Grey and purple, slightly banded, fine-grained
    Long Lane Quarries, Horderley, **Shropshire**
    Weight per cub. ft. 144 lbs.

88. DOWNTON STONE. *UPPER SILURIAN*
Greenish yellow, fine-grained, slightly micaceous
    Downton Castle Quarries, near Ludlow, **Shropshire**
    Weight per cub. ft. 119 lbs.

89. ONIBURY STONE. *UPPER SILURIAN*
Light yellow, fine-grained, calcareous
    Onibury Quarries, near Ludlow, **Shropshire**
    Weight per cub. ft. 147 lbs.

90. GREEN SLATE STONE. *LOWER SILURIAN*
Dark green, compact, fine-grained
    Honister Quarries, near Keswick, **Cumberland**
Presented by The Buttermere Green Slate Co., Keswick
    Chemical Composition:
SiO_2 52·34, Al_2O_3 11·05, Fe_2O_3 9·85,   MgO 6·10,   CaO 6·36,
H_2O 4·70, CO_2 4·51,   Other matter 5·80,   = 100·71
    Weight per cub. ft. 180 lbs.

91. ELTERWATER GREEN STONE. *LOWER SILURIAN*
Green, very fine-grained
    Elterwater Quarries, Ambleside, **Westmorland**
Presented by The Elterwater Green Slate Co., Ambleside
    Chemical Composition:
SiO_2 50·88, Al_2O_3 14·12, Fe_2O_3 9·96,   MgO 8·67,   CaO 8·72,
K_2O 0·88, CO_2 6·47,   = 99·70
    Weight per cub. ft. 176 lbs.
    Crushing strain per sqr. ft. 266·1 tons

92. GREEN SLATE STONE. *Lower Silurian*
Light green, medium-grained
  Tilberthwaite Quarries, Coniston, **Westmorland**
Presented by J. J. Thomas, Esq., Kendal
  Weight per cub. ft. 174·3 lbs.

93. PENRHYN LIGHT BLUE SLATE STONE.
    Cambrian
Light blue, fine-grained
  Penrhyn Quarries, near Bangor, **N. Wales**
Presented by Messrs Jones Bros. & Co., Garth
    Chemical Composition :
SiO_2 60·50, Al_2O_3 19·70, Fe_2O_3 7·83,   MgO 2·20,   CaO 1·12,
K_2O 3·18, Na_2O 2·20, H_2O 3·30,   =100·03
  Weight per cub. ft. 180 lbs.

94. PENRHYN DARK BLUE SLATE STONE.
    Cambrian
Dark blue, fine-grained
  Penrhyn Quarries, near Bangor, **N. Wales**
Presented by Messrs Jones Bros. & Co., Garth
  Weight per cub. ft. 180 lbs.

95. BLUE FLAG STONE. *Upper Silurian*
Dark blue-grey, medium-grained, siliceous mud-stone
  Swarthmoor Quarries, Settle, **Yorkshire**
Presented by Chr. Ralph, Esq., Settle
  Weight per cub. ft. 152 lbs.

96. IRISH FLAG STONE. *Cambrian*
Dark greenish grey, fine-grained
  Glaslacken Quarries, Newtownbarry, **Co. Wexford**
Presented by R. Hall Dare, Esq., Newtownbarry
  Weight per cub. ft. 167 lbs.

DEVONIAN AND OLD RED SANDSTONE.

97. STONEYCOMBE LIMESTONE. *Devonian*
*Pink with brown veins, rather crystalline, compact, admits of a
good polish*
    Stoneycombe Quarries, Ipplepen, **Devonshire**
Presented by The Stoneycombe Lime and Stone Co., Ipplepen
    Weight per cub. ft. 152 lbs.

98. TORQUAY LIMESTONE. *Devonian*
*Brown with pink veins, crystalline, compact, admits of a good
polish*
    Union Street Quarries, Torquay, **Devonshire**
Presented by Messrs Webber and Stedman, Torquay
    Weight per cub. ft. 169 lbs. Spec. grav. 2·71

98A. PICKWELL DOWN STONE. *Devonian*
Greyish pink, fine-grained sandstone
    South Down Quarries, Morthoe, **Devonshire**
Presented by Messrs Dart and Francis, Ltd., Crediton
    Weight per cub. ft. 155 lbs.

99. LYDE STONE. *Old Red Sandstone*
Drab, medium-grained, arenaceous
    Pipe and Lyde Quarries, **Herefordshire**
Presented by Messrs Heywood & Sons, Hereford
    Weight per cub. ft. 147 lbs.

100. GARNSTONE. *Old Red Sandstone*
Drab, medium-grained, arenaceous
    Weobley Quarries, **Herefordshire**
Presented by Messrs Heywood & Sons, Hereford
    Weight per cub. ft. 163 lbs.

101. LUSTON STONE. *Old Red Sandstone*
Drab and grey mottled, medium-grained, arenaceous
    Eye Quarries, **Herefordshire**
Presented by Messrs Heywood & Sons, Hereford
    Weight per cub. ft. 163 lbs.

102. WITHINGTON STONE. *OLD RED SANDSTONE*
Brownish grey mottled, coarse-grained sandstone
Withington Quarries, **Herefordshire**
Presented by Messrs Heywood & Sons, Hereford
Weight per cub. ft. 152 lbs.

103. ROSS STONE. *OLD RED SANDSTONE*
Deep chocolate-coloured, medium-grained, soft sandstone
Prospect Quarries, Ross, **Herefordshire**
Presented by Messrs Heywood & Sons, Hereford
Weight per cub. ft. 157 lbs.

104. TEMPLETON STONE. *OLD RED SANDSTONE*
Greenish drab, fine-grained sandstone
Templeton Quarries, **Pembrokeshire, S. Wales**
Presented by Messrs Beynon Bros., Ltd., Tenby
Weight per cub. ft. 140 lbs.

105. RED WILDERNESS STONE. *OLD RED SANDSTONE*
Dull red, medium-grained, micaceous, arenaceous
Mitcheldean Quarries, **Gloucestershire**
Presented by The Forest of Dean Stone Firms Co., Bristol
Chemical Composition:
SiO_2 88·70, Al_2O_3 3·25, Fe_2O_3 1·80, FeO 0·30, MnO 0·10,
MgO 0·11, CaO 2·90, Na_2O 0·31, H_2O 0·59, CO_2 1·94,
$=100·00$
Weight per cub. ft. 141 lbs.
Crushing strain per sqr. ft. 695 tons (Kirkaldy)

106. PEEL FREESTONE. *OLD RED SANDSTONE*
Dark purple, slightly streaked, fine-grained sandstone
Greg Malin Quarries, Peel, **Isle of Man**
Presented by S. H. Royston, Esq., Douglas
Weight per cub. ft. 171 lbs.

107. DODDINGTON HILL STONE. *OLD RED SANDSTONE*
Light purplish grey, fine-grained sandstone
Doddington Hill Quarries, Wooler, **Northumberland**
Presented by Messrs Emley & Sons, Ltd., Newcastle
Chemical Composition (Stevenson Macadam, Edinburgh):
SiO_2 97·44, Fe_2O_3 0·87, H_2O &c. 1·48, $MgCO_3$ 1·26,
$CaCO_3$ 0·12, $=101·17$
Weight per cub. ft. 154 lbs. Spec. grav. 2·471. Crushing strain
per sqr. ft. 476·64 tons (R. Stansfield, Edinburgh)

108. SWINTON STONE. *Old Red Sandstone*
Pinkish brown, fine-grained, micaceous sandstone
    Swinton Quarries, **Berwickshire**
Presented by John Steel, Esq., Greenlaw
    Weight per cub. ft. 152 lbs.

109. WHITSOME STONE. *Old Red Sandstone*
Light buff, medium-grained, micaceous sandstone
    Whitsome Newton Quarries, **Berwickshire**
Presented by John Steel, Esq., Greenlaw
    Weight per cub. ft. 148 lbs.

110. DUNBAR RED FREESTONE. *Old Red Sandstone*
Light red, medium-grained sandstone
    Broomhouse Quarries, Dunbar, **Haddingtonshire**
Presented by G. Cunningham, Esq., Dunbar
    Weight per cub. ft. 120 lbs.

111. AUCHINHEATH FREESTONE. *Old Red Sand-*
    stone
Light yellow with dark ferruginous specks, medium-grained
    sandstone
    Auchinheath Quarries, Hamilton, **Larnarkshire**
Presented by The Clydesdale Quarry Co., Glasgow
        Chemical Composition (Daniel Burns) :
SiO_2 99·22,   FeO 0·78,   = 100·00
    Weight per cub. ft. 148 lbs.
    Crushing strain per sqr. ft. 648 tons

112. AUCHINCARROCH RED FREESTONE. *Old*
    Red Sandstone
Bright red, medium-grained sandstone
    Auchincarroch Quarries, **Dumbartonshire**
Presented by The Auchincarroch Stone Co., Glasgow
    Weight per cub. ft. 152 lbs.

113. AUCHINCARROCH RED FREESTONE. *Old*
    Red Sandstone
Light red, medium-grained sandstone
    Auchincarroch Quarries, **Dumbartonshire**
Presented by The Auchincarroch Stone Co., Glasgow
    Weight per cub. ft. 152 lbs.

114. LAMBERKEN FREESTONE. *Old Red Sandstone*
Light greyish pink, medium-grained, micaceous sandstone
    Lamberken Quarries, Aberdalgie, **Perthshire**
Presented by Messrs J. and D. Beveridge, Perth
    Weight per cub. ft. 163 lbs.

115. WEST HALL ROCK. *Old Red Sandstone*
Brownish grey, streaked, coarse-grained sandstone
    West Hall Quarries, near Dundee, **Forfarshire**
Presented by Messrs Duncan Galloway & Co., Dundee
    Chemical Composition (Tatlock & Thomson, Glasgow):
SiO_2 69·01, Al_2O_3 13·59, Fe_2O_3 4·72, H_2O 1·34, $CaCO_3$ 5·17,
$MgCO_3$ 5·21,   = 99·04
Weight per cub. ft. 152 lbs.   Spec. grav. 2·48.   Absorp. 1·37 $^0/_0$
    Crushing strain per sqr. ft. 943·3 tons (Kirkaldy)

116. DUNTRUNE ROCK. *Old Red Sandstone*
Greenish grey, medium-grained sandstone
    West Hall Quarries, near Dundee, **Forfarshire**
Presented by Messrs Duncan Galloway & Co., Dundee
    Chemical Composition (Macdougald, Dundee):
SiO_2 76·23, Al_2O_3 & FeO 7·98,   MgO 2·92,   CaO 6·65,
H_2O &c. 5·50,   = 99·28
Weight per cub. ft. 161·25 lbs.   Spec. grav. 2·58.   Absorp. 0·65 $^0/_0$
    Crushing strain per sqr. ft. 1190·5 tons (Kirkaldy)

117. BLUE LIVER STONE. *Old Red Sandstone*
Blue-grey, fine-grained, slightly micaceous sandstone
    Carmyllie Quarries, near Arbroath, **Forfarshire**
Presented by Messrs Duncan Galloway & Co., Dundee
    Weight per cub. ft. 163 lbs.

118. BLUE LIVER STONE. *Old Red Sandstone*
Grey-blue, slightly banded, fine-grained, micaceous sandstone
    Carmyllie Quarries, near Arbroath, **Forfarshire**
Presented by Messrs Duncan Galloway & Co., Dundee
    Weight per cub. ft. 152 lbs.

119. DALGATY SANDSTONE. *Old Red Sandstone*
Light purple, medium-grained, highly micaceous
    Dalgaty Quarries, Turriff, **Aberdeenshire**
Presented by The Dalgaty Quarry Co., Turriff
    Weight per cub. ft. 127·5 lbs.

120. TARRADALE FREESTONE. *Old Red Sand-
stone*
Light pink, medium-grained, slightly micaceous sandstone
    Tarradale Quarries, **Inverness-shire**
Presented by A. E. Watson, Esq., Inverness
    Weight per cub. ft. 190 lbs.

121.   ROSEBRAE STONE (Upper bed). *OLD RED SANDSTONE*

Light straw-coloured, slightly banded, fine-grained sandstone
Rosebrae Quarries, **Elginshire**
Presented by W. Fraser, Esq., Elgin
Weight per cub. ft. 142·5 lbs.

122. NEWTON SANDSTONE. *OLD RED SANDSTONE*
Light pink, medium-grained
Newton Quarries, **Elginshire**
Presented by G. Geddes, Esq., Elgin
Weight per cub. ft. 133 lbs.

123. CAITHNESS FLAGSTONE. *OLD RED SANDSTONE*
Dark blue-grey, tough, compact
Castle Hill Quarries, Thurso, **Caithness-shire**
Presented by The Caithness Flagstone Co., Thurso
Chemical Composition :
SiO_2 69·45, Al_2O_3 & FeO_2O_3 10·50, K_2O & Na_2O 2·20,
$CaCO_3$ 10·65,   Organic matters 5·79,   = 98·80
Weight per cub. ft. 157 lbs.

124. FERSENES STONE. *OLD RED SANDSTONE*
Light drab, slightly streaked, fine-grained sandstone
Fersenes Quarries, near Wick, **Orkney Islands**
Presented by Messrs J. Hood & Son, Wick
Weight per cub. ft. 130 lbs.

CARBONIFEROUS.

125. HOPTON WOOD STONE. *Carboniferous Limestone*
Light drab, fine-grained, admits of a good polish
    Hopton Wood Quarries, Middleton, **Derbyshire**
Presented by Messrs J. Hodson & Son, Nottingham
    Weight per cub. ft. 158 lbs.
        Crushing strain per sqr. ft. 806 tons (Rivington)

126. STINK STONE. *Carboniferous Limestone*
Blue-grey, compact limestone, with pink veins
    Winsor's Hill Quarries, Shepton Mallet, **Somersetshire**
Presented by Dr Allen, St John's College, Cambridge

127. WESTLEIGH STONE. *Carboniferous Limestone*
Blue-grey, compact, crinoidal limestone, takes a polish
    Burlescombe Quarries, near Wellington, **Somersetshire**
Presented by The Westleigh Stone and Lime Co., Wellington
    Weight per cub. ft. 166 lbs.

128. RIBBLESDALE BLUE LIMESTONE. *Carbon-iferous Limestone*
Blue-grey, compact, crystalline
    Horton Quarries, near Settle, **Yorkshire**
Presented by John Delaney, Esq., Settle
    Chemical Composition (F. B. Last, Barrow):
SiO_2 0·96, Al_2O_3 0·05, Fe_2O_3 0·21,   MnO trace,   MgO 0·15,
CaO 54·70, H_2O &c. 0·30, CO_2 43·03, P_2O_5 0·17, SO_3 0·18,
= 99·75
    Weight per cub. ft. 170 lbs.

129. RIBBLESDALE WHITE LIMESTONE. *Carbon-iferous Limestone*
Light grey, compact, crystalline
    Horton Quarries, near Settle, **Yorkshire**
Presented by John Delaney, Esq., Settle
    Chemical Composition (F. B. Last, Barrow):
SiO_2 0·12, Al_2O_3 0·02, Fe_2O_3 0·13,   MnO trace,   MgO 0·10,
CaO 55·50, H_2O &c. 0·04, CO_2 43·80, P_2O_5 0·19, SO_3 0·01,
= 99·91
    Weight per cub. ft. 170 lbs.

130. HUMBLETON STONE. *Carboniferous Limestone*
Dark grey, compact, fine-grained limestone, with white veins,
admits of a fairly good polish
Humbleton Quarries, near Bolton Abbey, **Yorkshire**
Presented by Messrs J. Green & Son, Silsdon
Weight per cub. ft. 171 lbs.

131. SEDBUSK SANDSTONE. *Carboniferous Lime-*
stone
Light grey, fine-grained
Sedbusk Quarries, near Hawes, **Yorkshire**
Presented by the Earl of Wharncliffe, per E. Broughton, Esq.,
Hawes
Weight per cub. ft. 125·5 lbs.

132. STAG'S FELL SANDSTONE. *Carboniferous Lime-*
stone
Pinkish grey with brown ferruginous specks, fine-grained
Stag's Fell Quarries, Hawes, **Yorkshire**
Presented by the Earl of Wharncliffe, per E. Broughton, Esq.,
Hawes
Weight per cub. ft. 160·5 lbs.

133. BLACK PASTURE SANDSTONE. *Carboniferous*
Limestone
Light buff with small ferruginous specks, fine-grained
Black Pasture Quarries, Chollerford, **Northumberland**
Presented by Messrs Herbertson, Ltd., Galashiels
Chemical Composition (Greenwell, London):
SiO_2 96·13, Al_2O_3 2·05, Fe_2O_3 0·68, K_2O 0·30, H_2O &c. 0·81,
=99·97
Weight per cub. ft. 164 lbs. (Greenwell). Spec. grav. 2·63.
Absorp. 4·7 %. Crushing strain per sqr. ft. 673·8 tons
(Kirkaldy)

134. WOODBURN FREESTONE. *Carboniferous Lime-*
stone
Warm cream-coloured, medium-grained, micaceous sandstone
Woodburn Quarries, **Northumberland**
Presented by Messrs Wm. Benson & Son, Newcastle
Chemical Composition (Blake, Newcastle):
SiO_2 97·25, Al_2O_3 0·47, Fe_2O_3 0·62, MgO 0·27, CaO 0·15,
K_2O 0·07, H_2O 1·10, =99·93
Weight per cub. ft. 135 lbs. Absorp. 4·7 %
Crushing strain per sqr. ft. 573·8 tons (Beare)

135. PRUDHAM FREESTONE. *CARBONIFEROUS LIME-STONE*
Brown, slightly micaceous, coarse-grained sandstone
Prudham Quarries, **Northumberland**
Presented by Messrs Wm. Benson & Son, Newcastle
Chemical Composition (Blake, Newcastle):
SiO_2 86·40, Al_2O_3 6·45, Fe_2O_3 0·75, MgO 0·27, CaO 0·60,
K_2O 0·21, H_2O 5·25, = 99·93
Weight per cub. ft. 141 lbs. Spec. grav. 2·29. Absorp. 4·00 %
Crushing strain per sqr. ft. 455 tons.

136. DENWICK SANDSTONE. *CARBONIFEROUS LIME-STONE*
Greenish grey, fine-grained, micaceous
Denwick Quarries, near Alnwick, **Northumberland**
Presented by Messrs J. Green & Son, Alnwick
Weight per cub. ft. 142 lbs.

137. DENWICK SANDSTONE. *CARBONIFEROUS LIME-STONE*
Light brown, fine-grained
Denwick Quarries, near Alnwick, **Northumberland**
Presented by Messrs J. Green & Son, Alnwick
Weight per cub. ft. 142 lbs.

138. SPYLAW FREESTONE. *CARBONIFEROUS LIMESTONE*
Yellow, coarse-grained sandstone
Spylaw Quarries, near Bilton, **Northumberland**
Weight per cub. ft. 142·5 lbs.

139. GLANTON FREESTONE. *CARBONIFEROUS LIMESTONE*
Light straw-coloured, medium-grained sandstone
Glanton Quarries, **Northumberland**
Presented by T. M. Wilson, Esq., Glanton
Weight per cub. ft. 142 lbs.

140. PASTURE HILL STONE. *CARBONIFEROUS LIMESTONE*
Light cream-coloured, slightly banded, medium-grained sandstone
Pasture Hill Quarries, nr Chathill, **Northumberland**
Presented by The Doddington Hill Quarries Co., Chathill
Weight per cub. ft. 133 lbs.

141. FELL SANDSTONE. *CARBONIFEROUS LIMESTONE*
Light pink, medium-grained
King's Quarries, near Berwick, **Northumberland**
Presented by Messrs M. Gray & Son, Berwick-on-Tweed
Weight per cub. ft. 132 lbs.

142. WELSH LIMESTONE. *CARBONIFEROUS LIMESTONE*
Blue-grey, compact, veined with quartz
　　Town Quarry, Tenby, **Pembrokeshire**
Weight per cub. ft. 161 lbs.

143. WELSH LIMESTONE. *CARBONIFEROUS LIMESTONE*
Blue-grey, fine-grained, compact
　　Williamston Quarries, **Pembrokeshire**
Presented by Messrs Beynon Bros., Ltd., Tenby
Weight per cub. ft. 161 lbs.

144. PENMON STONE. *CARBONIFEROUS LIMESTONE*
Light grey, close-grained limestone
　　Beaumaris Quarries, Anglesey, **N. Wales**
Presented by The Penmon Park Quarries Co., Beaumaris
Weight per cub. ft. 171 lbs.

145. PENMON MARBLE STONE. *CARBONIFEROUS LIME-*
　　　　STONE
Blue-grey compact limestone
　　Beaumaris Quarries, Anglesey, **N. Wales**
Presented by The Penmon Park Quarries Co., Beaumaris
Weight per cub. ft. 166 lbs.

146. DINORBEN STONE (Lower Bed). *CARBONIFEROUS*
　　　　LIMESTONE
Dark blue-grey, compact limestone
　　Dinorben Quarries, **Anglesey**
Presented by The Mersey Dock and Harbour Board, Liverpool
Weight per cub. ft. 166 lbs.

147. DINORBEN STONE (Upper Bed). *CARBONIFEROUS*
　　　　LIMESTONE
Pink and brown mottled, fine, compact limestone
　　Dinorben Quarries, **Anglesey**
Presented by The Mersey Dock and Harbour Board, Liverpool
Weight per cub. ft. 166 lbs.

148. SCARLETT STONE. *CARBONIFEROUS LIMESTONE*
Dark blue-grey compact limestone, with narrow white veins,
　　admits of a fairly good polish
　　Scarlett Quarries, Castletown, **Isle of Man**
Presented by T. H. Royston, Esq., Douglas
Weight per cub. ft. 171 lbs.

149. POOLVASH STONE. *CARBONIFEROUS LIMESTONE*
Blue-black, compact, fine-grained limestone, admits of a fair polish
        Poolvash Quarries, **Isle of Man**
Presented by T. H. Royston, Esq., Douglas
        Weight per cub. ft. 180·5 lbs.

150. CRAIGLEITH STONE. *CALCIFEROUS SANDSTONE*
Delicate drab-coloured, fine, calcareous sandstone
        Craigleith Quarries, near **Edinburgh**
Presented by John Best, Esq., Edinburgh
        Chemical Composition (Middleton):
SiO_2 98·30, Al_2O_3 & Fe_2O_3 0·60, $CaCO_3$ 1·10, =100·00
Weight per cub. ft. 138 lbs. Spec. grav. 2·22 (Beare). Absorp. 3·61 $^0/_0$
        (Beare). Crushing strain per sqr. ft. 802 tons (Rivington)

151. CRAIGLEITH STONE. *CALCIFEROUS SANDSTONE*
Drab, fine-grained, calcareous sandstone
        Barnton Park Quarries, near **Edinburgh**
Presented by Messrs G. and R. Cousins, Edinburgh
        Chemical Composition (Bloxham):
SiO_2 96·95, Al_2O_3 & Fe_2O_3 2·30, MgO & CaO 0·52, H_2O 0·23,
=100·00
Weight per cub. ft. 152·6 lbs. Spec. grav. 2·443. Absorp. 3·6 $^0/_0$
        (Bloxham). Crushing strain per sqr. ft. 861·9 tons (*Geol.*
        Survey)

152. BLUE LIVER ROCK. *CALCIFEROUS SANDSTONE*
Blue-grey, medium-grained, tough sandstone
        Barnton Park Quarries, near **Edinburgh**
Presented by Messrs G. and R. Cousins, Edinburgh
        Weight per cub. ft. 152 lbs.

153. COMMON ROCK. *CALCIFEROUS SANDSTONE*
Light fawn-coloured sandstone, with black streaks
        Barnton Park Quarries, near **Edinburgh**
Presented by Messrs G. and R. Cousins, Edinburgh
        Weight per cub. ft. 152 lbs.

154. BLUE HAILES STONE. *CALCIFEROUS SANDSTONE*
Light drab, slightly micaceous, medium-grained sandstone
        Hailes Quarries, Slateford, **Edinburgh**
Presented by The Hailes Estate and Quarry Co., Edinburgh
        Chemical Composition (*Quarry* Analysis Dept.):
SiO_2 92·23, Al_2O_3 2·93, Fe_2O_3 0·01, FeO 1·71, MgO 0·19,
CaO 0·81, K_2O & Na_2O 0·18, H_2O &c. 1·73, =99·79
Weight per cub. ft. 149 lbs. Spec. grav. 2·55. Absorp. 4·70 $^0/_0$
        Crushing strain per sqr. ft. 459·7 tons (Beare)

w.   18

155. PINK HAILES STONE. *Calciferous Sandstone*
Light pink, fine-grained, micaceous sandstone
          Hailes Quarries, Slateford, near **Edinburgh**
Presented by The Hailes Estate and Quarry Co., Edinburgh
          Chemical Composition (*Quarry* Analysis Dept.) :
SiO_2 96·70, Al_2O_3 0·84, Fe_2O_3 1·46, FeO 0·25, CaO 0·36,
H_2O &c. 0·58, = 100·19
Weight per cub. ft. 142·5 lbs. Spec. grav. 2·60. Absorp. 5·3 %
          Crushing strain per sqr. ft. 511·3 tons (*Quarry*)

156. WHITE HAILES STONE. *Calciferous Sandstone*
Light drab, close-grained, fine sandstone
          Hailes Quarries, Slateford, near **Edinburgh**
Presented by The Hailes Estate and Quarry Co., Edinburgh
          Chemical Composition (*Quarry* Analysis Dept.) :
SiO_2 96·52, Al_2O_3 2·78, Fe_2O_3 0·05, FeO 0·19, CaO 0·41,
K_2O & Na_2O 0·27, H_2O &c. 0·31, = 100·33
Weight per cub. ft. 144 lbs. Spec. grav. 2·53. Absorp. 4 %
          Crushing strain per sqr. ft. 523·5 tons (*Quarry*)

157. TRANENT STONE. *Calciferous Sandstone*
Light cream, slightly banded, fine, calcareous sandstone
          Tranent Quarries, **Haddingtonshire**
Presented by Messrs A. Wilson & Son, Tranent
          Weight per cub. ft. 138 lbs.

158. PLEAN WHITE FREESTONE. *Upper Limestone,*
          Lower Carboniferous
Light drab, medium-grained, slightly micaceous sandstone
          Plean Quarries, near Bannockburn, **Stirlingshire**
Presented by Wm. Chalmers, Esq., Burntisland
Weight per cub. ft. 138·2 lbs. Spec. grav. 2·22. Absorp. 4·25 %
          Crushing strain per sqr. ft. 612·8 tons (Beare)

159. GIFNOCK LIVER ROCK. *Upper Limestone,*
          Lower Carboniferous
Light pinkish drab, fine-grained, micaceous sandstone
          Gifnock Quarries, **Renfrewshire**
Presented by Messrs Baird and Stevenson, Glasgow
          Weight per cub. ft. 143 lbs.
          Crushing strain per sqr. ft. 459·7 tons.

160. HUMBIE SANDSTONE. *Calciferous Sandstone*
Light yellow with brown ferruginous specks, medium-grained
          Humbie Quarries, Aberdour, **Fifeshire**
Presented by Thomas Locktie, Esq., Aberdour
          Weight per cub. ft. 136 lbs.

161. HUMBIE SANDSTONE. *CALCIFEROUS SANDSTONE*
Light drab, fine-grained
  Humbie Quarries, near Aberdour, **Fifeshire**
Presented by Thomas Locktie, Esq., Aberdour
  Weight per cub. ft. 136 lbs.

162. BURNTISLAND NEWBIGGEN FREESTONE.
        CALCIFEROUS SANDSTONE
Light yellow sandstone, with brown ferruginous specks
  Newbiggen Quarries, **Fifeshire**
Presented by Wm. Chalmers, Esq., Burntisland
  Weight per cub. ft. 133 lbs.

163. INVERKEITHING STONE. *CALCIFEROUS SAND-*
STONE
Light fawn-coloured, fine-grained sandstone
  Wall Dean Quarries, North Queensferry, **Fifeshire**
Presented by The Tilbury Contracting Co., Inverkeithing
  Weight per cub. ft. 158 lbs. Spec. grav. 2·54
  Crushing strain per sqr. ft. 723·4 tons (Kirkaldy)

164. CULLALOE STONE. *CALCIFEROUS SANDSTONE*
Light cream, fine-grained sandstone
  Cullaloe Quarries, Aberdour, **Fifeshire**
Presented by B. F. White, Esq., Pollokshields, Glasgow
  Chemical Composition (R. M. Clark, B.Sc., Glasgow):
SiO_2 99·17, Al_2O_3 0·11, Fe_2O_3 0·07, MgO 0·12, H_2O 0·49,
=99·96
Weight per cub. ft. 149·70 lbs. Spec. grav. 2·4. Absorp. 6·22 %

165. MILLSTONE MEADOW FREESTONE. *CAL-*
CIFEROUS SANDSTONE
Bright drab, fine-grained sandstone
  Millstone Meadow Quarries, near Fordell, **Fifeshire**
Presented by A. Wilson, Esq., Cowdenbeath, Fife
  Chemical Composition (J. Clark, Ph.D., Glasgow):
SiO_2 98·32, Al_2O_3 0·42, Fe_2O_3 0·30, H_2O 0·60, $MgCO_3$ 0·19,
=99·83
Weight per cub. ft. 145·60 lbs. Spec. grav. 2·344. Absorp. 6·95 %

166. IRISH LIMESTONE. *CARBONIFEROUS LIMESTONE*
Blue-grey, fine-grained
  Surra Quarries, near Birr, **King's County**
Presented by the Rt. Hon. Earl of Rosse, Birr Castle
  Weight per cub. ft. 180·5 lbs.

167. IRISH LIMESTONE. *CARBONIFEROUS LIMESTONE*
*Black, slightly banded, fine-grained, very compact, admits of a
    high polish*
Donaghcumper Quarries, **Co. Kildare**
*Presented by W. T. Kirkpatrick, Esq., J.P., Donaghcumper,
    Celbridge*
Weight per cub. ft. 171 lbs.

168. IRISH LIMESTONE. *CARBONIFEROUS LIMESTONE*
Drab, fine-grained, compact, fossiliferous, admits of a high polish
Carrick Quarries, Edenderry, **Co. Kildare**
Presented by E. S. Glanville, Esq., Dublin
Weight per cub. ft. 171 lbs.

169. IRISH LIMESTONE. *CARBONIFEROUS LIMESTONE*
Black and grey, mottled, fossiliferous, admits of a high polish
Three Castles Quarries, Kilkenny, **Co. Kilkenny**
Presented by E. S. Glanville, Esq., Dublin
Weight per cub. ft. 171 lbs.

170. IRISH LIMESTONE. *CARBONIFEROUS LIMESTONE*
Very dark blue-grey, close-grained
Royal Oak Quarries, Bagnalstown, **Co. Carlow**
Presented by A. Gordon, Esq., Inchicore, Dublin
Weight per cub. ft. 171 lbs.

171. IRISH LIMESTONE. *CARBONIFEROUS LIMESTONE*
Dark blue-grey, close-grained
Limestone Quarries, Foynes, **Co. Limerick**
Presented by A. Gordon, Esq., Inchicore, Dublin
Weight per cub. ft. 170 lbs.

172. IRISH LIMESTONE. *CARBONIFEROUS LIMESTONE*
Black, fine-grained
Rosbrien Quarries, **Co. Limerick**
Presented by M. D. Matthews, Esq., Limerick
Weight per cub. ft. 171 lbs.

173. IRISH LIMESTONE. *CARBONIFEROUS LIMESTONE*
Blue-black, fine-grained, compact
Corgrigg Quarries, **Co. Limerick**
Presented by The Lord Monteagle, K.P., London
Weight per cub. ft. 180·5 lbs.

174. IRISH LIMESTONE. *CARBONIFEROUS LIMESTONE*
Blue-black, fine-grained, compact
  Craglee Quarries, **Co. Limerick**
Presented by The Lord Monteagle, K.P., London
 Weight per cub. ft. 180·5 lbs.

175. IRISH LIMESTONE. *CARBONIFEROUS LIMESTONE*
Blue-black, fine-grained, compact
  Parkmoor Quarries, **Co. Limerick**
Presented by The Lord Monteagle, K.P., London
 Weight per cub. ft. 180·5 lbs.

176. IRISH LIMESTONE. *CARBONIFEROUS LIMESTONE*
Black, fine-grained, compact, takes a fair polish
  Meelin Quarries, **Co. Cork**
Presented by Major Aldworth, Newmarket Court, Co. Cork
 Weight per cub. ft. 176 lbs.

177. MOUNTCHARLES SANDSTONE. *CARBONIFEROUS*
   LIMESTONE
Light drab, fine-grained, slightly micaceous
  Drumkeelan Quarries, **Co. Donegal, Ireland**
Presented by Messrs G. A. Watson & Co., Liverpool
   Chemical Composition :
SiO_2 76·76, Al_2O_3 11·13, Fe_2O_3 0·50, FeO 0·15, MgO 0·59,
CaO 1·25, K_2O 3·72, Na_2O 2·47, Loss 2·95, =99·52
 Weight per cub. ft. 154 lbs. Spec. grav. 2·47
 Crushing strain per sqr. ft. 763 tons (Kirkaldy)

178. IRISH SANDSTONE. *LOWER CARBONIFEROUS*
Light brown, medium-grained
  Drumaroon Quarries, near Ballycastle, **Co. Antrim**
Presented by Jas. O'Connor, Esq., Ballycastle
 Weight per cub. ft. 133 lbs.

179. IRISH SANDSTONE. *LOWER CARBONIFEROUS*
Light red, medium-grained
  Drumaroon Quarries, near Ballycastle, **Co. Antrim**
Presented by Jas. O'Connor, Esq., Ballycastle
 Weight per cub. ft. 128 lbs.

180. GRINDELFORD STONE. *MILLSTONE GRIT*
Drab and brown, mottled, coarse-grained sandstone
  Bole Hill Quarries, Grindelford, **Derbyshire**
Presented by the Derwent Valley Water Board
   Chemical Composition (Earle, Hull) :
SiO_2 98·45, Al_2O_3 & Fe_2O_3 0·30, MgO 0·22, CaO 0·33,
H_2O 0·85, =100·15
 Weight per cub. ft. 145·5 lbs. Spec. grav. 2·33
 Crushing strain per sqr. ft. over 500 tons (Kirkaldy)

181. DARLEY DALE STONE. *Millstone Grit*
Light brown with ferruginous specks, medium-grained sandstone
         Peasenhurst Quarries, Darley Dale, **Derbyshire**
Presented by Messrs J. Hodson & Son, Nottingham
               Chemical Composition (Middleton):
SiO_2 96·4, Al_2O_3 & FeO 1·3, H_2O &c. 1·94, $CaCO_3$ 0·36,
=100·00
         Weight per cub. ft. 153 lbs.
         Crushing strain per sqr. ft. 455 tons (Rivington)

182. DARLEY DALE STONE. *Millstone Grit*
Light drab, compact, close-grained, micaceous sandstone
         Stancliffe Quarries, Darley Dale, **Derbyshire**
Presented by The Stancliffe Estate Co., Darley Dale
               Chemical Composition (Prof. Clowes):
SiO_2 96·40, Al_2O_3 & FeO 1·30, H_2O &c. 1·94,   =99·74
         Weight per cub. ft. 148 lbs.
         Crushing strain per sqr. ft. 670·3 tons (Kirkaldy)

183. HALL DALE STONE. *Millstone Grit*
Salmon-coloured, fine-grained sandstone
         Hall Dale Quarries, Darley Dale, **Derbyshire**
Presented by The Stancliffe Estate Co., Darley Dale
         Weight per cub. ft. 163 lbs.
         Crushing strain per sqr. ft. 534·1 tons (Kirkaldy)

184. CROMFORD STONE. *Millstone Grit*
Pink, coarse-grained sandstone
         Cromford Quarries, near Matlock, **Derbyshire**
Presented by The Stancliffe Estate Co., Darley Dale
         Weight per cub. ft. 163 lbs.
         Crushing strain per sqr. ft. 497·4 tons (Kirkaldy)

185. WHATSTANDWELL STONE. *Millstone Grit*
Pinkish brown, medium-grained sandstone
         Duke's Quarries, Whatstandwell, **Derbyshire**
Presented by Anthony Sims, Esq., Whatstandwell
         Weight per cub. ft. 144 lbs. (Rivington)
         Crushing strain per sqr. ft. 400 to 500 tons.

186. MATLOCK STONE. *Millstone Grit*
Yellow, fine-grained sandstone
         Oaks Quarries, Tansley, near Matlock, **Derbyshire**
Presented by Thomas Wragg, Esq., Loxley, near Sheffield
         Weight per cub. ft. 146 lbs.

187. POOR LOTS SANDSTONE. *Millstone Grit*
Rich brown, medium-grained
    Poor Lots Quarries, Tansley, near Matlock, **Derbyshire**
Presented by George Bodin, Esq., Matlock
Weight per cub. ft. 142·5 lbs.

188. BROWN FREESTONE. *Millstone Grit*
Yellow, medium-grained
    Brown Hill Quarries, Preston, **Lancashire**
Presented by The Tootal Height Quarry Co., Longridge, Preston
Weight per cub. ft. 146 lbs.

189. BROWN FREESTONE. *Millstone Grit*
Dark yellow, slightly banded, medium-grained sandstone
    Brown Hill Quarries, Preston, **Lancashire**
Presented by The Tootal Height Quarry Co., Longridge, Preston
Weight per cub. ft. 146·8 lbs.

190. LANCASTER FREESTONE. *Millstone Grit*
Yellow, medium-grained sandstone
    Scotch Quarries, Lancaster, **Lancashire**
Presented by Messrs J. Hatch & Sons, Lancaster
Weight per cub. ft. 147 lbs.

191. LONGRIDGE STONE. *Millstone Grit*
Yellow, medium-grained sandstone
    Longridge Quarries, near Preston, **Lancashire**
Presented by The Waring White Building Co., London
Weight per cub. ft. 163 lbs.

192. NIDDERDALE SANDSTONE. *Millstone Grit*
Light yellowish drab, slightly banded, fine-grained
    Scar Quarries, Middlesmoor, **Yorkshire**
Presented by W. A. Best, Esq., Angram
Weight per cub. ft. 127 lbs.

193. ILKLEY MOOR STONE. *Millstone Grit*
Yellow, very coarse-grained sandstone
    Hanging Stone Quarries, near Ilkley, **Yorkshire**
Presented by The Urban District Council, Ilkley.
Weight per cub. ft. 160 lbs.

194. ILKLEY MOOR STONE. *Millstone Grit*
Yellow, coarse-grained sandstone
    Hanging Stone Quarries, near Ilkley, **Yorkshire**
Presented by The Urban District Council, Ilkley
Weight per cub. ft. 160 lbs.

195. ILKLEY MOOR STONE. *MILLSTONE GRIT*
Yellow, fine-grained sandstone
     Hanging Stone Quarries, near Ilkley, **Yorkshire**
Presented by The Urban District Council, Ilkley
     Weight per cub. ft. 160 lbs.

196. BRAMLEY FALL GRIT. *MILLSTONE GRIT*
Light brown, mottled, coarse-grained sandstone
     Horsforth Quarries, near Leeds, **Yorkshire**
Presented by Messrs Pawson Bros., Morley
Weight per cub. ft. 132 lbs.   Spec. grav. 2·12.   Absorp. 3·70 %
     (Beare).   Crushing strain per sqr. ft. 389 tons (Rennie)

197. BRAMLEY FALL STONE. *MILLSTONE GRIT*
Light brown, mottled, coarse-grained sandstone
     Horsforth Quarries, Horsforth, **Yorkshire**
Presented by Messrs B. Whitaker & Son, Horsforth, Leeds
         Chemical Composition (Fairley, Leeds):
SiO_2 96·58, Al_2O_3 1·10, Fe_2O_3 0·34,   MgO & CaO 0·49,
H_2O 1·49,   =100·00
Weight per cub. ft. 162·3 lbs.   Spec. grav. 2·628.   Absorp. 3·70 %
     Crushing strain per sqr. ft. 552·2 tons.

198. BRAMLEY FALL STONE. *MILLSTONE GRIT*
Drab, fine-grained sandstone
     Pool Bank Quarries, near Otley, **Yorkshire**
Presented by Messrs B. Whitaker & Son, Horsforth, Leeds
     Weight per cub. ft. 142 lbs.
     Crushing strain per sqr. ft. 238 tons (Middleton)

199. DUNHOUSE STONE. *MILLSTONE GRIT*
Dark brown, fine-grained sandstone
     Dunhouse Quarries, near Barnard Castle, **Durham**
Presented by George Thompson, Esq., Barnard Castle
     Weight per cub. ft. 149 lbs.

200. HEWORTH BURN BLUE STONE. *MILLSTONE GRIT*
Blue-grey, fine-grained, slightly micaceous sandstone
     Heworth Burn Quarries, Felling, **Durham**
Presented by Messrs Tate Brown & Co., Felling
     Weight per cub. ft. 160 lbs.
     Crushing strain per sqr. ft. 664 tons (Durham College)

201. HEWORTH BURN BROWN STONE. *MILLSTONE GRIT*
Light brown, speckled with ferruginous spots, fine-grained sandstone
     Heworth Burn Quarries, Felling, **Durham**
Presented by Messrs Tate Brown & Co., Felling
     Weight per cub. ft. 160 lbs.
     Crushing strain per sqr. ft. 651 tons (Durham College)

202. KENTON COARSE FREESTONE. *Millstone Grit*
Light drab, medium-grained, slightly micaceous sandstone
Kenton Quarries, Newcastle, **Northumberland**
Presented by Messrs Taylor Bros., Newcastle
Weight per cub. ft. 138 lbs.

203. KENTON FINE FREESTONE. *Millstone Grit*
Light brown, fine-grained sandstone
Kenton Quarries, Newcastle, **Northumberland**
Presented by Messrs Taylor Bros., Newcastle
Weight per cub. ft. 142·5 lbs.

204. STOCKSFIELD FREESTONE. *Millstone Grit*
*Light yellow, slightly speckled with ferruginous spots, medium-
        grained, slightly micaceous sandstone*
Bearl Quarries, Stocksfield, **Northumberland**
Presented by Messrs R. Patterson & Son, Newcastle
Weight per cub. ft. 152 lbs.

205. BIRLING FREESTONE. *Millstone Grit*
Dark pink (puce), medium-grained sandstone
Birling Quarries, Warkworth, **Northumberland**
Presented by Messrs R. Moore & Sons, Warkworth
Weight per cub. ft. 138 lbs.

206. BIRLING FREESTONE. *Millstone Grit*
Yellow, medium-grained sandstone
Birling Quarries, Warkworth, **Northumberland**
Presented by Messrs R. Moore & Sons, Warkworth
Weight per cub. ft. 142·5 lbs.

207. BRAEHEAD FREESTONE. *Millstone Grit,
        Upper Carboniferous* (of Scotland)
Yellow, slightly banded, coarse-grained sandstone
Braehead Quarries, Fauldhouse, **Linlithgowshire**
Presented by Messrs Wm. Forrest & Co., Edinburgh
Weight per cub. ft. 133 lbs.

208. OLD BRIGHTON FREESTONE. *Millstone Grit,
        Upper Carboniferous* (of Scotland)
*Light straw-coloured with ferruginous specks, medium-grained
        sandstone*
Old Brighton Quarries, Polmont Station, **Stirlingshire**
Presented by Messrs Wm. Forrest & Co., Edinburgh
Weight per cub. ft. 142·5 lbs.

209. SHAMROCK STONE. *Millstone Grit*
Dark blue-grey, very fine, compact sandstone, admits of a polish
        Doonagore Quarries, **Co. Clare, Ireland**
Presented by Messrs G. A. Watson & Co., Liverpool
                Chemical Composition :
SiO_2 84·90, Al_2O_3 6·60, Fe_2O_3 3·60,   MnO 0·65,   MgO 1·26,
CaO 0·90, Na_2O 0·39, H_2O 1·70,   =100·00
        Weight per cub. ft. 167 lbs.   Spec. grav. 2·7
        Crushing strain per sqr. ft. 1902 tons (Kirkaldy)

209A. HARD YORK FLAGSTONE. *Coal Measures*
Yellow and brown, very distinctly banded, medium-grained sandstone
        Thornton Quarries, Bradford, **Yorkshire**
Presented by Messrs Pawson Bros., Morley
        Weight per cub. ft. 133 lbs.

**209B. SELF-FACED HARD BRADFORD FLAG-
        STONE.** *Coal Measures*
Yellow and brown, distinctly banded, medium-grained sandstone
        Thornton Quarries, Bradford, **Yorkshire**
Presented by Messrs Pawson Bros., Morley
        Weight per cub. ft. 133 lbs.

210. HARD BLUE YORK STONE. *Coal Measures*
Fawn-coloured, very fine-grained sandstone
        Arthwith Quarries, Bradford, **Yorkshire**
Presented by Messrs John Shackleton & Son, Goole
        Weight per cub. ft. 168 lbs.
        Crushing strain per sqr. ft. 700 tons.

211. HARD BROWN YORK STONE. *Coal Measures*
Light brown, fine-grained, micaceous sandstone
        Arthwith Quarries, Bradford, **Yorkshire**
Presented by Messrs John Shackleton & Son, Goole
        Weight per cub. ft. 168 lbs.
        Crushing strain per sqr. ft. 600 tons.

212. BIRSTALL STONE. *Coal Measures*
Drab, very fine-grained sandstone
        Birstall Quarries, Bradford, **Yorkshire**
Presented by Messrs W. Pickard & Son, Handsworth, Sheffield
        Weight per cub. ft. 171 lbs.

213. BLUE NELT STONE. *Coal Measures*
Greyish drab with rusty specks, medium-grained sandstone
        Heaton Park Quarries, Bradford, **Yorkshire**
Presented by Messrs A. Kendal & Sons, Shipley
        Weight per cub. ft. 130 lbs.

214. HOWLEY PARK BROWN YORK STONE.
COAL MEASURES
Light brown, ferruginous, fine-grained sandstone
Howley Park Quarries, Morley, **Yorkshire**
Presented by Messrs Pawson Bros., Morley
Weight per cub. ft. 140·3 lbs.   Spec. grav. 2·25.   Absorp. 4·90 %
Crushing strain per sqr. ft. 466·7 tons (Beare)

215. SILEX HARD YORK STONE. *COAL MEASURES*
Drab, fine-grained, slightly micaceous sandstone
Tuck Royd Quarries, near Halifax, **Yorkshire**
Presented by Messrs Brooks, Ltd., London
Chemical Composition :
SiO_2 97·83, Al_2O_3 & Fe_2O_3 1·17,   CaO 1·00,   = 100·00
Weight per cub. ft. 160 lbs.   Absorp. 1·02 %
Crushing strain per sqr. ft. 1100 tons.

216. YORKSHIRE FREESTONE. *COAL MEASURES*
Drab, fine-grained sandstone
Shibden Head Quarries, near Halifax, **Yorkshire**
Presented by James Dalton, Esq., Leeds
Weight per cub. ft. 157 lbs.

217. HARD YORK STONE. *COAL MEASURES*
Light brown, medium-grained sandstone
Morton Quarries, Bradford, **Yorkshire**
Presented by Messrs A. Kendal & Sons, Shipley
Weight per cub. ft. 152 lbs.

218. RINGBY BLOCK STONE. *COAL MEASURES*
Dark yellow, slightly banded, fine-grained sandstone
North Bridge Quarries, Halifax, **Yorkshire**
Presented by Messrs Jaggers Bros., Ambler Thorn, Halifax
Weight per cub. ft. 190 lbs.
Crushing strain per sqr. ft. 600 tons.

219. HARD YORK FLAGSTONE. *COAL MEASURES*
Yellowish drab, fine-grained sandstone
Shibden Head Quarries, near Halifax, **Yorkshire**
Presented by Messrs Dalton and Higgins, Halifax
Weight per cub. ft. 155 lbs.
Crushing strain per sqr. ft. 900 tons (Kirkaldy)

220. CROSLAND HILL HARD YORK STONE.
COAL MEASURES
Yellow, medium-grained, slightly micaceous sandstone
  Crosland Hill Quarries, Huddersfield, **Yorkshire**
Presented by Messrs Graham and Jessop, Moldgreen, Huddersfield
  Chemical Composition :
SiO₂ 93·30, Al₂O₃ 0·44, Fe₂O₃ 2·85, MgO 1·32, CaO 1·46,
H₂O 0·33, =99·70
  Weight per cub. ft. 163 lbs. Spec. grav. 2·60.

221. BROWN SHIPLEY STONE. *COAL MEASURES*
Light brown, slightly banded, fine-grained sandstone
  Shipley Quarries, near Huddersfield, **Yorkshire**
Presented by Messrs Wm. Stephenson & Son, Shipley
  Weight per cub. ft. 157 lbs.

222. BLUE SHIPLEY STONE. *COAL MEASURES*
Light blue-grey, fine-grained sandstone
  Shipley Quarries, near Huddersfield, **Yorkshire**
Presented by Messrs Wm. Stephenson & Son, Shipley
  Weight per cub. ft. 157 lbs.

223. BOLTON WOOD STONE. *COAL MEASURES*
Light greenish brown fine-grained sandstone
  Bolton Wood Quarries, Bradford, **Yorkshire**
Presented by Messrs Pawson Bros., Morley
  Weight per cub. ft. 146 lbs.

224. ROBIN HOOD STONE. *COAL MEASURES*
Blue-grey, fine-grained, slightly micaceous sandstone
  Rothwell Haigh Quarries, near Leeds, **Yorkshire**
Presented by Messrs Pawson Bros., Morley
  Weight per cub. ft. 145 lbs. Spec. grav. 2·32. Absorp. 5·41 %
  Crushing strain per sqr. ft. 574 tons (Rivington)

225. PARK SPRING STONE. *COAL MEASURES*
Yellowish drab, micaceous, medium-grained sandstone
  Park Spring Quarries, near Leeds, **Yorkshire**
Presented by Messrs Pawson Bros., Morley
  Weight per cub. ft. 151 lbs.
  Crushing strain per sqr. ft. 487 tons (Rivington)

226. ROBIN HOOD BLUE YORK STONE. *Coal*
Measures
Light bluish drab, fine-grained sandstone
Thorp Quarries, near Leeds, **Yorkshire**
Presented by Messrs Pawson Bros., Morley
Weight per cub. ft. 144·6 lbs. Spec. grav. 2·32. Absorp. 3·90 %
Crushing strain per sqr. ft. 574 tons (Beare)

227. HARD BLUE YORK STONE. *Coal Measures*
Light blue-grey, very fine-grained sandstone
Carlinghoe Quarries, near Batley, **Yorkshire**
Presented by Messrs Pawson Bros., Morley
Weight per cub. ft. 172 lbs.

228. SCOUT HARD YORK STONE. *Coal Measures*
Yellow with brown specks, micaceous, medium-grained sandstone
Eccleshills Quarries, **Yorkshire**
Presented by Messrs Pawson Bros., Morley
Weight per cub. ft. 172 lbs.

229. HARD YORK BROWN STONE. *Coal Measures*
Greenish brown, medium-grained, highly micaceous sandstone
Wycliffe Quarries, Shipley, **Yorkshire**
Presented by Messrs A. Kendal & Sons, Shipley
Weight per cub. ft. 152 lbs.

230. SOFT BROWN YORK STONE. *Coal Measures*
Brown, medium-grained, slightly micaceous sandstone
Bracken Hill Quarries, Pontefract, **Yorkshire**
Presented by Messrs Pawson Bros., Morley
Weight per cub. ft. 133 lbs.

231. HANDSWORTH STONE. *Coal Measures*
Grey with brown ferruginous specks, fine-grained sandstone
Handsworth Quarries, Sheffield, **Yorkshire**
Presented by Messrs W. Pickard & Son, Handsworth
Weight per cub. ft. 171 lbs.

232. GREENMORE STONE. *Coal Measures*
Dark grey, very fine-grained sandstone
Greenmore Quarries, near Sheffield, **Yorkshire**
Presented by Messrs W. Pickard & Son, Handsworth
Weight per cub. ft. 163 lbs.

233. BRISTOL PENNANT SANDSTONE. *COAL MEASURES*
Blue-grey, fine-grained
          Fish Pond Quarries, Bristol, **Gloucestershire**
Presented by The Hard Stones Firms, Ltd., Bristol
          Chemical Composition (E. H. Cook, Bristol):
SiO_2 72·44, Al_2O_3 & FeO 15·78,   CaO 2·03, K_2O & Na_2O 8·01,
H_2O 0·50, CO_2 1·24,   =100·00
          Weight per cub ft. 172 lbs.
          Crushing strain per sqr. ft. 1001 tons (Kirkaldy)

234. GREY FOREST OF DEAN STONE (*PENNANT GRIT*). *COAL MEASURES*
Drab-grey, medium-grained, slightly micaceous sandstone
          Forest of Dean Quarries, nr Lydney, **Gloucestershire**
Presented by The Forest of Dean Stone Firms, Ltd., Bristol
          Chemical Composition :
SiO_2 80·16, Al_2O_3 14·40, Fe_2O_3 1·65,   MgO 0·24,   CaO 2·55,
$CaSO_4$ 1·00,   =100·00
          Weight per cub. ft. 149 lbs.
          Crushing strain per sqr. ft. 569 tons (Kirkaldy)

235. BLUE FOREST OF DEAN STONE (*PENNANT GRIT*). *COAL MEASURES*
Dark blue-grey, compact, hard sandstone
          Forest of Dean Quarries, nr Lydney, **Gloucestershire**
Presented by The Forest of Dean Stone Firms, Ltd., Bristol
          Weight per cub. ft. 151·4 lbs.   Absorp. 2·71 %
          Crushing strain per sqr. ft. 631 tons (Kirkaldy)

236. RED PENNANT STONE. *COAL MEASURES*
Pinkish brown, fine-grained
          Huckford Quarries, Winterbourne, **Gloucestershire**
Presented by The United Stone Firms, Ltd., Bristol
          Weight per cub. ft. 171 lbs.

237. ABERCARNE STONE. *COAL MEASURES*
Blue-grey, fine-grained sandstone
          Mynyddyslwyn Quarries, **Monmouthshire**
Presented by The Ebbw Vale Iron and Steel Co., Ebbw Vale
          Weight per cub. ft. 152 lbs.

238. TINTERN ABBEY STONE. *Coal Measures*
Dark purple-brown, fine-grained sandstone
Tintern Quarries, **Monmouthshire**
Presented by Messrs Turner & Son, Cardiff
Chemical Composition (Sir A. Church):
SiO_2 78·20, Al_2O_3 & Fe_2O_3 11·54, $CaCO_3$ 5·62,   CaO 0·44,
MgO 0·18, K_2O & Na_2O 2·28, H_2O 0·33,   =98·59
Weight per cub. ft. 155 lbs.   Absorp. 1·48 %.   Crushing
strain per sqr. ft. 850 tons (Burstall and Monkhouse).

239. PONTYPRIDD STONE. *Coal Measures*
Blue-grey, hard, fine-grained, micaceous sandstone
Craig-y-Herg Quarries, Pontypridd, **S. Wales**
Presented by Messrs Mackay and Davies, Cardiff
Chemical Composition:
SiO_2 83·15, Al_2O_3 8·10, Fe_2O_3 4·54,   MgO 0·68, K_2O & Na_2O 1·78,
=98·25
Weight per cub. ft. 160 lbs.
Crushing strain per sqr. ft. 1617 tons.

240. PENNANT SANDSTONE. *Coal Measures*
Blue-grey, hard, close-grained
Llwyndu Quarries, near Glais, **S. Wales**
Presented by Messrs J. Evans & Co., Glais
Weight per cub. ft. 160·5 lbs.

241. BRYN TEG BUFF FREESTONE. *Coal Measures*
Buff-coloured, fine-grained sandstone
Bryn Teg Quarries, Broughton, Denbigh, **N. Wales**
Presented by Messrs Davies Bros., Wrexham
Weight per cub. ft. 158 lbs.

242. CEFN STONE. *Coal Measures*
Drab, fine-grained sandstone
Ruabon Quarries, near Denbigh, **N. Wales**
Presented by Messrs H. Dennis & Co., Ruabon
Weight per cub. ft. 155 lbs.

243. PEASENHURST STONE. *Coal Measures*
Yellowish drab, medium-grained sandstone
Peasenhurst Quarries, **Derbyshire**
Presented by The Peasenhurst Quarries Co., Stamford
Weight per cub. ft. 140 lbs.
Crushing strain 420 tons per sqr. ft.

244. TURF MOOR STONE. *Coal Measures*
Light greenish grey, fine-grained sandstone
    Worthsthorne Quarries, near Burnley, **Lancashire**
Presented by Messrs Smith Bros. (Burnley), Ltd.
    Weight per cub. ft. 171 lbs.

245. BURNLEY BLUE STONE. *Coal Measures*
Light blue-grey, fine-grained sandstone
    Habergham Quarries, near Burnley, **Lancashire**
Presented by Messrs Smith Bros. (Burnley), Ltd.
    Weight per cub. ft. 147 lbs.

246. APPLEY BRIDGE STONE. *Coal Measures*
Light grey, fine-grained sandstone
    Appley Road Quarries, Appley Bridge, Wigan, **Lancashire**
Presented by Messrs Walter Martland, Ltd., Burscough Bridge
    Weight per cub. ft. 152 lbs.

247. WINDY NOOK FREESTONE. *Coal Measures*
Amber yellow, medium-grained sandstone
    Windy Nook Quarries, near Gateshead, **Durham**
Presented by Messrs R. Kell & Co., Newcastle-on-Tyne
    Weight per cub. ft. 160 lbs.

248. PENSHAW STONE. *Coal Measures*
Light brown, coarse-grained sandstone
    Penshaw Quarries, **Durham**
Presented by Messrs J. and W. Lowry, Newcastle-on-Tyne
    Chemical Composition (G. P. Lishman) :
SiO_2 85·00, Al_2O_3 6·09, Fe_2O_3 0·51, K_2O 4·05, Na_2O 1·02,
H_2O &c. 2·31, $CaCO_3$ 0·04, $MgCO_3$ 0·62, =99·64
    Weight per cub. ft. 133 lbs.

249. BURRADON FIRESTONE. *Coal Measures*
Light yellow, medium-grained, slightly micaceous sandstone
    Burradon Quarries, Killingworth, **Northumberland**
Presented by Messrs R. Patterson & Son, Newcastle
    Weight per cub. ft. 142 lbs.

250. ROUND CLOSE FREESTONE. *Coal Measures*
Light pinkish yellow, medium-grained
    Castle Quarries, Whitehaven, **Cumberland**
Presented by Messrs Walker Bros., Cockermouth
    Chemical Composition (W. Allen, Workington) :
SiO_2 62·44, Al_2O_3 19·23, Fe_2O_3 4·43, MgO 4·47, CaO 5·20,
H_2O 4·21, =99·98
    Weight per cub. ft. 165 lbs.

251. AUCHINLEA SANDSTONE. *Upper Carboniferous*
Drab, slightly banded, medium-grained, micaceous
Auchinlea Quarries, **Lanarkshire**
Presented by J. R. Taylor, Esq., Cleland
Weight per cub. ft. 128·9 lbs. Spec. grav. 2·06 (Beare). Absorp.
6.90 %. Crushing strain per sqr. ft. 203·6 tons (Beare)

252. AUCHINLEA SANDSTONE. *Upper Carboniferous*
Drab, with brown ferruginous specks, medium-grained
Auchinlea Quarries, **Lanarkshire**
Presented by J. R. Taylor, Esq., Cleland
Chemical Composition (Tatlock, Glasgow) :
SiO_2 90·17, Al_2O_3 6·62, FeO 1·26, MgO 0·31, CaO 0·54,
K_2O 0·90, Na_2O 0·20, = 100·00
Weight per cub. ft. 132 lbs. Absorp. 6·90 %
Crushing strain per sqr. ft. 203·6 tons (Beare)

253. OVER WOOD LIVER ROCK. *Upper Carboniferous*
Dark drab, medium-grained sandstone
Over Wood Quarries, Stone House, **Lanarkshire**
Presented by Messrs Baird and Stevenson, Glasgow
Weight per cub. ft. 147 lbs.

254. ALLOA FREESTONE. *Upper Carboniferous*
Light pinkish grey, medium-grained sandstone
Devon Quarries, near Alloa, **Clackmannanshire**
Presented by Messrs G. and R. Cousins, Edinburgh
Weight per cub. ft. 152 lbs.

255. SANDS FREESTONE. *Upper Carboniferous*
Light drab, fine-grained sandstone
Sands Quarries, Kincardine, **Kincardineshire**
Presented by Messrs R. McAlpine & Son, Glasgow
Weight per cub. ft. 133 lbs.

256. SCARES LIVER ROCK. *Upper Carboniferous*
Warm drab, medium-grained, slightly micaceous sandstone
Scares Quarries, Old Cummock, **Ayrshire**
Presented by Messrs Baird and Stevenson, Glasgow
Weight per cub. ft. 152 lbs.

257. AYRSHIRE FREESTONE. *Upper Carboniferous*
Light straw-coloured, with small brown specks, slightly micaceous
sandstone
Sevenacres Quarries, near Irvine, **Ayrshire**
Presented by Messrs Maclachlan & Son, Irvine
Weight per cub. ft. 128 lbs.

PERMIAN.

258. RED ROCK. *Permian*
Warm red, with light particles, coarse-grained
      Blatchcombe Quarries, Paignton, **Devonshire**
Presented by Jos. Pollard, Esq., Paignton
    Weight per cub. ft. 166 lbs.

259. NESSCLIFFE STONE. *Permian*
Dull brick red, fine-grained sandstone
      Nesscliffe Quarries, Great Ness, **Shropshire**
Presented by The Earl of Bradford, Oswestry
    Weight per cub. ft. 152 lbs.

260. RED MANSFIELD STONE. *Permian*
Light red, fine-grained, calcareous sandstone
      Lindley Quarries, Mansfield **Nottinghamshire**
Presented by Messrs Lindley & Son, Nottingham
            Chemical Composition :
SiO_2 49·4, Al_2O_3 & Fe_2O_3 3·2, H_2O &c. 4·8, $CaCO_3$ 26·5,
$MgCO_3$ 16·1, = 100·00
    Weight per cub. ft. 143·2 lbs. Spec. grav. 2·30 (Beare).
    Absorp. 4·58 %. Crushing strain per sqr. ft. 591·9 tons
    (Beare)

261. WHITE MANSFIELD STONE. *Permian*
Light yellow, fine-grained, calcareous sandstone
      Lindley Quarries, Mansfield, **Nottinghamshire**
Presented by Messrs Lindley & Son, Nottingham
            Chemical Composition :
SiO_2 51·40, Al_2O_3 & Fe_2O_3 1·32, H_2O &c. 2·08, $CaCO_3$ 26·50,
$MgCO_3$ 17·98, = 99·28
Weight per cub. ft. 140·1 lbs. Spec. grav. 2·25. Absorp. 5·01 %
    Crushing strain per sqr. ft. 461·7 tons (Beare)

262. RED MANSFIELD STONE. *PERMIAN*
Warm red, fine-grained, calcareous sandstone
Sills Quarries, Mansfield, **Nottinghamshire**
Presented by Wm. Sills, Esq., Mansfield
Chemical Composition :
SiO_2 49·4, Al_2O_3 & Fe_2O_3 3·2, H_2O &c. 4·8, $CaCO_3$ 26·5,
$MgCO_3$ 16·1, = 100·00
Weight per cub. ft. 133 lbs.
Crushing strain per sqr. ft. 609 tons

263. WHITE MANSFIELD STONE. *PERMIAN*
Warm yellow, with small ferruginous spots, fine-grained calcareous sandstone
Sills Quarries, Mansfield, **Nottinghamshire**
Presented by Wm. Sills, Esq., Mansfield
Chemical Composition :
SiO_2 51·40, Al_2O_3 & Fe_2O_3 1·32, H_2O &c. 2·08, $CaCO_3$ 26·50,
$MgCO_3$ 17·98, = 99·28
Weight per cub. ft. 140 lbs.
Crushing strain per sqr. ft. 462 tons

264. ABENBURY PINK FREESTONE. *PERMIAN*
Purplish brown, fine-grained sandstone
Abenbury Quarries, Wrexham, **N. Wales**
Presented by Messrs Davies Bros., Wrexham
Weight per cub. ft. 160 lbs.

265. REMINGTON STONE. *PERMIAN*
Bright red, coarse-grained, micaceous sandstone
Remington Quarries, near Penrith, **Cumberland**
Presented by W. Grisenthwaite, Esq., Penrith
Weight per cub. ft. 152 lbs.

266. LAZONBY RED STONE. *PERMIAN*
Light terra-cotta, medium-grained sandstone
Lazonby Fell Quarries, near Lazonby, **Cumberland**
Presented by Henry Graves, Esq., Lazonby
Weight per cub. ft. 151·5 lbs.

267. LAZONBY WHITE STONE. *PERMIAN*
Light yellowish pink, medium-grained sandstone
Lazonby Fell Quarries, near Lazonby, **Cumberland**
Presented by Henry Graves, Esq., Lazonby
Weight per cub. ft. 151·5 lbs.

268. RED MOAT STONE. *PERMIAN*
Dull red, very fine-grained sandstone
  Moat Quarries, near Netherby, **Cumberland**
Presented by Messrs Herbertsons, Ltd., Galashiels
  Weight per cub. ft. 123 lbs. Spec. grav. 1·97
  Crushing strain per sqr. ft. 274·4 tons.

269. HALF WAY WELL STONE. *PERMIAN*
Bright red, mottled, coarse-grained sandstone
  Half Way Well Quarries, near Penrith, **Cumberland**
Presented by W. Grisenthwaite, Esq., Penrith
  Weight per cub. ft. 152 lbs.

270. DE WHELPDALE STONE. *PERMIAN*
Light red, medium-grained, micaceous sandstone
  De Whelpdale Quarries, Boscar, **Cumberland**
Presented by W. Grisenthwaite, Esq., Penrith
  Weight per cub. ft. 133 lbs.

271. HEIGHTS STONE. *PERMIAN*
Light buff, with brown specks, medium-grained sandstone
  Heights Quarries, near Appleby, **Westmorland**
Presented by Messrs Lindsay and Robinson, Appleby
  Weight per cub. ft. 142·5 lbs.

272. BULWELL STONE. *MAGNESIAN LIMESTONE*
Light pink, medium-grained limestone
  Bulwell Quarries, **Nottinghamshire**
Presented by Messrs Wilkinson & Son, Bulwell
  Weight per cub. ft. 142 lbs.

273. MANSFIELD YELLOW STONE. *MAGNESIAN*
    LIMESTONE
Orange yellow, very fine-grained limestone, capable of taking a
  polish
  Lindley Wood Quarries, Mansfield, **Nottinghamshire**
Presented by Messrs Lindley & Son, Mansfield
    Chemical Composition :
SiO_2 3·70, H_2O &c. 2·05, $CaCO_3$ 51·65, $MgCO_3$ 42·60,
= 100·00
Weight per cub. ft. 145·4 lbs. Spec. grav. 2·33. Absorp. 4·62 %
  Crushing strain per sqr. ft. 577·4 tons (Beare)

274. MANSFIELD WOODHOUSE STONE. *MAGNESIAN*
LIMESTONE
Warm yellow, very fine-grained, compact limestone
Mansfield Quarries, **Nottinghamshire**
Presented by Wm. Sills, Esq., Mansfield
Chemical Composition :
SiO_2 & H_2O 1·38,   CaO 74·32, $MgCO_3$ 24·30,   =100·00
Weight per cub. ft. 145 lbs.
Crushing strain per sqr. ft. 577 tons (Rivington)

275. BOLSOVER MOOR STONE. *MAGNESIAN LIMESTONE*
Warm yellowish brown, very fine-grained limestone
Bolsover Moor Quarries, **Derbyshire**
Presented by J. Day, Esq., Cambridge
Chemical Composition (Rivington) :
SiO_2 3·60, Al_2O_3 & Fe_2O_3 1·80, H_2O &c. 3·30, $CaCO_3$ 51·10,
$MgCO_3$ 40·20,   =100·00
Weight per cub. ft. 152 lbs. (Rivington)
Crushing strain per sqr. ft. 484 tons (Royal Com.)

276. ANSTON STONE. *MAGNESIAN LIMESTONE*
Light cream, fine-grained limestone
Kiveton Park Quarries, near Sheffield, **Yorkshire**
Presented by Messrs J. Turner & Son, Sheffield
Chemical Composition :
SiO_2 1·00, Fe_2O_3 1·30, CH_2O_3 & H_2O 45·60,   MgO 21·40,
CaO 30·00, H_2PO_4 0·07,   =99·37
Weight per cub. ft. 134 lbs.   Spec. grav. 2·12.   Absorp. 7·50 %
Crushing strain per sqr. ft. 833·1 tons (Kirkaldy)

277. HUDDLESTONE STONE. *MAGNESIAN LIMESTONE*
Light cream, fine-grained, semi-crystalline limestone
Huddlestone Quarries, **Yorkshire**
Presented by Messrs W. H. Newton & Co., Leeds
Chemical Composition (Rivington) :
SiO_2 2·53, Al_2O_3 & Fe_2O_3 0·30, H_2O &c. 1·61, $CaCO_3$ 54·19,
$MgCO_3$ 41·37,   =100·00
Weight per cub. ft. 138 lbs.
Crushing strain per sqr. ft. 278 tons (Rivington)

278. ROCHE ABBEY STONE. *MAGNESIAN LIMESTONE*
Cream-coloured, with brown ferruginous specks, fine-grained limestone
Bawtry Quarries, **Yorkshire**
Presented by R. Rollit, Esq., Worksop
Chemical Composition (Rivington) :
SiO_2 0·80, Al_2O_3 & Fe_2O_3 0·70, H_2O &c. 0·60, $CaCO_3$ 57·5,
$MgCO_3$ 39·4,   =99·0
Weight per cub. ft. 139 lbs.
Crushing strain per sqr. ft. 250 tons (Rivington)

279. BRACKENHILL STONE. *MAGNESIAN LIMESTONE*
Light yellow, with brown specks, fine-grained limestone
        Brackenhill Quarries, near Pontefract, **Yorkshire**
Presented by Messrs Hardaker & Son, Ackworth
        Weight per cub. ft. 138 lbs.

280. ACKWORTH STONE. *MAGNESIAN LIMESTONE*
Light straw-coloured, fine-grained limestone
        Brackenwell Quarries, near Pontefract, **Yorkshire**
Presented by Messrs Camplin Bros., Ackworth
        Weight per cub. ft. 140·7 lbs.   Absorp. 5·00 %
        Crushing strain per sqr. ft. 389·1 tons (Beare)

281. TADCASTER STONE. *MAGNESIAN LIMESTONE*
Rich dark cream, very fine-grained, compact limestone
        Smaws Quarries, near Tadcaster, **Yorkshire**
Presented by Samuel Smith, Esq., Tadcaster
        Weight per cub. ft. 123·5 lbs.

282. SOUTH ELMSALL STONE. *MAGNESIAN LIMESTONE*
Yellow, compact, fine-grained limestone
        South Elmsall Quarries, near Doncaster, **Yorkshire**
Presented by Messrs J. Hinchcliffe & Son, South Elmsall
        Weight per cub. ft. 138 lbs.

283. MARSDEN LIMESTONE. *MAGNESIAN LIMESTONE*
Light yellowish brown, slightly cellular, crystalline
        Marsden Quarries, South Shields, **Durham**
Presented by The Harton Coal Co., Ltd., South Shields
        Chemical Composition (Pattinson):
SiO_2 0·28, Al_2O_3 0·04, Fe_2O_3 0·06, H_2O 0·07, $CaCO_3$ 98·31,
$MgCO_3$ 1·24, H_2PO_4 0·02,   S 0·02,   =100·04
        Weight per cub. ft. 150 lbs.

TRIAS.

284. RAINHILL STONE. *Pebble Beds, Bunter*
Light terra-cotta, medium-grained sandstone
    Mill Quarries, Rainhill, **Lancashire**
Presented by Messrs T. Welsby & Son, Rainhill
    Weight per cub. ft. 133 lbs.
    Crushing strain per sqr. ft. 351 tons (Liverpool University)

285. RAINHILL RED SANDSTONE. *Pebble Beds, Bunter*
Dull red, medium-grained
    Sandstone Quarries, Rainhill, **Lancashire**
Presented by Messrs Morrison & Sons, Wavertree, Liverpool
    Weight per cub. ft. 160 lbs.   Crushing strain per sqr. ft.
    351 lbs.   (Watkinson, Liverpool University)

286. WOOLTON STONE. *Lower Pebble Beds, Bunter*
Dull red, fine-grained sandstone
    Stone Quarries, Woolton, **Lancashire**
Presented by Messrs J. Morrison & Sons, Wavertree, Liverpool
    Weight per cub. ft. 160 lbs.   Crushing strain per sqr. ft.
    400 lbs. (Watkinson, Liverpool University)

287. ST BEES RED FREESTONE. *Bunter*
Bright red, slightly banded, fine-grained sandstone
    Bankend Quarries, near Bigrigg, **Cumberland**
Presented by H. T. Doloughan, Esq., Bigrigg
    Chemical Composition (R. H. Hellon, Whitehaven):
SiO_2 83·37, Al_2O_3 7·17, Fe_2O_3 1·77, CH_2O_3 0·98,
MgO 0·59, CaO 0·59, K_2O 3·73, H_2O 1·80, $=100\cdot00$
    Weight per cub. ft. 142 lbs.
    Crushing strain per sqr. ft. 393·1 tons (H. Stanger)

287A. ST BEES HEAD FREESTONE. BUNTER
Dark red, fine-grained, micaceous sandstone
Sandwith Quarries, near Whitehaven, **Cumberland**
Presented by The Sandwith Quarries Co., Whitehaven
Chemical Composition (R. H. Hellon, County Analyst,
Whitehaven):
SiO_2 85·20, Al_2O_3 7·00, Fe_2O_3 2·00, CH_2O_3 1·00,
MgO 0·30, CaO 0·10, K_2O & Na_2O 3·30, H_2O 1·10,
=100·00
Weight per cub. ft. 134·6 lbs.
Crushing strain per sqr. ft. 506·9 tons (Kirkaldy)

288. ASPATRIA STONE. BUNTER
Chocolate red, fine-grained sandstone
Aspatria Quarries, **Cumberland**
Presented by H. Graves, Esq., Aspatria
Weight per cub. ft. 125 lbs. Spec. grav. 1·98. Absorp. 8·50 %
Crushing strain per sqr. ft. 759·7 tons (Kirkaldy)

289. RED SHAWK STONE. BUNTER
Dark dull red, medium-grained sandstone
Shawk Quarries, near Thursby, **Cumberland**
Presented by James McKay, Esq., Thursby
Weight per cub. ft. 133 lbs.

290. MOTTLED SHAWK STONE. BUNTER
Pink and yellow mottled, medium-grained sandstone
Shawk Quarries, near Thursby, **Cumberland**
Presented by James McKay, Esq., Thursby
Weight per cub. ft. 138 lbs.

291. MORLEY MOOR STONE. KEUPER
Light drab, medium-grained, slightly micaceous sandstone
Morley Quarries, Morley, **Derbyshire**
Presented by Messrs S. Seal & Sons, Derby
Weight per cub. ft. 123·5 lbs.

292. MORLEY MOOR STONE. KEUPER
Chocolate-coloured, with brown specks, micaceous sandstone
Morley Quarries, Morley, **Derbyshire**
Presented by Messrs S. Seal & Sons, Derby
Weight per cub. ft. 123·5 lbs.

293. HORSLEY CASTLE STONE. *Keuper*
Drab, medium-grained, micaceous sandstone
    Horsley Castle Quarries, near Coxbench, **Derbyshire**
Presented by Messrs Slater & Co., Derby
    Weight per cub. ft. 138 lbs.

294. RED HOLLINGTON STONE. *Keuper*
Bright red, even fine-grained sandstone
    Beggars Well Quarries, Rocester, **Staffordshire**
Presented by Messrs Stanton and Bettany, Rocester
    Weight per cub. ft. 133 lbs.

295. FINE WHITE HOLLINGTON STONE. *Keuper*
Light pinkish drab, medium-grained, micaceous sandstone
    Hollington Quarries, Rocester, **Staffordshire**
Presented by Messrs Stanton and Bettany, Rocester
    Weight per cub. ft. 133 lbs.

296. SECONDS WHITE HOLLINGTON STONE.
    Keuper
Light pinkish drab, coarse-grained, micaceous sandstone
    Hollington Quarries, Rocester, **Staffordshire**
Presented by Messrs Stanton and Bettany, Rocester
    Weight per cub. ft. 133 lbs.

297. MOTTLED HOLLINGTON STONE. *Keuper*
Red and drab, mottled and banded, even-grained sandstone
    Beggars Well Quarries, Rocester, **Staffordshire**
Presented by Messrs Stanton and Bettany, Rocester
    Weight per cub. ft. 133 lbs.

298. RED HOLLINGTON STONE. *Keuper*
Dull red, fine-grained sandstone
    Hollington Quarries, Stoke-on-Trent, **Staffordshire**
Presented by Messrs J. Stevenson & Co., Hollington
    Weight per cub. ft. 138 lbs.

299. WHITE HOLLINGTON STONE. *Keuper*
Light dove-coloured, fine-grained sandstone
    Hollington Quarries, Stoke-on-Trent, **Staffordshire**
Presented by Messrs J. Stevenson & Co., Hollington
    Weight per cub. ft. 138 lbs.

300. MOTTLED HOLLINGTON STONE. *KEUPER*
Drab and red mottled, fine-grained, slightly micaceous sandstone
    Hollington Quarries, Stoke-on-Trent, **Staffordshire**
Presented by Messrs J. Stevenson & Co., Hollington
    Weight per cub. ft. 138 lbs.
    Crushing strain per sqr. ft. 289 tons (Kirkaldy)

301. WHITE HOLLINGTON STONE. *KEUPER*
Light dove-coloured, medium-grained sandstone
    Hollington Quarries, Stoke-on-Trent, **Staffordshire**
Presented by J. Day, Esq., Cambridge
    Weight per cub. ft. 138 lbs.

302. PENKRIDGE STONE. *KEUPER*
Rich red and drab, fine-grained, micaceous sandstone
    Penkridge Quarries, **Staffordshire**
Presented by Frank Sprenger, Esq., Penkridge
    Weight per cub: ft. 133 lbs.

303. STANTON STONE. *KEUPER*
Light pink and drab, medium-grained sandstone
    Stanton Quarries, near Leek, **Staffordshire**
Presented by Messrs P. Ford & Son, Uttoxeter
    Weight per cub. ft. 166 lbs.

304. BROMSGROVE STONE. *KEUPER*
Terra-cotta, fine-grained sandstone
    Lineage and Hill Top Quarries, Bromsgrove,
                                    Worcestershire
Presented by C. L. Willcox, Esq., Bromsgrove
    Chemical Composition (A. B. Hill, Birmingham):
SiO_2 83·19, Al_2O_3 6·70, Fe_2O_3 1·94, CH_2O_3 0·90, MgO 0·92,
CaO 3·18, H_2O 0·68, Loss 2·49, = 100·00
    Weight per cub. ft. 140 lbs.
    Crushing strain per sqr. ft. 245 tons (Kennedy)

305. BROMSGROVE STONE. *KEUPER*
Drab, fine, even-grained sandstone
    Lineage and Hill Top Quarries, Bromsgrove,
                                    Worcestershire
Presented by C. L. Willcox, Esq., Bromsgrove
    Weight per cub. ft. 140 lbs.

306. BROMSGROVE STONE. *KEUPER*
Yellow and reddish brown, banded, fine-grained sandstone
    Lineage and Hill Top Quarries, Bromsgrove,
                                    Worcestershire
Presented by C. L. Willcox, Esq., Bromsgrove
    Weight per cub. ft. 137 lbs.

307. BROMSGROVE STONE. *Keuper*
Light yellowish brown, fine-grained sandstone
Lineage and Hill Top Quarries, Bromsgrove,
Worcestershire
Presented by C. L. Willcox, Esq., Bromsgrove
Weight per cub. ft. 137 lbs.

308. RED GRINSHILL SANDSTONE. *Keuper*
Bright red, fine-grained
Bridge and Cureton Quarries, Grinshill, **Shropshire**
Presented by The Grinshill Stone Quarries Co., Grinshill
Weight per cub. ft. 123·5 lbs.

309. WHITE GRINSHILL SANDSTONE. *Keuper*
Light drab, with small brown specks, fine-grained
Bridge and Cureton Quarries, Grinshill, **Shropshire**
Presented by The Grinshill Stone Quarries Co., Grinshill
Chemical Composition :
SiO_2 95·46, Al_2O_3 1·17, Fe_2O_3 0·87, $CaCO_3$ 0·61, $MgCO_3$ 0·69,
H_2O & loss, 1·20, $= 100·00$
Weight per cub. ft. 122·5 lbs. Spec. grav. 1·96. Absorp. 7·80 %
Crushing strain por sqr. ft. 332 tons (Kirkaldy)

310. HARMER HILL SANDSTONE. *Keuper*
Red terra-cotta coloured, medium-grained
Harmer Hill Quarries, **Shropshire**
Weight per cub. ft. 128 lbs.

311. RUNCORN RED SANDSTONE. *Keuper*
Dull red, medium-grained
Sandstone Quarries, Runcorn, **Cheshire**
Presented by Messrs J. Morrison & Sons, Wavertree, Liverpool
Weight per cub. ft. 160 lbs. Crushing strain per sqr. ft.
254 tons (Watkinson, Liverpool University)

312. WESTON RED SANDSTONE. *Keuper*
Bright terra-cotta, medium-grained
South Quarries, Runcorn, **Cheshire**
Presented by Messrs Orme and Muntz, Weston
Weight per cub. ft. 160 lbs.
Crushing strain per sqr. ft. 240 tons (Heenan and Froude)

313. WESTON FLECKED SANDSTONE. *Keuper*
Drab and red mottled, medium-grained
South Quarries, Runcorn, **Cheshire**
Presented by Messrs Orme and Muntz, Weston
Weight per cub. ft. 160 lbs.
Crushing strain per sqr. ft. 430 tons (Heenan and Froude)

314. STORETON STONE. *KEUPER*
Pinkish buff, medium-grained sandstone
    Higher Bebington Quarries, Wirral, **Cheshire**
Presented by Chas. Wells, Esq., Bootle
              Chemical Composition :
SiO_2 95·48, Al_2O_3 3·06, FeO 0·42, = 98·96
    Weight per cub. ft. 139 lbs.

315. ALDERLEY EDGE STONE. *KEUPER*
Purplish red, fine-grained, slightly micaceous sandstone
    Alderley Edge Quarries, **Cheshire**
Presented by G. E. Mills, Esq., Alderley Edge
    Weight per cub. ft. 160 lbs.

316. ALDERLEY EDGE STONE. *KEUPER*
Yellow, fine-grained, slightly micaceous sandstone
    Alderley Edge Quarries, **Cheshire**
Presented by G. E. Mills, Esq., Alderley Edge
    Weight per cub. ft. 160 lbs.

317. GREEN QUERELLA STONE. *RHÆTIC*
Light green, fine-grained sandstone
    Bridgend Quarries, **Glamorgan, S. Wales**
Presented by The Forest of Dean Stone Firms Co., Bristol
           Chemical Composition (J. S. Merry) :
SiO_2 91·20, Al_2O_3 3·35, Fe_2O_3 0·58, MgO 0·25, CaO 1·23,
H_2O 2·57, Loss 0·82, = 100·00
Weight per cub. ft. 136 lbs. Spec. grav. 2·39. Absorp. 4·5 %
    Crushing strain per sqr. ft. 438·4 tons (Kirkaldy)

318. WHITE QUERELLA STONE. *RHÆTIC*
White, with slight green tint, fine-grained sandstone
    Bridgend Quarries, **Glamorgan, S. Wales**
Presented by The Forest of Dean Stone Firms Co., Bristol
Weight per cub. ft. 142 lbs. Spec. grav. 2·23. Absorp. 6 %
    Crushing strain per sqr. ft. 546·4 tons (Kirkaldy)

319. DRAYCOTT STONE. *TRIAS*
Grey and red, coarse-grained sandstone
    Draycott Quarries, near Weston-super-Mare,
                                Somersetshire
    Weight per cub. ft. 161 lbs.

320. GREENBRAE STONE. *TRIAS*
Drab, medium-grained sandstone
    Greenbrae Quarries, Hopeman, **Elginshire**
Presented by Messrs J. and L. Anderson, Hopeman
    Weight per cub. ft. 119 lbs.

321. CORNCOCKLE STONE. *TRIAS*
Light terra-cotta, fine-grained sandstone
Corncockle Quarries, **Dumfriesshire**
Presented by Messrs Wm. Benson & Son, Newcastle-on-Tyne
Chemical Composition (Blake):
SiO_2 94·75, Al_2O_3 2·81, Fe_2O_3 0·09, MgO 0·27, CaO 0·20,
K_2O, HC_2O 0·08, H_2O 1·77, = 99·97
Weight per cub. ft. 132 lbs. Spec. grav. 2·12. Absorp. 4·57 %
Crushing strain per sqr. ft. 383·8 tons (London University)

322. CLOSEBURN RED FREESTONE. *TRIAS*
Warm red, fine-grained, slightly micaceous sandstone
Closeburn Quarries, Thornhill, **Dumfriesshire**
Presented by The Scottish Stone Quarries, Ltd., Closeburn
Chemical Composition (A. Greenwell):
SiO_2 91·86, Al_2O_3 4·50, Fe_2O_3 0·97, FeO 0·07, MgO 0·33,
CaO 0·30, K_2O 2·50, Na_2O trace, Loss 2·50, = 103·03
Weight per cub. ft. 123·5 lbs.
Crushing strain per sqr. ft. 478 tons (Beare)

323. CLOSEBURN RED FREESTONE. *TRIAS*
Light red, fine-grained, slightly micaceous sandstone
Closeburn Quarries, near Thornhill, **Dumfriesshire**
Presented by The Scottish Stone Quarries, Ltd., Closeburn
Weight per cub. ft. 123·5 lbs.

324. GATELAW BRIDGE RED FREESTONE. *TRIAS*
Bright red, with small rusty specks, medium-grained sandstone
Gatelaw Bridge Quarries, nr Thornhill, **Dumfriesshire**
Presented by Marcus Bain, Esq., Mauchline
Chemical Composition:
SiO_2 97·70, Al_2O_3 0·75, Fe_2O_3 0·55, MgO 0·10, CaO 0·10,
= 99·20
Weight per cub. ft. 129·5 lbs. Spec. grav. 2·07. Absorp. 5·81 %
Crushing strain per sqr. ft. 495 tons (Beare)

325. CORSEHILL RED STONE. *TRIAS*
Warm red, fine-grained, slightly micaceous sandstone
Stone Quarries, near Annan, **Dumfriesshire**
Presented by Messrs J. Murray & Sons, Annan
Chemical Composition:
SiO_2 95·24, Al_2O_3 0·50, Fe_2O_3 1·40, $CaCO_3$ 1·40, $MgCO_3$, 1·23
= 99·77
Weight per cub. ft. 141 lbs. Spec. grav. 2·262
Crushing strain per sqr. ft. 635 tons (Kirkaldy)

326. CORSEHILL RED STONE (Top Bed). *TRIAS*
Light terra-cotta, medium-grained sandstone
         Stone Quarries, near Annan, **Dumfriesshire**
Presented by Messrs Pawson Bros., Morley
Weight per cub. ft. 130·4 lbs.   Spec. grav. 2·24.   Absorp. 7·94 %
         Crushing strain per sqr. ft. 579 tons (Beare)

327. ANNANLEA RED STONE. *TRIAS*
Pink, fine-grained, slightly micaceous sandstone
         Kirtle Bridge Quarries, **Dumfriesshire**
Presented by The Symington Quarries, Ltd., Coatbridge
         Weight per cub. ft. 133 lbs.

328. ANNANLEA WHITE STONE. *TRIAS*
Light buff, fine-grained, slightly micaceous sandstone
         Kirtle Bridge Quarries, **Dumfriesshire**
Presented by The Symington Quarries, Ltd., Coatbridge
         Weight per cub. ft. 133 lbs.

329. RED DUMFRIES STONE. *TRIAS*
Light terra-cotta, medium-grained, slightly micaceous sandstone
         Dumfries Quarries, **Dumfriesshire**
Presented by Messrs Pawson Bros., Morley
         Weight per cub. ft. 152 lbs.

330. OLD LOCHARBRIGGS FREESTONE. *TRIAS*
Light red, medium-grained sandstone
         Locharbriggs Quarries, **Dumfriesshire**
Presented by Messrs Baird and Stevenson, Glasgow
Weight per cub. ft. 143 lbs.   Spec. grav. 2·29.   Absorp. 4·06 %
         Crushing strain per sqr. ft. 394·6 tons (Beare)

331. NEW LOCHARBRIGGS FREESTONE. *TRIAS*
Light red, medium-grained sandstone
         New Locharbriggs Quarries, **Dumfriesshire**
Presented by Messrs Wm. Forrest & Co., Edinburgh
         Weight per cub. ft. 143·5 lbs.

332. BALLOCHMYLE RED STONE. *TRIAS*
Bright terra-cotta, with ferruginous specks, fine-grained sandstone
         Ballochmyle Quarries, Mauchline, **Ayrshire**
Presented by Marcus Bain, Esq., Mauchline
                    Chemical Composition :
SiO_2 96·36, Al_2O_3 1·13, Fe_2O_3 0·65, $CaCO_3$ 0·19, $MgCO_3$ 0·42,
= 98·75
         Weight per cub. ft. 129 lbs.   Absorp. 6·77 %
         Crushing strain per sqr ft. 326 tons

333. BARSKIMNIEG FREESTONE. *TRIAS*
Terra-cotta, medium-grained
 Barskimnieg Quarries, Mauchline, **Ayrshire**
Presented by Messrs Baird and Stevenson, Glasgow

334. SCRABO SANDSTONE. *KEUPER*
Light pink, fine-grained
 Scrabo Quarries, **Co. Down**
Presented by Wm. Gill, Esq., Newtownards, Co. Down
 Weight per cub. ft. 138 lbs.

335. SCRABO SANDSTONE. *KEUPER*
Light pink and yellow banded, fine-grained
 Scrabo Quarries, **Co. Down**
Presented by Wm. Gill, Esq., Newtownards, Co. Down
 Weight per cub. ft. 152 lbs.

336. SCRABO SANDSTONE. *KEUPER*
Light yellow banded, fine-grained
 Scrabo Quarries, **Co. Down**
Presented by Wm. Gill, Esq., Newtownards, Co. Down
 Weight per cub. ft. 138 lbs.

JURASSIC.

337. KEINTON STONE. *Lias*
Blue-grey, hard, compact limestone
    Keinton Mandeville Quarries, **Somersetshire**
Presented by The Hard Stone Firms Co., Bristol
    Weight per cub. ft. 160 lbs.

338. KEINTON STONE. *Lias*
Dark blue-grey, fine-grained, crystalline limestone
    Seaton Mandeville Quarries, **Somersetshire**
Presented by Messrs Kerridge and Shaw, Cambridge
    Weight per cub. ft. 157 lbs.

339. BASTARD FREESTONE (Grey Bed). *Upper Lias*
Light cream, medium-grained limestone
    Bath Road Quarries, Shepton Mallet, **Somersetshire**
Presented by Dr Allen, St John's College, Cambridge
    Weight per cub. ft. 133 lbs.

340. BASTARD FREESTONE (White Bed). *Upper Lias*
Light cream, medium-grained, siliceous limestone
    Bath Road Quarries, Shepton Mallet, **Somersetshire**
Presented by Dr Allen, St John's College, Cambridge
    Weight per cub. ft. 133 lbs.

341. BASTARD FREESTONE (Shelly Bed). *Lias*
Light cream, highly cellular, shelly limestone
    Bath Road Quarries, Shepton Mallet, **Somersetshire**
Presented by Dr Allen, St John's College, Cambridge
    Weight per cub. ft. 133 lbs.

342. BASTARD FREESTONE. *LIAS*
Light cream, fine-grained limestone
    Bath Road Quarries, Shepton Mallet, **Somersetshire**
Presented by Dr Allen, St John's College, Cambridge
    Weight per cub. ft. 138 lbs.

343. BLUE LIAS STONE. *LIAS*
Dark blue-black, fine, compact limestone
    Marchalee Elm Quarries, Street, **Somersetshire**
Presented by Messrs F. Seymour & Son, Street
    Weight per cub. ft. 161·5 lbs.

344. HORNTON STONE. *LIAS*
Brown and green, fine-grained limestone
    Hornton Quarries, near Banbury, **Oxfordshire**
Presented by Jos. Stanley, Esq., Hornton, Banbury
    Weight per cub. ft. 112 lbs.

345. HAM HILL STONE (Yellow Bed). *INFERIOR OOLITE*
Light brown, ferruginous, shelly limestone
    Ham Hill Quarries, Norton, near Yeovil,
                             Somersetshire
Presented by The Ham Hill and Doulting Stone Co., Yeovil
    Chemical Composition (D. Forbes, F.R.S.):
SiO_2 4·00, FeO 2·00, H_2O 5·00, $CaCO_3$ 52·00, $MgPO_3$ 37·00,
= 100·00
    Weight per cub. ft. 136 lbs. Spec. grav. 2·18
    Crushing strain per sqr. ft. 207 tons (Kirkaldy)

346. HAM HILL STONE (Grey Bed). *INFERIOR OOLITE*
Brown, ferruginous, hard, coarse-grained limestone
    Ham Hill Quarries, Norton, near Yeovil,
                             Somersetshire
Presented by The Ham Hill and Doulting Stone Co., Yeovil
Weight per cub. ft. 141·6 lbs. (Rivington). Spec. grav. 2·26
    Crushing strain per sqr. ft. 259 tons (Royal Com.)

347. DOULTING STONE (Chelynch Beds). *INFERIOR OOLITE*
Rich cream, coarse-grained, crystalline limestone
    St Andrew's Quarries, Doulting, near Shepton Mallet,
                             Somersetshire
Presented by The Ham Hill and Doulting Stone Co., Yeovil
Weight per cub. ft. 150 lbs. Spec. grav. 2·41. Absorp. 3·36 %
    Crushing strain per sqr. ft. 211·6 tons (Kirkaldy)

348. DOULTING STONE (Fine Beds). *INFERIOR OOLITE*
Rich cream, fine-grained, crystalline limestone
St Andrew's Quarries, Doulting, near Shepton Mallet,
Somersetshire
Presented by The Ham Hill and Doulting Stone Co., Yeovil
Chemical Composition (D. Forbes, F.R.S.):
SiO_2 2·04, Al_2O_3 0·79, FeO 0·85, H_2O 0·32, $CaCO_3$ 95·89,
$MgCO_3$ 0·11, = 100·00
Weight per cub. ft. 125 lbs.   Spec. grav. 2·01.   Absorp. 10·87 %
Crushing strain per sqr. ft. 103·9 tons (Beare)

349. CARY HILL STONE. *INFERIOR OOLITE*
Yellow, medium-grained limestone
Cary Hill Quarries, Cary, **Somersetshire**
Presented by Dr Allen, St John's College, Cambridge
Weight per cub. ft. 138 lbs.

350. PAINSWICK STONE. *INFERIOR OOLITE*
Light cream, fine-grained, granular limestone
Painswick Hill Quarries, **Gloucestershire**
Presented by The Painswick Hill Stone Co., Painswick
Weight per cub. ft. 145 lbs.

351. NAILSWORTH STONE. *INFERIOR OOLITE*
Light cream, fine-grained, granular limestone
Ball's Green Quarries, Nailsworth, **Gloucestershire**
Presented by The United Stone Firms, Ltd., Bristol
Weight per cub. ft. 142 lbs.

352. CHIPPING CAMPDEN STONE. *INFERIOR OOLITE*
Light cream, fine, compact, siliceous limestone
Chipping Campden Quarries, Campden,
Gloucestershire
Presented by W. Plested, Esq., Chipping Campden
Weight per cub. ft. 153·3 lbs. Spec. grav. 2·47

353. GUITING STONE. *INFERIOR OOLITE*
Rich orange colour, medium-grained limestone
Tally Ho Quarries, Guiting, **Gloucestershire**
Presented by The Taynton and Guiting Quarries, Ltd., Cheltenham
Weight per cub. ft. 123·5 lbs.

354. MILL GRIT WEATHER STONE. *INFERIOR OOLITE*
Dark cream, coarse-grained, granular limestone
  The Quarries, Leckhampton, **Gloucestershire**
Presented by The Leckhampton Quarries Co., Leckhampton
  Chemical Composition (A. Greenwell, F.G.S.):
SiO_2 0·91, Al_2O_3 0·54, Fe_2O_3 0·23, CH_2O_3 40·11,   FeO 0·26,
MgO 0·51,   CaO 55·04, K_2O & Na_2O 0·29, H_2O 3·07,   = 100·96
Weight per cub. ft. 133 lbs.   Spec. grav. 2·33.

355. LECKHAMPTON STONE. *INFERIOR OOLITE*
Light cream, medium-grained limestone
  The Quarries, Leckhampton, **Gloucestershire**
Presented by The Leckhampton Quarries Co., Leckhampton
  Weight per cub. ft. 133 lbs.

356. BARNACK STONE. *INFERIOR OOLITE, LINCOLNSHIRE*
  LIMESTONE
Dark cream, medium-grained, shelly limestone
  *Old Barnack Quarries, **Northamptonshire***
Presented by Lord de Ramsey, Ramsey Abbey
  Weight per cub. ft. 133 lbs.

357. BARNACK STONE. *INFERIOR OOLITE, LINCOLNSHIRE*
  LIMESTONE
Dark cream, medium-grained, shelly limestone
  *Old Barnack Quarries, **Northamptonshire***
  Weight per cub. ft. 135 lbs.
  Crushing strain per sqr. ft. 114 tons (Royal Com.)

358. BARNACK STONE. *INFERIOR OOLITE, LINCOLNSHIRE*
  LIMESTONE
Cream, medium-grained, shelly limestone
  *Old Barnack Quarries, **Northamptonshire***
Presented by J. Lewis, Esq., Ely
  Chemical Composition:
Fe_2O_3 1·30, H_2O & loss 1·50, $CaCO_3$ 93·40, $MgCO_3$ 3·80,   = 100·00

359. BARNACK STONE. *INFERIOR OOLITE, LINCOLNSHIRE*
  LIMESTONE
Dark cream, medium-grained, shelly limestone
  *Old Barnack Quarries, **Northamptonshire***

* This specimen was procured from the buildings mentioned in
the descriptive notes. (See p. 170.)

360. BARNACK STONE. *INFERIOR OOLITE, LINCOLNSHIRE*
LIMESTONE
Dark yellow, coarse, shelly limestone
    *Old Barnack Quarries, **Northamptonshire**
Presented by the Duke of Bedford, Thorney Abbey
    Weight per cub. ft. 137 lbs.

361. WELDON STONE. *INFERIOR OOLITE, LINCOLNSHIRE*
LIMESTONE
Pinkish brown, coarse, cellular, shelly limestone
    Weldon Quarries, Kettering, **Northamptonshire**
Presented by John Rook, Esq., Corby
        Chemical Composition (Prof. Attfield):
SiO₂ 0·08,   Al₂O₃ 0·28,   Fe₂O₃ 0·61,   H₂O &c. 1·13,
CaCO₃ 94·35,   MgCO₃ 3·55,   = 100·00
    Weight per cub. ft. 150 lbs.   Spec. grav. 2·42

362. WANSFORD STONE. *INFERIOR OOLITE, LINCOLN-*
SHIRE LIMESTONE
Dark brown, very coarse, siliceous limestone
    Wansford Quarries, **Northamptonshire**
Presented by Hugh Williams, Esq., Thorney
    Weight per cub. ft. 128 lbs.

363. GLENDON STONE. *INFERIOR OOLITE, LINCOLNSHIRE*
LIMESTONE
Dark reddish brown, fine-grained, slightly micaceous limestone
    Glendon Quarries, near Kettering, **Northamptonshire**
Presented by James Pain, Esq., Kettering
    Weight per cub. ft. 133 lbs.

364. KING'S CLIFFE STONE. *INFERIOR OOLITE, LINCOLN-*
SHIRE LIMESTONE
Light yellow, medium-grained, shelly limestone
    *Old King's Cliffe Quarries, **Northamptonshire**
Presented by J. Rook, Esq., Weldon
    Weight per cub. ft. 152 lbs.

365. KETTON STONE. *INFERIOR OOLITE, LINCOLNSHIRE*
LIMESTONE
Light pinkish brown, medium-grained, granular limestone
    *Old Ketton Quarries, Stamford, **Rutland**
Presented by Messrs Kerridge and Shaw, Cambridge
    Weight per cub. ft. 133 lbs.

    * This specimen was procured from the buildings mentioned in
the descriptive notes. (See p. 172.)

366. KETTON STONE. *INFERIOR OOLITE, LINCOLNSHIRE LIMESTONE*

Yellowish brown, medium-grained, granular limestone
Ketton Quarries, Stamford, **Rutland**
Presented by Messrs Kerridge and Shaw, Cambridge
Chemical Composition (Middleton):
Fe_2O_3 0·90, H_2O & loss 2·83, $CaCO_3$ 92·17, $MgCO_3$ 4·10, = 100·00
Weight per cub. ft. 156·7 lbs. Spec. grav. 2·50 (Beare)

367. EDITHWESTON STONE. *INFERIOR OOLITE, LINCOLN-SHIRE LIMESTONE*

Light yellow, fine-grained, granular limestone
Edithweston Quarries, Stamford, **Rutland**
Presented by Messrs Woodford and Wing, Stamford
Weight per cub. ft. 142 lbs.

368. CASTERTON STONE. *INFERIOR OOLITE, LINCOLN-SHIRE LIMESTONE*

Dark cream, coarse-grained limestone
Casterton Quarries, Stamford, **Rutland**
Weight per cub. ft. 129 lbs.

369. CLIPSHAM STONE. *INFERIOR OOLITE, LINCOLNSHIRE LIMESTONE*

Cream, coarse-grained, shelly limestone
Clipsham Old Quarries, near Oakham, **Rutland**
Presented by M. Medwell, Esq., Oakham
Chemical Composition (Blount):
SiO_2 & insoluble residue 0·84, Al_2O_3 & Fe_2O_3 0·82, MgO 0·53, H_2O 0·02, $CaCO_3$ 97·50, = 99·71
Weight per cub. ft. 150 lbs. Spec. grav. 2·427
Crushing strain per sqr. ft. 291·6 tons (Kirkaldy)

370. ANCASTER STONE (Weather Bed). *INFERIOR OOLITE, LINCOLNSHIRE LIMESTONE*

Brown and grey, mottled, crystalline, coarse-grained limestone
Ancaster Quarries, near Grantham, **Lincolnshire**
Presented by Messrs Kerridge and Shaw, Cambridge
Chemical Composition (Prof. Attfield):
SiO_2 0·11, Al_2O_3 0·32, Fe_2O_3 1·48, H_2O 0·20, $CaCO_3$ 95·99, $MgCO_3$ 1·90, = 100·00
Weight per cub. ft. 156 lbs. Spec. grav. 2·50. Absorp. 2·42 %
Crushing strain per sqr. ft. 552·6 tons (Beare)

371. ANCASTER STONE (Free Bed). *INFERIOR OOLITE,*
LINCOLNSHIRE LIMESTONE
Cream, fine-grained limestone
    Ancaster Quarries, near Grantham, **Lincolnshire**
Presented by The Thompson's Ancaster Quarries Co., Grantham
        Chemical Composition (Prof. Attfield):
SiO_2 0·09, Al_2O_3 0·43, Fe_2O_3 0·78, H_2O 0·21, $CaCO_3$ 96·34,
$MgCO_3$ 2·15, = 100·00
Weight per. cub. ft. 135·3 lbs. Spec. grav. 2·18. Absorp. 6·27 %
    Crushing strain per sqr. ft. 218·6 tons (Kirkaldy)

372. ANCASTER FREESTONE (Weather Bed). *INFERIOR*
OOLITE. LINCOLNSHIRE LIMESTONE
Brown and yellow, crystalline, coarse-grained limestone
    Lindley Quarries, Ancaster, **Lincolnshire**
Presented by Messrs Lindley & Son, Ancaster
Weight per cub. ft. 156·3 lbs. Spec. grav. 2·50. Absorp. 2·42 %
    Crushing strain per sqr. ft. 552·6 tons (Beare)

373. ANCASTER FREESTONE (Free Bed). *INFERIOR*
OOLITE, LINCOLNSHIRE LIMESTONE
Cream, fine-grained, granular limestone
    Lindley Quarries, Ancaster, **Lincolnshire**
Presented by Messrs Lindley & Son, Ancaster
      Chemical Composition (Daniels and Wheatstone):
Al_2O_3 & Fe_2O_3 0·80, H_2O & loss 2·71, $CaCO_3$ 93·59,
$MgCO_3$ 2·90, = 100·00
Weight per cub. ft. 140·4 lbs. Spec. grav. 2·25. Absorp. 6·27 %
    Crushing strain per sqr. ft. 184 tons (Beare)

374. LINCOLN FREESTONE. *INFERIOR OOLITE, LINCOLN-*
SHIRE LIMESTONE
Dark cream, fine-grained, slightly shelly limestone
    * Old Quarries, near Lincoln, **Lincolnshire**
Presented by R. A. MacBride, Esq., Lincoln
    Weight per cub. ft. 142 lbs.

375. HAYDOR STONE. *INFERIOR OOLITE, LINCOLNSHIRE*
LIMESTONE
Dark cream, fine-grained, granular limestone
    Haydor Quarries, near Grantham, **Lincolnshire**
Presented by The Thompson's Ancaster Quarries Co., Grantham
    Weight per cub. ft. 133 lbs.

    * This specimen was procured from the buildings mentioned in
the descriptive notes. (See p. 174.)

376. AISLABY FREESTONE. *INFERIOR OOLITE*
Light yellow, medium-grained, calcareous sandstone
        Swerthow Quarries, near Aislaby, **Yorkshire**
Presented by J. Blythman, Esq., Aislaby
        Weight per cub. ft. 127 lbs.

377. ST ALDHELM BOX GROUND STONE. *GREAT*
        OOLITE
Cream, coarse-grained, granular limestone
        Box Ground Quarries, near Bath, **Wiltshire**
Presented by The Bath Stone Firms Co., Bath
        Chemical Composition (Middleton):
Al_2O_3 & Fe_2O_3 1·20, H_2O & loss 1·78, $CaCO_3$ 94·52,
$MgCO_3$ 2·50, =100·00
Weight per cub. ft. 129 lbs.   Spec. grav. 2·07.   Absorp. 7·75 %
        Crushing strain per sqr. ft. 107 tons (Rivington)

378. CORSHAM DOWN STONE. *GREAT OOLITE*
Light cream, fine-grained, granular limestone
        Corsham Down Quarries, near Bath, **Wiltshire**
Presented by The Bath Stone Firms Co., Bath
        Chemical Composition:
Al_3O_3 & Fe_2O_3 1·20, H_2O & loss 1·78, $CaCO_3$ 94·52,
$MgCO_3$ 2·50, =100·00
        Weight per cub. ft. 129 lbs.   Absorp. 11·18 %
        Crushing strain per sqr. ft. 128 tons (Beare)

379. HARTHAM PARK STONE. *GREAT OOLITE*
Dark cream, fine-grained, granular limestone
        Hartham Park Quarries, near Bath, **Wiltshire**
Presented by The Bath Stone Firms Co., Bath
        Weight per cub. ft. 123½ lbs.

380. FARLEIGH DOWN STONE. *GREAT OOLITE*
Rich cream, fine-grained, granular limestone
        Farleigh Down Quarries, near Bath, **Wiltshire**
Presented by The Bath Stone Firms Co., Bath
Weight per cub. ft. 120 lbs.   Spec. grav. 1·93.   Absorp. 13·13 %
        Crushing strain per sqr. ft. 62 tons (Beare)

381. CORSHAM STONE (Blue Beds). *GREAT OOLITE*
Light grey, fine-grained, granular limestone
        Corsham Quarries, near Bath, **Wiltshire**
Presented by Messrs Yockney & Co., Box Tunnel, Bath
        Weight per cub. ft. 128 lbs.

382. STOKE GROUND STONE. *Great Oolite*
Light brown, medium-grained, granular limestone
    Stoke Ground Quarries, near Bath, **Somersetshire**
Presented by The Bath Stone Firms Co., Bath
Weight per cub. ft. 126 lbs.   Spec. grav. 2·02.   Absorp. 10·52 %
    Crushing strain per sqr. ft. 107 tons (Beare)

383. COMBE DOWN STONE. *Great Oolite*
Cream, medium-grained, granular limestone
    Combe Down Quarries, near Bath, **Somersetshire**
Presented by The Bath Stone Firms Co., Bath
        Chemical Composition (Middleton):
Al_2O_3 & Fe_2O_3 1·20, H_2O &c. 1·78, $CaCO_3$ 94·52,
$MgCO_3$ 2·50,   = 100·00
Weight per cub. ft. 128 lbs.   Spec. grav. 2·06.   Absorp. 6·00 %
    Crushing strain per sqr. ft. 151 tons (Rivington)

384. MONKS PARK STONE. *Great Oolite*
Cream, fine-grained, granular limestone
    Monks Park Quarries, near Bath, **Somersetshire**
Presented by The Bath Stone Firms Co., Bath
        Chemical Composition (Prof. J. Attfield):
SiO_2 1·00, Al_2O_3 & Fe_2O_3 1·60, $CaCO_3$ 97·20, $MgCO_3$ 0·20,
= 100·00
Weight per cub. ft. 137 lbs.   Spec. grav. 2·19.   Absorp. 7·51 %
    Crushing strain per sqr. ft. 223·5 tons (Kirkaldy)

385. TAYNTON STONE. *Great Oolite*
Brown, with light streaks, coarse, shelly limestone
    Taynton Quarries, Wychwood, **Oxfordshire**
Presented by The Taynton and Guiting Quarries, Ltd., Cheltenham
        Chemical Composition:
SiO_2 1·36, Al_2O_3 1·05, Fe_2O_3 0·69, H_2O 0·95,
$CaCO_3$ 94·20, $MgCO_3$ 1·20, $CaPO_3$ 0·26,   = 99·71
Weight per cub. ft. 136 lbs.   Spec. grav. 2·091

386. HEADINGTON STONE (Hard).   *Middle Oolite*
Dark cream, slightly fossiliferous
    Headington Quarries, **Oxfordshire**
Presented by Robt. Edwards, Esq., Oxford
    Weight per cub. ft. 160 lbs.

387. HEADINGTON STONE (Soft).   *Middle Oolite*
Light cream, cellular, medium-grained
    Headington Quarries, **Oxfordshire**
Presented by Robt. Edwards, Esq., Oxford
    Weight per cub. ft. 120 lbs.

388. PORTLAND STONE (Whit Bed). *Upper Oolite*
Cream, fine-grained, granular limestone
    Weston Quarries, Portland, **Dorsetshire**
Presented by The Bath Stone Firms Co., Bath
    Weight per cub. ft. 132 lbs. Absorp. 7·51 %
    Crushing strain per sqr. ft. 205 tons (Rivington)

389. PORTLAND STONE (Base Bed). *Upper Oolite*
Cream, medium-grained, shelly limestone
    Weston Quarries, Portland, **Dorsetshire**
Presented by The Bath Stone Firms Co., Bath
    Chemical Composition (Daniel and Wheatstone):
SiO_2 1·20, Al_2O_3 & Fe_2O_3 0·50, H_2O & loss 1·94,
$CaCO_3$ 95·16, $MgCO_3$ 1·20, = 100·00
    Weight per cub. ft. 137 lbs.
    Crushing strain per sqr. ft. 287 tons (Rivington)

390. PORTLAND STONE (Whit Bed). *Upper Oolite*
Light dove-coloured, fine-grained, granular limestone
    Inmosthay Quarries, Portland, **Dorsetshire**
Presented by F. J. Barnes, Esq., Isle of Portland
    Chemical Composition :
SiO_2 3·19, Al_2O_3 & Fe_2O_3 0·27, CH_2O_3 42·00, MgO 0·65,
CaO 53·69, H_2O 0·10, H_2SO_4 0·10, = 100·00
Weight per cub. ft. 132·3 lbs. Spec. grav. 2·12. Absorp. 7·51 %
    Crushing strain per sqr. ft. 204·7 tons (Beare)

391. PORTLAND STONE (Base Bed). *Upper Oolite*
Very light dove-coloured, fine-grained, granular limestone
    Inmosthay Quarries, Portland, **Dorsetshire**
Presented by F. J. Barnes, Esq., Isle of Portland
    Chemical Composition :
SiO_2 3·19, Al_2O_3 & Fe_2O_3 0·27, CH_2O_3 42·00, MgO 0·65,
CaO 53·69, H_2O 0·10, H_2SO_4 0·10, = 100·00
Weight per cub. ft. 137·6 lbs. Spec. grav. 2·20. Absorp. 6·84 %
    Crushing strain per sqr. ft. 287·0 tons (Beare)

392. PORTLAND STONE (Roche Bed). *Upper Oolite*
Light dove-coloured, highly cellular, limestone
    Inmosthay Quarries, Portland, **Dorsetshire**
Presented by F. J. Barnes, Esq., Isle of Portland
    Chemical Composition :
SiO_2 3·19, Al_2O_3 & Fe_2O_3 0·27, CH_2O_3 42·00, MgO 0·65,
CaO 53·69, H_2O 0·10, H_2SO_4 0·10, = 100·00
    Weight per cub. ft. 142·5 lbs.

393. CHILMARK STONE (Pinney Bed). *UPPER OOLITE*
Yellowish brown, siliciferous, limestone
       Chilmark Quarries, Tisbury, **Wiltshire**
Presented by T. P. Lilly, Esq., Gillingham
                 Chemical Composition :
SiO_2 10·4, Al_2O_3 & Fe_2O_3 2·0, H_2O &c. 4·2, $CaCO_3$ 79·0,
$MgCO_3$ 3·7,   =99·30
       Weight per cub. ft. 135 lbs.   Spec. grav. 2·481
          Crushing strain per sqr. ft. 136·6 tons (Kirkaldy)

394. PURBECK CLIFF STONE. *UPPER OOLITE*
Cream, fine-grained limestone
          Seacombe Quarries, near Swanage, **Dorsetshire**
Presented by Messrs Burt and Burt, Swanage
       Weight per cub. ft. 169·2 lbs. (Rivington)

395. PURBECK FREESTONE. *UPPER OOLITE*
Light cream, fine-grained sandstone
          Langton Quarries, near Swanage, **Dorsetshire**
Presented by Messrs Burt and Burt, Swanage
       Weight per cub. ft. 150 lbs. (Rivington)

CRETACEOUS.

396. WEALDEN SANDSTONE. *Lower Cretaceous*
Light yellow, with brown streaks, fine-grained
    Selsfield Quarries, East Grinstead, **Sussex**
Presented by R. Gunter, Esq., East Grinstead, Sussex
    Weight per cub. ft. 150 lbs. Absorp. 10·34 %
      Crushing strain per sqr. ft. 279·2 tons (Kirkaldy)

397. HENLEY HILL STONE. *Lower Cretaceous*
Light yellow, medium-grained, shelly, arenaceous
    North Heath Quarries, Midhurst, **Sussex**
Presented by the Earl of Egmont, Cowley, Midhurst
    Weight per cub. ft. 133 lbs.

398. HARROW STONE. *Lower Cretaceous*
Yellow and brown, mottled, fine-grained, arenaceous
    Harrow Quarries, Hastings, **Sussex**
Presented by Charles Hughes, Esq., St Leonards
    Weight per cub. ft. 123·5 lbs.

399. ARDINGLY SANDSTONE. *Lower Cretaceous*
Yellow, fine-grained
    Firewood Quarries, near Ardingly, **Sussex**
Presented by Lord Cowdray, London
    Weight per cub. ft. 114 lbs.

400. SUSSEX MARBLE. *Lower Cretaceous*
Greenish grey, highly fossiliferous, compact, calcareous, admits
    of a polish
    Goringler Quarries, near Grinstead, **Sussex**
Presented by Messrs Norman and Burt, Burgess Hill
    Weight per cub. ft. 171 lbs.

401. KENTISH RAG. *Lower Greensand, Cretaceous*
Blue-grey, compact, crystalline, calcareous
            Borough Green Quarries, Seven Oaks, **Kent**
Presented by Messrs Chittenden and Simmons, West Malling
                    Chemical Composition (Woodward):
Al$_2$O$_3$ 6·50,   Fe$_2$O$_3$ 0·50,   H$_2$O 0·40,   CaCO$_3$ 92·60,   = 100·00
      Weight per cub. ft. 166·6 lbs. (Rivington)

402. CARSTONE. *Lower Greensand, Cretaceous*
Deep orange brown, medium-grained, ferruginous
            Snettisham Quarries, near Hunstanton, **Norfolk**
      Weight per cub. ft. 152 lbs.

403. CARSTONE. *Lower Greensand, Cretaceous*
Deep orange brown, fine-grained, ferruginous grit
            Snettisham Quarries, near Hunstanton, **Norfolk**
      Weight per cub. ft. 161·5 lbs.

404. ST BONIFACE STONE. *Upper Greensand,*
            Cretaceous
Light grey, very fine-grained, calcareous
            St Boniface Quarries, Ventnor, **Isle of Wight**
Presented by The St Boniface Stone Quarries Co., Ventnor
      Weight per cub. ft. 114 lbs.

405. DENNER HILL STONE. *Upper Greensand,*
            Cretaceous
Light grey, medium-grained, arenaceous
            Denner Hill Quarries, Walters Ash, **Buckinghamshire**
Presented by Messrs T. Bristow Bros., Speen, Princes Risborough
      Weight per cub. ft. 142 lbs.

406. HEATH STONE. *Upper Greensand, Cretaceous*
Light yellowish grey, medium-grained, arenaceous
            Wycombe Quarries, Walters Ash, **Buckinghamshire**
Presented by Messrs T. Bristow Bros., Speen, Princes Risborough
      Weight per cub. ft. 142 lbs.

407. MERSTHAM FIRE STONE. *Upper Greensand,*
            Cretaceous
Light whitish grey, fine-grained, calcareous
            Merstham Quarries, near Red Hill, **Surrey**
Presented by J. S. Peters, Esq., Merstham
      Weight per cub. ft. 114 lbs.

408. REIGATE STONE. *UPPER GREENSAND, CRETACEOUS*
Greenish grey, fine-grained, calcareous
Reigate Quarries, **Surrey**
Presented by T. Carruthers, Esq., Reigate
Weight per cub. ft. 103 lbs.

409. BARGATE STONE. *UPPER GREENSAND, CRETACEOUS*
Dark brownish grey, hard crystalline, arenaceous
Milford Quarries, Witley, **Surrey**
Presented by Messrs F. Milton & Sons, Witley
Weight per cub. ft. 171 lbs.

410. WINDRUSH STONE. *UPPER GREENSAND, CRETACEOUS*
Light cream, medium-grained, calcareous
Windrush Quarries, near Farringdon, **Hampshire**
Presented by J. Day, Esq., Cambridge
Weight per cub. ft. 135·9 lbs.

411. RED CHALK. *UPPER CRETACEOUS*
Red, nodular, ferruginous limestone
Sea Cliffs at Old Hunstanton, **Norfolk**
Chemical Composition (R. A. Berry, F.I.C.)
SiO_2 9·18, Al_2O_3 1·94, Fe_2O_3 4·62, MnO 1·30, CaO 2·14,
$CaCO_3$ 74·0, $MgCO_3$ 1·93, Fe_2PO_4 2·44, Ca_3PO_4 1·65, $= 99·20$
Weight per cub. ft. 151 lbs.

412. CLUNCH. *LOWER CHALK, CRETACEOUS*
White, compact, fine-grained, chalky
Quarries, Cherryhinton, **Cambridgeshire**
Presented by A. B. Day, Esq., Cambridge
Weight per cub. ft. 133 lbs.
Crushing strain per sqr. ft. 72 tons.

413. CLUNCH. *LOWER CHALK, CRETACEOUS*
White, compact, fine-grained, chalky
From Old Wall of First Court, Christ's College, **Cambridge**
Presented by Messrs Prime & Co., Cambridge
Weight per cub. ft. 135 lbs.

414. TOTTERNHOE STONE. *LOWER CHALK, CRETACEOUS*
Greenish white, calcareous, fine-grained
Totternhoe Quarries, **Bedfordshire**
Presented by The Totternhoe Lime Co., Totternhoe
Weight per cub. ft. 116 lbs.
Crushing strain per sqr. ft. 124 tons (Rivington)

415. BEER STONE. *MIDDLE CHALK, CRETACEOUS*
White, fine-grained, compact, calcareous
        Beer Quarries, Seaton, **Devonshire**
Presented by The Beer Stone Co., Seaton
        Weight per cub. ft. 131·7 lbs. (Rivington)
        Crushing strain per sqr. ft. 151·2 tons (Kirkaldy)

416. CHALK STONE. *LOWER CHALK, CRETACEOUS*
White, soft, chalky, fine-grained
        Chalk Quarries, near Devizes, **Wiltshire**
Presented by Henry Ash, Esq., Devizes
        Weight per cub. ft. 123·5 lbs.

417. KNAPPED FLINT. *UPPER CHALK, CRETACEOUS*
Blue-black, hard, very compact, conchoidal fracture
        Chalk Quarries, Greenhithe, **Kent**
Presented by C. H. Watson, Esq., Stone Castle, Greenhithe

418. KNAPPED AND DRESSED FLINT. *UPPER
        CHALK, CRETACEOUS*
Fawn-coloured, hard, compact
        From buttress of Covehithe Church, **Suffolk**
Presented by Professor T. McKenny Hughes, Cambridge

418A. KNAPPED AND DRESSED FLINT. *UPPER
        CHALK, CRETACEOUS*
Dark blue-black, hard, compact
        From ruins of Covehithe Church, **Suffolk**
Presented by Professor T. McKenny Hughes, Cambridge

PLEISTOCENE AND RECENT.

419. SAUNTON SANDSTONE. *Post Tertiary*
Light brown, ferruginous, coarse-grained
Saunton Sands Caves, **North Devonshire**
Weight per cub. ft. 147 lbs.

COLONIAL AND FOREIGN
BUILDING STONES

IGNEOUS ROCKS (PLUTONIC).

425. AUSTRIAN GRANITE. *MUSCOVITE-BIOTITE-GRANITE*
Grey, coarse, porphyritic, large white felspar crystals
   Gmünd Quarries, **Lower Austria**
Weight per cub. ft. 166 lbs. Spec. grav. 2·67. Absorp. 0·94 %
   Crushing strain per sqr. ft. 946·5 tons

426. AUSTRIAN GRANITE. *MUSCOVITE-BIOTITE-GRANITE*
Grey, fine-grained
   St Oswald Quarries, near Freistadt, **Upper Austria**
Weight per cub. ft. 166 lbs. Spec. grav. 2·67. Absorp. 0·26 %
   Crushing strain per sqr. ft. 1512·5 tons

427. AUSTRIAN GRANITE. *HORNBLENDE-BIOTITE-*
       GRANITE
Light grey, fine-grained
   Mauthausen Quarries, **Upper Austria**
Weight per cub. ft. 165 lbs. Spec. grav. 2·65. Absorp. 0·62 %
   Crushing strain per sqr. ft. 1418·5 tons

428. AUSTRIAN GRANITE. *MUSCOVITE-BIOTITE-GRANITE*
Grey, medium-grained
   Mrakotin Quarries, **Moravia**
  Weight per cub. ft. 160 lbs. Spec. grav. 2·57
   Crushing strain per sqr. ft. 1415 tons

429. AUSTRIAN GRANITE. *QUARTZ-DIORITE*
Dark grey, medium-grained
   Granite Quarries, near Pilsen, **Hungary**
Presented by Messrs Kirkpatrick Bros., Manchester
   Weight per cub. ft. 170 lbs.

430. DANISH GRANITE. *Hornblende-Granite*
Pinkish grey, coarse-grained
    Mosselökke Quarries, near Allinge, **Island of Bornholm**
Presented by Messrs H. and J. Larsen, Copenhagen
    Weight per cub. ft. 165 lbs. Spec. grav. 2·65
    Crushing strain per sqr. ft. 2052 tons

431. DANISH GRANITE. *Hornblende-Granite*
Dark grey with pink tinge, coarse-grained
    Klintelökke Quarries, near Allinge, **Island of Bornholm**
Presented by Messrs H. and J. Larsen, Copenhagen
    Weight per cub. ft. 165 lbs.

432. DANISH GRANITE. *Diorite*
Dark grey, nearly black, medium-grained
    Klippegaard Quarries, near Rönne, **Island of Bornholm**
Presented by Messrs H. and J. Larsen, Copenhagen
    Weight per cub. ft. 176 lbs.

433. DANISH GRANITE. *Hornblende-Granite (Gneissic)*
Pink, medium-grained
    Bjergbakken Quarries, near Rönne, **Island of Bornholm**
Presented by Messrs H. and J. Larsen, Copenhagen
    Weight per cub. ft. 176 lbs.

434. DANISH GRANITE. *Hornblende-Granite*
Pink, coarse-grained
    Helvedsbakkerne Quarries, near Nexö,
                          Island of Bornholm
Presented by Messrs H. and J. Larsen, Copenhagen
    Weight per cub. ft. 170 lbs.

435. DANISH GRANITE. *Granite-Porphyry (with*
                  Hornblende and Biotite).
Dark grey with blue veins
    Paradisbakkerne Quarries, near Svanike,
                          Island of Bornholm
Presented by Messrs H. and J. Larsen, Copenhagen
    Weight per cub. ft. 171 lbs.

436. FRENCH GRANITE. *Muscovite-Biotite-Granite*
Light grey, medium-grained
    Chatelard Quarries, **Haute Savoie**
    Weight per cub. ft. 166 lbs.

437. FRENCH GRANITE. *Biotite-Hornblende-Granite*
Pinkish grey mottled, medium-grained
　　　Raon l'Étape Quarries, **Vosges**
Presented by H. Vienne, Esq., Consolere
　　Weight per cub. ft. 168 lbs.
　　Crushing strain per sqr. ft. 1735 tons.

438. FRENCH GRANITE. *Muscovite-Biotite-Granite*
Light grey, coarse-grained, porphyritic
　　　Vagney Quarries, **Vosges**
Presented by H. Vienne, Esq., Consolere
　　Weight per cub. ft. 177 lbs.

439. FRENCH GRANITE. *Biotite-Granite (some Muscovite)*
Grey, medium-grained
　　　Bécon Quarries, **Maine et Loire**
Presented by Messrs G. Larivière & Co., Angers
　　Weight per cub. ft. 176 lbs.

440. FRENCH GRANITE. *Biotite-Granite*
Brownish grey, medium-grained
　　　Bécon Quarries, **Maine et Loire**
Presented by Messrs G. Larivière & Co., Angers
　　Weight per cub. ft. 166 lbs.

441. FRENCH GRANITE. *Biotite-Granite*
Bright red, medium-grained
　　　Vezins Quarries, **Maine et Loire**
Presented by Messrs G. Larivière & Co., Angers
　　Weight per cub. ft. 152 lbs.

442. FRENCH GRANITE. *Hornblende-Biotite-Granite*
Dark pinkish grey, medium-grained
　　　Vezins Quarries, **Maine et Loire**
Presented by Messrs G. Larivière & Co., Angers
　　Weight per cub. ft. 169 lbs.

443. FRENCH GRANITE. *Muscovite-Biotite-Granite*
Light purplish grey, medium-grained
　　　Torfou Quarries, **Maine et Loire**
Presented by Messrs G. Larivière & Co., Angers
　　Weight per cub. ft. 157 lbs.

444. FRENCH GRANITE. *BIOTITE-GRANITE*
Brown, medium-grained
  Fougères Quarries, **Ille et Vilaine**
Presented by Messrs G. Larivière & Co., Angers
  Weight per cub. ft. 166 lbs.

445. FRENCH GRANITE. *BIOTITE-GRANITE*
Dark grey, medium-grained
  Fougères Quarries, **Ille et Vilaine**
Presented by Messrs G. Larivière & Co., Angers
  Weight per cub. ft. 176 lbs.

446. FRENCH GRANITE. *HORNBLENDE-GRANITE*
Light pink, coarse-grained, with large felspathic crystals
  Laber Quarries, near Brest, **Finistère**
Presented by L. Omnes, Esq., Brest
  Weight per cub. ft. about 160 lbs.

447. FRENCH GRANITE. *MUSCOVITE-BIOTITE-GRANITE*
Blue-grey, medium-grained
  Montjoie Quarries, near Vire, **Manche**
Presented by H. Vienne, Esq., Consolere
  Weight per cub. ft. 171 lbs.
  Crushing strain per sqr. ft. 847·8 tons

448. CORSICAN GRANITE. *PORPHYRY, EPIDOTIC?*
Green and red mottled, coarse-grained
  Granite Quarries, near **Ajaccio**
  Weight per cub. ft. 175 lbs.

449. GERMAN GRANITE. *HORNBLENDE-BIOTITE-GRANITE*
Light grey, medium-grained
  Ratschken Quarries, near Schmöllen, **Saxony**
Presented by Messrs C. Sparmann & Co., Dresden
  Weight per cub. ft. 166 lbs.

450. GERMAN GRANITE. *HORNBLENDE-BIOTITE-GRANITE*
Dark grey, medium-grained
  Grund Quarries, near Schmöllen, **Saxony**
Presented by Messrs C. Sparmann & Co., Dresden
  Weight per cub. ft. 170·5 lbs.

451. GERMAN GRANITE. *Hornblende-Biotite-Granite*
Light grey, coarse-grained
Sandberg Quarries, near Kamenz, **Saxony**
Presented by Messrs C. Sparmann & Co., Dresden
Weight per cub. ft. 171 lbs.

452. GERMAN GRANITE. *Hornblende-Biotite-Granite*
Light grey, fine-grained
Horka Quarries, near Neschwitz, **Saxony**
Presented by Messrs C. Sparmann & Co., Dresden
Weight per cub. ft. 171 lbs.

453. GERMAN GRANITE. *Muscovite-Biotite-Granite*
Light grey, medium-grained
Alpirsbach Quarries, **Würtemberg**
Weight per cub. ft. 170 lbs.

454. ITALIAN GRANITE. *Hornblende-Biotite-Granite*
Light grey and pink, medium-grained
Baveno Quarries, **Lake Maggiore**
Presented by Messrs A. Guttridge & Co., London
Chemical Composition (Bunsen) :
SiO_2 74·82, Al_2O_3 16·14, Fe_2O_3 1·52,   MgO 0·47,   CaO 1·68,
K_2O 3·55, Na_2O 6·12,   =104·30
Weight per cub. ft. 170 lbs.

455. ITALIAN GRANITE. *Muscovite-Biotite-Granite*
Very light grey, nearly white, medium-grained
Alzo Quarries, **Piedmont**
Presented by Messrs C. and G. Fratelli Peverilli, Turin
Weight per cub. ft. 170 lbs.

456. ITALIAN GRANITE. *Hornblende-Biotite-Granite*
Light grey with pink tinge, medium-grained
Montorfano Quarries, near Gravellona, **Piedmont**
Presented by Messrs A. Guttridge & Co., London
Weight per cub. ft. 165 lbs.

457. ITALIAN GRANITE. *Hornblende-Biotite-Granite*
Dark brown with white specks, medium-grained
Black Granite Quarries, **Piedmont**
Weight per cub. ft. 166 lbs.

458. NORWEGIAN GRANITE. *LARVIKITE (AUGITE-SYENITE)*
Blue-grey, coarse-grained, iridescent
    Granite Quarries, near **Larvik**
Presented by Messrs Kirkpatrick Bros., Manchester
        Chemical Composition (Meriau):
SiO_2 58·88, Al_2O_3 20·30, Fe_2O_3 3·63,   FeO 2·58,   MgO 0·79,
CaO 3·03, K_2O 4·50, Na_2O 5·73, H_2O 1·01, P_2O_5 0·54,
= 100·99
    Weight per cub. ft. 166 lbs.

459. NORWEGIAN GRANITE. *LARVIKITE (AUGITE-SYENITE)*
Dark grey with green tinge, coarse-grained, iridescent
    Granite Quarries, near **Larvik**
Presented by Messrs Kirkpatrick Bros., Manchester
    Weight per cub. ft. 163 lbs.

460. NORWEGIAN GRANITE. *LARVIKITE (AUGITE-SYENITE)*
Brown with blue tinge, coarse-grained, iridescent
    Granite Quarries, near **Larvik**
Presented by Messrs Kirkpatrick Bros., Manchester
    Weight per cub. ft. 160 lbs.

461. NORWEGIAN GRANITE. *LARVIKITE (AUGITE-SYENITE)*
Dark grey, coarse-grained, iridescent
    Granite Quarries, **Larvik**
Presented by De Forende Stenhuggerin, Christiania
    Weight per cub. ft. 161 lbs.

462. NORWEGIAN GRANITE. *LARVIKITE (AUGITE-SYENITE)*
Dark grey, coarse-grained, iridescent
    Tjylling Quarries, near **Larvik**
Presented by De Forende Stenhuggerin, Christiania
    Weight per cub. ft. 157 lbs.

463. NORWEGIAN GRANITE. *BIOTITE-GRANITE*
Light grey, fine-grained
    Bakka Quarries, **Frederikshald**
Presented by De Forende Stenhuggerin, Christiania
    Weight per cub. ft. 163 lbs.

464. NORWEGIAN GRANITE. *Nepheline-Syenite*
Pink, medium-grained
Kgogoln Quarries, near **Frederiksstadt**
Presented by De Forende Stenhuggerin, Christiania
Weight per cub. ft. 163 lbs.

465. NORWEGIAN GRANITE. *Tönsbergite (altered*
Larvikite)
Dark brown, medium-grained
Granite Quarries, **Nölbrod**
Presented by De Forende Stenhuggerin, Christiania
Weight per cub. ft. 166 lbs.

466. NORWEGIAN GRANITE. *Muscovite-Biotite-*
Granite
Light grey, medium-grained
Hov Quarries, near **Christiania**
Presented by Sir John Jackson, LL.D., London
Weight per cub. ft. 171 lbs.

467. NORWEGIAN GRANITE. *Biotite-Granite*
Pinkish grey, medium-grained
Ulleberg Quarries, near **Christiania**
Presented by Sir John Jackson, LL.D., London
Weight per cub. ft. 166 lbs.

468. NORWEGIAN GRANITE. *Nordmarkite (Quartz-*
Syenite)
Light brownish red, medium-grained
Granite Quarries, near **Christiania**
Presented by H. A. Paulsen, Esq., Christiania
Weight per cub. ft. 176 lbs.

469. NORWEGIAN GRANITE. *Biotite-Granite*
Grey with buff tinge, fine-grained
Lilholt Quarries, near **Frederikshald**
Presented by Messrs Blichfeldt & Co., London
Chemical Composition :
SiO_2 70·78, TiO_2 0·11, Al_2O_3 15·27, Fe_2O_3 2·01, FeO 1·00,
MgO 0·97, CaO 1·82, K_2O 3·59, Na_2O 3·57, H_2O & loss 0·88,
=100·00
Weight per cub. ft. 166·5 lbs. Spec. grav. 2·67. Absorp. 0·34 %
Crushing strain per sqr. ft. 1870 tons.

470. NORWEGIAN GRANITE. *Biotite-Granite*
Pinkish grey, very fine-grained
Ystehede Quarries, near **Frederikshald**
Presented by Messrs Brooks, Ltd., London
Weight per cub. ft. 171 lbs.
Crushing strain per sqr. ft. 2144 tons (Grey)

471. PORTUGUESE GRANITE. *MUSCOVITE-BIOTITE-GRANITE*
Very light pinkish brown and white, medium-grained
Curraes Quarries, near **Oporto**
Presented by The Booth Steamship Co., Ltd., Liverpool
Weight per cub. ft. 162 lbs.

472. PORTUGUESE GRANITE. *MUSCOVITE-BIOTITE-GRANITE*
Light brown and white, coarse-grained, porphyritic
Monte Pedral Quarries, near **Oporto**
Presented by The Booth Steamship Co., Ltd., Liverpool
Weight per cub. ft. 161 lbs.

473. PORTUGUESE GRANITE. *MUSCOVITE-BIOTITE-GRANITE*
Light brown and white, medium-grained
Milheiroz Quarries, near **Oporto**
Presented by The Booth Steamship Co., Ltd., Liverpool
Weight per cub. ft. 162 lbs.

474. PORTUGUESE GRANITE. *MUSCOVITE-BIOTITE-GRANITE*
Light brownish grey, medium-grained
St Gens Quarries, near **Oporto**
Presented by The Booth Steamship Co., Ltd., Liverpool
Weight per cub. ft. 160 lbs.

475. PORTUGUESE GRANITE. *MUSCOVITE-BIOTITE-GRANITE*
Light brownish grey, medium-grained
St Gens Quarries, near **Oporto**
Presented by The Booth Steamship Co., Ltd., Liverpool
Weight per cub. ft. 162 lbs.

476. RUSSIAN GRANITE. *BIOTITE-GRANITE*
Dark grey, fine-grained
Gamla Carlby Quarries, near **Kalajoki**
Weight per cub. ft. 170 lbs.

477. RUSSIAN GRANITE. *DIORITE*
Dark blue-grey, fine-grained
Ruskeala Quarries, Sordowala, **Finland**
Presented by Aktiebolaget Granit, Helsingfors
Weight per cub. ft. 175 lbs.

478. RUSSIAN GRANITE. *AMPHIBOLITE*
Black, very fine-grained
    Svartä Quarries, Ekanäs, **Finland**
Presented by Aktiebolaget Granit, Helsingfors
Weight per cub. ft. 175 lbs.

479. RUSSIAN GRANITE. *MUSCOVITE-BIOTITE-GRANITE*
    (pale variety)
Light blue-grey, fine-grained
    Hiitis Quarries, near Hangö, **Finland**
Presented by Aktiebolaget Granit, Helsingfors
Weight per cub. ft. 171 lbs.

480. RUSSIAN GRANITE. *BIOTITE-GRANITE*
Dark grey, medium-grained
    Hiitis Quarries, near Hangö, **Finland**
Presented by Aktiebolaget Granit, Helsingfors
Weight per cub. ft. 170 lbs.

481. RUSSIAN GRANITE. *DIORITE*
Dark grey, nearly black, medium-grained
    Simola Quarries, Willmanstvand, **Finland**
Presented by Aktiebolaget Granit, Helsingfors
Weight per cub. ft. 171 lbs.

482. RUSSIAN GRANITE. *BIOTITE-GRANITE*
Warm red, fine-grained
    Hangö Quarries, Hangö, **Finland**
Presented by Aktiebolaget Granit, Helsingfors
Weight per cub. ft. 176 lbs.

483. RUSSIAN GRANITE. *HORNBLENDE-GNEISS*
Brown and red mottled, coarse-grained
    Kokar Quarries, Mariehamn, **Finland**
Presented by Aktiebolaget Granit, Helsingfors
Weight per cub. ft. 171 lbs.

484. RUSSIAN GRANITE. *(BIOTITE)-HORNBLENDE-*
    GRANITE
Warm red, coarse-grained
    Bjärnå Quarries, near Ekanäs, **Finland**
Presented by Aktiebolaget Granit, Helsingfors
Weight per cub. ft. 171 lbs.

332   IGNEOUS ROCKS (PLUTONIC)

485. RUSSIAN GRANITE. *AMPHIBOLITE*
Very dark grey, nearly black, fine-grained
    Nurmijarvi Quarries, near Helsingfors, **Finland**
Presented by Aktiebolaget Granit, Helsingfors
    Weight per cub. ft. 176 lbs.

486. RUSSIAN GRANITE. *OLIVINE-GABBRO*
Dark grey, medium-grained
    Mariehamn Quarries, Mariehamn, **Finland**
Presented by Aktiebolaget Granit, Helsingfors
    Weight per cub. ft. 170 lbs.

487. RUSSIAN GRANITE. *BIOTITE-HORNBLENDE-GRANITE*
Dark red, medium-grained
    Vehmo Quarries, near Nystad, **Finland**
Presented by Aktiebolaget Granit, Helsingfors
    Weight per cub. ft. 176 lbs.

488. RUSSIAN GRANITE. *BIOTITE-GRANITE*
Warm red, coarse-grained
    Tofsalo Quarries, near Nystad, **Finland**
Presented by Aktiebolaget Granit, Helsingfors
    Weight per cub. ft. 176 lbs.

489. RUSSIAN GRANITE. *BIOTITE-GRANITE*
Dark red, medium-grained
    Quarries on Island near Helsingfors, **Finland**
Presented by G. A. List, Esq., Moscow
    Weight per cub. ft. 176 lbs.

490. RUSSIAN GRANITE. *HORNBLENDE-GRANITE*
Dark red and black, medium-grained
    Quarries near Antrea, **Finland**
Presented by G. A. List, Esq., Moscow
    Weight per cub. ft. 176 lbs.

491. RUSSIAN GRANITE. *BIOTITE-GRANITE (Rapakiwi Granite)*
Dark red, coarse-grained
    Granite Quarries, near Wyborg, **Finland**
Presented by G. A. List, Esq., Moscow
    Weight per cub. ft. 171 lbs.

492. RUSSIAN GRANITE. *BIOTITE-GRANITE*
Red, fine-grained
  Quarries on Island in **Lake Ladoga**
Presented by G. A. List, Esq., Moscow
  Weight per cub. ft. 171 lbs.

493. RUSSIAN GRANITE.   *QUARTZ-MICA-DIORITE*
  (*TONALITE*)
Dark grey mottled, medium-grained
  Quarries near **Lake Ladoga**
Presented by G. A. List, Esq., Moscow
  Weight per cub. ft. 180 lbs.

494. RUSSIAN GRANITE. *BIOTITE-GRANITE*
Light brownish grey, medium-grained
  Granite Quarries, Govt. **Ekaterinoslav**
Presented by G. A. List, Esq., Moscow
  Weight per cub. ft. 171 lbs.

495. RUSSIAN GRANITE. *ANALCIME-DOLERITE*
Grey, medium-grained
  Granite Quarries, Kursebi, Kutais, **Caucasus**
  Weight per cub. ft. 161·5 lbs.

496. SWEDISH GRANITE. *BIOTITE-GRANITE*
Dark red, coarse-grained
  Granite Quarries, **Grafversfors**
  Weight per cub. ft. 176 lbs.

497. SWEDISH GRANITE. *HORNBLENDE-BIOTITE-GRANITE*
Darkish pink, medium-grained
  Lamo Quarries, near **Strömstad**
Presented by De Forende Stenhuggerin, Christiania
  Weight per cub. ft. 166 lbs.

498. SWEDISH GRANITE. *HORNBLENDE-BIOTITE-GRANITE*
Dark pinkish grey, coarse-grained
  Reso Quarries, near **Strömstad**
Presented by De Forende Stenhuggerin, Christiania
  Weight per cub. ft. 171 lbs.

499. SWEDISH GRANITE. *BIOTITE-GRANITE*
Light grey, fine-grained
  Rabbaisbede Quarries, near **Uddevalla**
Presented by De Forende Stenhuggerin, Christiania
  Weight per cub. ft. 166 lbs.

500. SWEDISH GRANITE. *MUSCOVITE-BIOTITE-GRANITE*
Light grey, fine-grained
    Buar Quarries, near **Strömstad**
Presented by De Forende Stenhuggerin, Christiania
    Weight per cub. ft. 163 lbs.

501. SWEDISH GRANITE. *AMPHIBOLITE*
Black, fine-grained
    Granite Quarries, **Smalund**
Presented by De Forende Stenhuggerin, Christiania
    Weight per cub. ft. 120 lbs.

502. SWEDISH GRANITE. *EPIDIORITE*
Black, very fine-grained
    Lilla Bokhult Quarries, Herrestad, **Kärda**
Presented by W. D. Makinson, Esq., Kärda
    Weight per cub. ft. 210 lbs.

503. SWEDISH GRANITE. *OLIVINE-GABBRO*
Very dark grey, nearly black, medium-grained
    Bjoerkelund Hill Quarries, Herrestad, **Kärda**
Presented by W. D. Makinson, Esq., Kärda
    Weight per cub. ft. 188 lbs.

504. SWEDISH GRANITE. *OLIVINE-GABBRO*
Very dark grey, nearly black, coarse-grained
    Sjoehag Hill Quarries, Herrestad, **Kärda**
Presented by W. D. Makinson, Esq., Kärda
    Weight per cub. ft. 188 lbs.

505. SWEDISH GRANITE. *BIOTITE-GRANITE*
Dark red, coarse-grained
    Granite Quarries, **Oscarhamn**
Presented by Messrs Kirkpatrick Bros., Manchester
    Weight per cub. ft. 176 lbs.

506. SWISS GRANITE. *BIOTITE-GRANITE*
Light grey, coarse-grained
    Monthey Quarries, **Valais**
            Chemical Composition (Zirkel) :
SiO_2 71·41, Al_2O_3 14·45, Fe_2O_3 2·58, MgO 1·11, CaO 2·49,
K_2O 2·77, Na_2O 1·25, H_2O 1·25, =97·31
    Weight per cub. ft. 171 lbs.

507. SWISS GRANITE. *MUSCOVITE-BIOTITE-GNEISS*
Light grey, coarse-grained
  Biasca Quarries, **Ticino**
  Weight per cub. ft. 162·5 lbs.

508. EGYPTIAN GRANITE. *HORNBLENDE-BIOTITE-GRANITE*
Warm red, coarse-grained
  Syene Quarries, **Aswan**
Presented by Sir John Aird & Co., London
  Weight per cub. ft. 171 lbs. Spec. grav. 2·731

509. EGYPTIAN GRANITE. *HORNBLENDE-BIOTITE-GRANITE*
Warm dark red, coarse-grained
  Syene Quarries, **Aswan**
Presented by The Egyptian Government
  Weight per cub. ft. 171 lbs. Spec. grav. 2·731

510. EGYPTIAN GRANITE. *HORNBLENDE-BIOTITE-GRANITE*
Light red, coarse-grained
  Syene Quarries, **Aswan**
Presented by The Egyptian Government
      Chemical Composition (Scheerer):
  SiO_2 69·95, TiO_2 0·95, Al_2O_3 13·32, FeO 4·90, MgO 0·66, CaO 1·79, K_2O 3·47, Na_2O 3·31, H_2O 1·27, = 99·62
  Weight per cub. ft. 171 lbs. Spec. grav. 2·731

511. EGYPTIAN GRANITE. *HORNBLENDE-BIOTITE-GRANITE*
Light grey, fine-grained
  Syene Quarries, **Aswan**
Presented by The Egyptian Government
  Weight per cub. ft. 172 lbs.

512. EGYPTIAN GRANITE. *DIORITE*
Dark blue-grey, fine-grained
  Syene Quarries, **Aswan**
Presented by The Egyptian Government
  Weight per cub. ft. 171 lbs.

513. CENTRAL AFRICAN GRANITE. *HORNBLENDE-GRANITE*
Dark pinkish grey, coarse-grained
   Lakoma Island Quarries, **Lake Nyasa**
Presented by The Universities' Mission, Central Africa
   Weight per cub. ft. 161 lbs.

514. WEST AFRICAN GRANITE. *GARNETIFEROUS GNEISS*
Dark grey, garnetiferous, medium-grained
   Gneiss Quarries, Akusi, **Gold Coast**
Presented by The Gold Coast Colony, Accra
   Weight per cub. ft. 151 lbs.

515. WEST AFRICAN GRANITE. *BIOTITE-GRANITE*
Light grey, fine-grained
   Abbontiakoon Quarries, near Tarqua, **Gold Coast**
Presented by The Gold Coast Colony, Accra
   Weight per cub. ft. 152 lbs.

516. WEST AFRICAN GRANITE. *HORNBLENDIC GNEISS*
Light grey, fine-grained
   Abeokuta Quarries, Lagos, **South Nigeria**
Presented by Bishop Tugwell, Lagos
   Weight per cub. ft. 165 lbs.

517. WEST AFRICAN GRANITE
Dark grey, coarse-grained
   Lengue Quarries, Benguela, **Angola**
Presented by H. G. Mackie, Esq., H.B.M. Consul, Loanda
   Weight per cub. ft. 160 lbs.

518. WEST AFRICAN GRANITE
Pinkish grey, medium-grained
   Lengue Quarries, Benguela, **Angola**
Presented by H. G. Mackie, Esq., H.B.M. Consul, Loanda
   Weight per cub. ft. 158 lbs.

519. EAST AFRICAN GRANITE. *MUSCOVITE-BIOTITE-GNEISS*
Yellowish pink, slightly banded, coarse-grained
   Amatongas Quarries, Beira, **Portuguese Territory**
Presented by The Companhia de Mozambique, London
   Weight per cub. ft. 152 lbs.

520. EAST AFRICAN GRANITE. *MUSCOVITE-BIOTITE-*
GNEISS
Yellowish pink, medium-grained
  Amatongas Quarries, Beira, **Portuguese Territory**
Presented by The Companhia de Moçambique, London
  Weight per cub. ft. 152 lbs.

521. EAST AFRICAN GRANITE. *MUSCOVITE-BIOTITE-*
GNEISS
Yellowish pink and black, fine-grained, banded
  Amatongas Quarries, Beira, **Portuguese Territory**
Presented by The Companhia de Moçambique, London
  Weight per cub. ft. 152 lbs.

522. EAST AFRICAN GRANITE. *MUSCOVITE-BIOTITE-*
GNEISS
Yellowish pink and black, fine-grained, banded
  Amatongas Quarries, Beira, **Portuguese Territory**
Presented by The Companhia de Moçambique, London
  Weight per cub. ft. 152 lbs.

523. SOUTH AFRICAN GRANITE. *HORNBLENDE-*
SYENITE
Dark red, medium-grained
  Pyramid Quarries, near Pretoria, **Transvaal**
Presented by Frank Watson, Esq., Game Pass, Natal
  Weight per cub. ft. 166 lbs.

524. SOUTH AFRICAN GRANITE. *HORNBLENDE-*
SYENITE
Dark grey, medium-grained
  Waterval North Quarries, near Pretoria, **Transvaal**
Presented by Frank Watson, Esq., Game Pass, Natal
  Weight per cub. ft. 170 lbs.

525. SOUTH AFRICAN GRANITE. *HORNBLENDE-*
GRANITE
Light grey, fine-grained
  Wittkopje Quarries, near Johannesburg, **Transvaal**
Presented by Weightman and Avery, Johannesburg
  Weight per cub. ft. 170 lbs.

526. SOUTH AFRICAN GRANITE. *HORNBLENDE-GRANITE*
Grey with green tinge, medium-grained
Hyde Park Quarries, near Johannesburg, **Transvaal**

527. SOUTH AFRICAN GRANITE. *BIOTITE-GRANITE*
Greyish buff, fine-grained
Lake Quarries, Craighall, near Johannesburg, **Transvaal**

528. SOUTH AFRICAN GRANITE. *QUARTZ-DIORITE*
Dark grey with pink tinge, medium-grained
Craighall Quarries, near Johannesburg, **Transvaal**

529. SOUTH AFRICAN GRANITE. *GABBRO*
Dark green, medium-grained
Newlands Quarries, near Johannesburg, **Transvaal**

530. SOUTH AFRICAN GRANITE. *GABBRO*
Grey and green, medium-grained
Newlands Quarries, near Johannesburg, **Transvaal**

531. SOUTH AFRICAN GRANITE. *BIOTITE-HORNBLENDE-GRANITE*
Light grey with green tinge, medium-grained
Allens Quarries, Paarl, **Cape of Good Hope**
Weight per cub. ft. 171 lbs.

532. SOUTH AFRICAN GRANITE. *BIOTITE-GRANITE*
Light grey, coarse-grained, porphyritic
Higgo Quarries, near Cape Town, **Cape of Good Hope**
Weight per cub. ft. 152 lbs.

534. SOUTH AFRICAN GRANITE. *HORNBLENDE-GRANITE*
Warm red, medium-grained
Park Rynie Quarries, Port Shepstone, **Natal**
Presented by Messrs Cornelius and Hollis, Durban
Weight per cub. ft. 163 lbs.

535. SOUTH AFRICAN GRANITE. *HORNBLENDE-GRANITE*
Grey with pink tinge, medium-grained
    Rifle Butts Kopje Quarries, Matopos, **Rhodesia**
Presented by The British South Africa Co., London
    Weight per cub. ft. 171 lbs.

536. INDIAN GRANITE. *HORNBLENDE-GRANITE*
Dark purple, medium-grained
    Musapet Quarries, near Hyderabad, **Deccan**
Presented by P. G. Messent, Esq., M.Inst.C.E., Bombay
    Weight per cub. ft. 186 lbs.

537. INDIAN GRANITE. *HORNBLENDE-GRANITE*
Light purple, medium-grained
    Musapet Quarries, near Hyderabad, **Deccan**
Presented by P. G. Messent, Esq., M.Inst.C.E., Bombay
    Weight per cub. ft. 176 lbs.

538. INDIAN GRANITE. *BIOTITE-GRANITE*
Light grey, medium-grained
    Sarakki Quarries, near **Bangalore**
Presented by Dr W. F. Smeeth, Bangalore
    Weight per cub. ft. 158 lbs.

539. INDIAN GRANITE. *GRANITE*
Dark red, medium-grained
    Uttenhalli Quarries, Chamundi, **Mysore**
Presented by Dr W. F. Smeeth, Bangalore
    Weight per cub. ft. 153 lbs.

540. INDIAN GRANITE. *GRANITE*
Light pink with green tint, medium-grained
    Chamundi Hills Quarries, near **Mysore**
Presented by Dr W. F. Smeeth, Bangalore

541. INDIAN BLACK TRAP. *MICA-TALC-CHLORITE-ROCK*
Blue-black, fine-grained
    Kadéhalli Quarries, District of **Tumkur**
Presented by Dr W. F. Smeeth, Bangalore
    Weight per cub. ft. 174 lbs.

542. INDIAN BRONZITE. *PERIDOTITE (Serpentinized)*
Blue-black, fine-grained
    Bronzite Quarries, near **Mysore**
Presented by Dr W. F. Smeeth, Bangalore
    Weight per cub. ft. 174 lbs.

543. INDIAN GRANITE. *PYROXENE-GNEISS*
Dark greenish brown, medium-grained
    Granite Quarries, near Cochin, **Malabar Coast**
Presented by Messrs Aspinwall & Co., Cochin
    Weight per cub. ft. 161·5 lbs.

544. CEYLON GRANITE. *NODULAR BIOTITE-GRANITE*
Dark brown, medium-grained
    Granite Quarries, near **Colombo**
    Weight per cub. ft. 186 lbs.

545. CEYLON GRANITE. *GRANULITE*
Dark green and black, slightly foliated
    Mahara Quarries, **Western Province**
Presented by J. H. Bostock, Esq., Colombo
    Weight per cub. ft. 166 lbs.

546. CEYLON GRANITE. *BIOTITE-GNEISS*
Grey and pink, slightly banded, medium-grained
    Mahara Quarries, **Western Province**
Presented by J. H. Bostock, Esq., Colombo
    Weight per cub. ft. 171 lbs.

547. CEYLON GRANITE. *GARNETIFEROUS-GNEISS*
Dark grey and pink, medium-grained
    Mahara Quarries, **Western Province**
Presented by J. H. Bostock, Esq., Colombo
    Weight per cub. ft. 171 lbs.

548. CEYLON GRANITE. *PYROXENE-GNEISS (probably*
               Charnockite)
Pink and grey, medium-grained
    Mahara Quarries, **Western Province**
Presented by J. H. Bostock, Esq., Colombo
    Weight per cub. ft. 171 lbs.

549. CEYLON GRANITE. *Biotite-Hornblende-Granite*
Yellowish grey, coarse-grained
    Granite Quarries, near **Galle**
Presented by J. R. Black, Esq., Galle
    Weight per cub. ft. 161 lbs.

550. CEYLON GRANITE. *Biotite-Granite (with Gneissic*
                character)
Light pink, foliated, fine-grained
    Granite Quarries, near **Galle**
Presented by J. R. Black, Esq., Galle
    Weight per cub. ft. 162 lbs.

551. CEYLON GRANITE. *Biotite-Granite (with Gneissic*
                character)
Greenish white, coarse-grained
    Granite Quarries, near **Galle**
Presented by J. R. Black, Esq., Galle
    Weight per cub. ft. 162 lbs.

552. CEYLON GRANITE. *Biotite-Granite (with Gneissic*
                character)
Greenish grey, coarse-grained
    Granite Quarries, near **Galle**
Presented by J. R. Black, Esq., Galle
    Weight per cub. ft. 162 lbs.

553. CEYLON GRANITE. *Hypersthene-Biotite-Granite*
Very dark greenish grey, fine-grained
    Granite Quarries, near **Galle**
Presented by J. R. Black, Esq., Galle
    Weight per cub. ft. 162 lbs.

554. CEYLON GRANITE. *Biotite-Granite*
Light greenish grey, coarse-grained
    Granite Quarries, near **Galle**
Presented by J. R. Black, Esq., Galle
    Weight per cub. ft. 162 lbs.

555. CEYLON GRANITE. *Enstatite-Biotite-Granite*
                (Charnockite)
Dark green and black, medium-grained
    Boralasgamawa Quarries, **Western Province**
Presented by The Mineral Survey, Colombo
    Weight per cub. ft. 170 lbs.

556. SINGAPORE GRANITE. *Quartz-Monzonite*
Light grey, medium-grained
     Pulo Obin Quarries, **Singapore**
Presented by Messrs Swan and Maclaren, Singapore
     Weight per cub. ft. 175 lbs.

557. SINGAPORE GRANITE. *Quartz-Diorite*
Light grey, medium grained
     Granite Quarries, **Singapore**
     Weight per cub. ft. 171 lbs.

558. SINGAPORE GRANITE. *Biotite-Granite*
Light brownish grey, medium-grained
     8th-Mile Quarries, Bukit Timah, **Singapore**
Presented by Messrs Swan and Maclaren, Singapore
     Weight per cub. ft. 170 lbs.

559. SINGAPORE GRANITE. *Biotite-Granite*
Very light grey, medium-grained
     Bungalow Quarries, Bukit Timah, **Singapore**
Presented by Messrs Swan and Maclaren, Singapore
     Weight per cubic ft. 161·5 lbs.

560. CHINESE GRANITE. *Biotite-Granite*
Light pink and grey, medium-grained
     Kowloon Quarries, near **Hong Kong**
Presented by Messrs Leigh and Orange, Hong Kong
     Weight per cub. ft. 166 lbs.

561. CHINESE GRANITE. *Biotite-Hornblende-Granite*
Light grey and pink, medium-grained
     Kowloon Quarries, near **Hong Kong**
Presented by Messrs Leigh and Orange, Hong Kong
     Weight per cub. ft. 166 lbs.

562. CHINESE GRANITE. *Biotite-Hornblende-Granite*
Light grey, medium-grained
     Kowloon Quarries, near **Hong Kong**
Presented by Messrs Leigh and Orange, Hong Kong
     Weight per cub. ft. 166 lbs.

563. CHINESE GRANITE. *Biotite-Granite*
Light pink and grey, medium-grained
    Mo-Tao Quarries, near **Soochow**
Presented by A. E. Algar, Esq., Shanghai
    Weight per cub. ft. 156 lbs.

564. CHINESE GRANITE. *Granite*
Light brown with black patches, medium-grained
    Granite Quarries, near **Ning-po**
Presented by A. E. Algar, Esq., Shanghai
    Weight per cub. ft. 154 lbs.

565. JAPANESE GRANITE. *Hornblende-Granite*
Light pinkish grey, medium-grained
    Rokkosan Quarries, near **Kobe**
Presented by Messrs Cameron & Co., Ltd., Kobe
    Weight per cub. ft. 166 lbs.

566. AUSTRALIAN GRANITE. *Hornblende-Granite*
Red and green, medium-grained
    Gabo Island Quarries, **New South Wales**
Presented by The Technological Museum, Sydney
    Weight per cub. ft. 171 lbs.

567. AUSTRALIAN GRANITE. *Biotite-Granite*
Brownish red, medium-grained
    Barren Jack Quarries, **New South Wales**
Presented by The Technological Museum, Sydney
    Weight per cub. ft. 176 lbs.

568. AUSTRALIAN GRANITE. *Hornblende-Granite*
Black and white, medium-grained
    Moruya Quarries, **New South Wales**
Presented by The Technological Museum, Sydney
    Weight per cub. ft. 176 lbs.

569. AUSTRALIAN GRANITE. *Hornblende-Granite*
Pink and grey porphyritic, large felspar crystals
    Tenterfield Quarries, **New South Wales**
Presented by The Technological Museum, Sydney
    Weight per cub. ft. 171 lbs.

570. AUSTRALIAN SYENITE. *SYENITE*
Dark olive-green, fine-grained
      Bowral Quarries, The Gib, **New South Wales**
Presented by The Technological Museum, Sydney
      Chemical Composition (J. C. H. Mingage) :
SiO_2 57·14, Al_2O_3 16·13, Fe_2O_3 4·69, FeO 4·00, MnO trace,
MgO 0·63, CaO 3·44, K_2O 5·07, Na_2O 4·87, H_2O 2·20,
CO_2 1·42, P_2O_5 0·25, NaCl 0·04, SO_3 0·30, = 100·18
      Weight per cub. ft. 171 lbs.

571. AUSTRALIAN GRANITE. *HORNBLENDE-GRANITE*
      (*Epidotic*)
Pinkish brown, medium-grained
      Cleveland Quarries, near Townsville,
                             North Queensland
Presented by The Government of Queensland
      Weight per cub. ft. 171 lbs.

572. AUSTRALIAN GRANITE. *BIOTITE-GRANITE*
Pink and black, medium-grained
      Magnetic Island Quarries, near Townsville,
                             North Queensland
Presented by The Government of Queensland
      Weight per cub. ft. 164 lbs.

573. AUSTRALIAN GRANITE. *BIOTITE-GRANITE (with*
      Pyrites)
Light grey, coarse-grained
      Ennoggera Quarries, 5 miles W. of **Brisbane**
Presented by The Government of Queensland
      Weight per cub. ft. 164 lbs.

574. AUSTRALIAN GRANITE. *QUARTZ-DIORITE*
Darkish grey, medium-grained
      Mount Crosby Quarries, 16 miles W. of **Brisbane**
Presented by The Government of Queensland
      Weight per cub. ft. 176 lbs.

575. AUSTRALIAN GRANITE. *HORNBLENDE-GRANITE*
Warm red, fine-grained
      Granite Quarries, Gabo Island, **Victoria**
      Weight per cub. ft. 180·5 lbs.

576. AUSTRALIAN GRANITE. *HORNBLENDE-GRANITE*
Light grey, medium-grained
Granite Quarries, Harcourt, **Victoria**
Weight per cub. ft. 180·5 lbs.

577. AUSTRALIAN GRANITE. *HORNBLENDE-GRANITE*
Chocolate-coloured, coarse-grained
Murray Bridge Quarries, near Adelaide,
South Australia
Weight per cub. ft. 171 lbs.

578. AUSTRALIAN GRANITE. *BIOTITE-GRANITE*
Light pinkish grey, medium-grained, porphyritic
Kellerberrin Quarries, **Western Australia**
Presented by The Government of Western Australia
Weight per cub. ft. 171 lbs.

579. AUSTRALIAN GRANITE. *BIOTITE-GRANITE*
Light grey, fine-grained
Meckering Quarries, **Western Australia**
Presented by The Government of Western Australia
Weight per cub. ft. 175·7 lbs.

580. NEW ZEALAND GRANITE. *QUARTZ-DIORITE*
Light grey, medium-grained
Coromandel Quarries, near **Auckland**
Presented by R. C. Hooper, Esq., Ponsonby, Auckland
Weight per cub. ft. 175·5 lbs.

581. NEW ZEALAND GRANITE. *DIORITE*
Dark blue-grey, medium-grained
Granite Quarries, Ruapuke Island, **Otago**
Weight per cub. ft. 178 lbs.

582. NORTH AMERICAN GRANITE. *BIOTITE-GRANITE*
Light grey, very fine-grained
White Granite Quarries, Waterford,
Connecticut, U.S.A.
Presented by Messrs Booth Bros. and The Hurricane Isle Granite Co., New York
Chemical Composition (Ricketts and Banks, N. York):
SiO_2 68·11, Al_2O_3 14·28, FeO 2·63, MgO 0·68, CaO 1·86, K_2O 5·46, Na_2O 6·57, S 0·34, = 99·93
Weight per cub. ft. 166 lbs.
Crushing strain per sqr. ft. 1537 tons

583. NORTH AMERICAN GRANITE. *Biotite-Granite*
Light pink, coarse-grained
  Norcross Quarries, Stoney Creek, Bradford,
                Connecticut, U.S.A.
Presented by The Norcross Brothers Co., Worcester, Mass.
       Chemical Composition (L. P. Kinnicutt) :
SiO_2 72·73, $Al_2O_3 + Fe_2O_3$ 16·95, MgO trace, CaO 1·05,
K_2O 8·15, Na_2O 0·90, Loss 0·22, =100·00
  Weight per cub. ft. 171 lbs.
  Crushing strain per sqr. ft. 1443 tons without crushing

584. NORTH AMERICAN GRANITE. *Muscovite-Granite*
Light grey, fine-grained
  Stone Mountain Quarries, **Georgia, U.S.A.**
Presented by Messrs Venable Bros., Atlanta
       Chemical Composition (Prof. McCaudless) :
SiO_2 74·50, Al_2O_3 14·66, Fe_2O_3 0·56, MnO 0·11, MgO 0·35,
CaO 1·25, K_2O 4·80, Na_2O 2·95, H_2O 0·90, =100·08
  Weight per cub. ft. 168 lbs. Spec. grav. 2·686
  Crushing strain per sqr. ft. 1808 tons

585. NORTH AMERICAN GRANITE. *Biotite-Granite*
Pink and grey mottled, coarse-grained
  Palmer Quarries, Vinalhaven, **Maine, U.S.A.**
Presented by The Bodwell Granite Co., Vinalhaven
       Chemical Composition (Ricketts and Brown) :
SiO_2 70·94, Al_2O_3 15·68, FeO 2·29, MnO 0·13, MgO 0·19, CaO 1·23,
K_2O 5·54, Na_2O 3·58, H_2O & Loss 0·37, S 0·05, =100·00
  Weight per cub. ft. 166·3. Spec. grav. 2·66

586. NORTH AMERICAN GRANITE. *Biotite-Granite*
Light fawn-coloured, mottled, coarse-grained
  Sands Quarries, Vinalhaven, **Maine, U.S.A.**
Presented by The Bodwell Granite Co., Vinalhaven
  Weight per cub. ft. 165 lbs. Spec. grav. 2·63

587. NORTH AMERICAN GRANITE. *Hornblende-Granite*
Pink and grey, medium-grained
  Hurricane Isle Quarries, Vinalhaven, **Maine, U.S.A.**
Presented by Messrs Booth Bros. and The Hurricane Isle Granite Co., New York
       Chemical Composition (Ricketts and Banks, N. York) :
SiO_2 70·94, Al_2O_3 15·68, FeO 2·29, MnO 0·13, MgO 0·19,
CaO 1·23, K_2O 5·54, Na_2O 3·58, CO_2 0·13, S 0·05,
Loss 0·37, =100·00
  Weight per cub. ft. 176 lbs.
  Crushing strain per sqr. ft. 1259 tons (Woolson)

588. NORTH AMERICAN GRANITE. *Biotite-Granite*

Light pink and grey, medium-grained

High Island Quarries, Muscle Ridge Plantation,

Maine, U.S.A.

Presented by Messrs Wm. Gray & Sons, Philadelphia

Chemical Composition (I. F. Kemp, Columbia University, New York):

SiO_2 74·54, Al_2O_3 13·30, Fe_2O_3 0·92, FeO 0·79, MnO 0·51, MgO 0·009, CaO 1·26, Na_2O 3·69, K_2O 5·01, S 0·038, = 100·067

Weight per cub. ft. 167 lbs. Spec. grav. 2·64

Crushing strain per sqr. ft. 1989 tons

589. NORTH AMERICAN GRANITE. *Biotite-Granite*

Light pinkish grey, medium-grained

High Island Quarries, Muscle Ridge Plantation,

Maine, U.S.A.

Presented by Messrs. Wm. Gray & Sons, Philadelphia

Chemical Composition (J. F. Kemp, Columbia University, New York):

SiO_2 74·54, Al_2O_3 13·30, Fe_2O_3 0·92, FeO 0·79, MnO 0·51, MgO 0·009, CaO 1·26, Na_2O 3·69, K_2O 5·01, S 0·038, = 100·067

Weight per cub. ft. 167 lbs. Spec. grav. 2·64

Crushing strain per sqr. ft. 1989 tons

590. NORTH AMERICAN GRANITE. *Quartz-Monzonite*

Grey, coarse-grained

Spruce Head Quarries, Spruce Head Island,

Maine, U.S.A.

Presented by The Bodwell Granite Co., Vinalhaven

Weight per cub. ft. 171 lbs. Spec. grav. 2·75

591. NORTH AMERICAN GRANITE. *Biotite-Hornblende-Granite*

Pink and white, coarse-grained

Hardwood Island Quarry, Jonesport,

Hardwood Island, **Maine, U.S.A.**

Presented by The Rockport Granite Co., Rockport

Weight per cub. ft. 166 lbs. Spec. grav. 2·76

Crushing strain per sqr. ft. 1543 tons

592. NORTH AMERICAN GRANITE. *Muscovite-Biotite-Granite*

Light grey, fine-grained
    Waldoboro Quarries, Waldoboro, **Maine, U.S.A.**
Presented by Messrs Booth Bros. and The Hurricane Isle Granite Co., New York
    Chemical Composition (Ricketts and Banks, N. York):
SiO_2 73·48, Al_2O_3 15·26, FeO 1·42, MnO 0·10, MgO 0·09, CaO 0·88, K_2O 5·66, Na_2O 3·12, S trace, =100·01
    Weight per cub. ft. 171 lbs.
    Crushing strain per sqr. ft. 1485·5 tons (Woolson)

593. NORTH AMERICAN GRANITE. *Muscovite-Biotite-Granite*

Light grey, fine-grained
    Lincolnville Quarries, **Maine, U.S.A.**
Presented by The Bodwell Granite Co., Vinalhaven
    Weight per cub. ft. 171 lbs.

594. NORTH AMERICAN GRANITE. *Muscovite-Biotite-Granite*

Light grey, fine-grained
    Hallowell Granite Quarries, **Maine, U.S.A.**
Presented by The Hallowell Granite Co., Hallowell
    Weight per cub. ft. 176 lbs. Absorp. 0·370 %
    Crushing strain per sqr. ft. 2654·3 tons (Buckley)

595. NORTH AMERICAN GRANITE. *Biotite-Muscovite-Granite*

Light grey, fine-grained
    Long Cove Quarries, Tenant Harbour, **Maine, U.S.A.**
Presented by Messrs Booth Bros. and The Hurricane Isle Granite Co., New York
    Weight per cub. ft. 176 lbs. Crushing strain per sqr. ft. 1414 tons (Columbia School of Maine)

596. NORTH AMERICAN GRANITE. *Biotite-Granite*
Light red, coarse-grained
    Jonesborough Quarries, Washington Co., **Maine, U.S.A.**
Presented by The Bodwell Granite Co., Vinalhaven
    Chemical Composition (Ricketts and Banks, N. York):
SiO_2 72·97, Al_2O_3 14·63, FeO 1·73, MnO 0·10, MgO 0·27, CaO 1·48, K_2O 5·18, Na_2O 3·28, H_2O & Loss 0·33, S 0·03, =100·00
    Weight per cub. ft. 166 lbs.

597. NORTH AMERICAN GRANITE. *Olivine-Gabbro*
Dark blue-grey, fine-grained
  Pleasant River Quarries, Addison, **Maine, U.S.A.**
Presented by The Pleasant River Granite Co., Addison
  Weight per cub. ft. 184 lbs.
  Crushing strain per sqr. ft. 1441 tons

598. NORTH AMERICAN GRANITE. *Biotite-Granite*
Pinkish grey, medium-grained
  Granite Quarries, Woodstock, **Maryland, U.S.A.**
    Chemical Composition (W. F. Hillebrand):
SiO_2 71·79, Al_2O_3 15·00, Fe_2O_3 0·77, FeO 1·12, MgO 0·51,
CaO 2·50, K_2O 4·75, Na_2O 3·09, H_2O 0·64, = 100·17
  Weight per cub. ft. 161·5 lbs.
  Crushing strain per sqr. ft. 1359 tons

599. NORTH AMERICAN GRANITE. *Riebeckite-Granite*
Grey, coarse-grained
  Granite Quarries, Quincy, **Massachusetts, U.S.A.**
  Weight per cub. ft. 166·2 lbs. Spec. grav. 2·66
  Crushing strain per sqr. ft. 1141 tons

600. NORTH AMERICAN GRANITE. *Hornblende-Granite*
Light grey, medium-grained
  Rockport Quarries, Cape Ann, **Massachusetts, U.S.A.**
Presented by The Rockport Granite Co., Rockport
      Chemical Composition :
SiO_2 74·70, Al_2O_3 13·71, FeO 2·37, MnO 0·05, CaO 0·65,
K_2O 4·28, Na_2O 4·22, = 99·98
  Weight per cub. ft. 166 lbs. Spec. grav. 2·76
  Crushing strain per sqr. ft. 2121 tons

601. NORTH AMERICAN GRANITE. *Hornblende-Granite*
Greenish grey, medium-grained
  Gloucester Quarries, Cape Ann,
                    Massachusetts, U.S.A.
Presented by The Rockport Granite Co., Rockport
      Chemical Composition :
SiO_2 74·70, Al_2O_3 13·71, FeO 2·37, MnO 0·05, CaO 0·65,
K_2O 4·28, Na_2O 4·22, = 99·98
  Weight per cub. ft. 166 lbs. Spec. grav. 2·76
  Crushing strain per sqr. ft. 2121 tons

602. NORTH AMERICAN GRANITE. *HORNBLENDE-*
BIOTITE-GNEISS
Pink and black, coarse-grained
    Granite Quarries, Milford, **Massachusetts, U.S.A.**
Chemical Composition (Talbott, *Geol. Survey U.S.A.*) :
SiO_2 76·95, Al_2O_3 11·15, Fe_2O_3 0·80, K_2O 5·03, Na_2O 5·60,
H_2O & Loss 0·20, =99·73
    Weight per cub. ft. 176 lbs.
    Crushing strain per sqr. ft. 1985 tons

603. NORTH AMERICAN GRANITE. *MUSCOVITE-*
BIOTITE-GRANITE
Grey, medium-grained
    Granite Quarries, Concord, **New Hampshire, U.S.A.**
    Weight per cub. ft. 171 lbs.

604. NORTH AMERICAN GRANITE. *MUSCOVITE-*
BIOTITE-GRANITE
Light grey, fine-grained
    Rattlesnake Hill Quarries, Concord,
                    New Hampshire, U.S.A.
Presented by The New England Granite Co., Westerly R.I.
    Chemical Composition (Howard) :
SiO_2 70·20, Al_2O_3 17·74, Fe_2O_3 2·26, CaO 1·15, K_2O 3·52,
Na_2O 5·15, =100·02
    Weight per cub. ft. 165 lbs.

605. NORTH AMERICAN GRANITE. *HORNBLENDE-*
BIOTITE-GRANITE
Light grey, medium-grained
    Lovejoy Granite Quarries, S. Milford,
                    New Hampshire, U.S.A.
Presented by The Lovejoy Granite Co., Milford
    Weight per cub. ft. 171 lbs.

606. NORTH AMERICAN GRANITE. *BIOTITE-GRANITE*
Pink mottled with light yellowish green, coarse-grained
    Pompton Junction Quarries, **New Jersey, U.S.A.**
Presented by Dr G. Merrill, Washington
    Chemical Composition (R. B. Gage, Chemist of *Geol. Survey*) :
SiO_2 71·91, TiO_2 0·03, Al_2O_3 15·71, Fe_2O_3 0·21, FeO 0·13,
MnO 0·02, MgO 0·03, CaO 0·70, BaO 0·14, K_2O 8·60,
Na_2O 2·61, H_2O 0·27, CO_2 0·22, P_2O_5 0·11, SO_3 0·05,
=100·74
    Weight per cub. ft. 161 lbs.

607. NORTH AMERICAN GRANITE. *HORNBLENDE-BIOTITE-GRANITE*
Light grey, medium-grained
    Mount Airy Quarries, **North Carolina, U.S.A.**
Presented by The North Carolina Granite Corporation, Mount Airy
    Weight per cub. ft. 165·5 lbs.

608. NORTH AMERICAN GRANITE. *BIOTITE-GRANITE*
Pink and white, coarse-grained
    Granite Quarries, Salisbury, **North Carolina, U.S.A.**
    Weight per cub. ft. 171 lbs.

609. NORTH AMERICAN GRANITE. *HORNBLENDE-GRANITE*
Medium grey, fine-grained
    Anderson Quarries, Winnsborough,
                    South Carolina, U.S.A.
Presented by The Winnsboro Granite Corporation, Rion, S. Carolina
    Chemical Composition (Booth, Garrett and Blair):
SiO_2 69·74, Al_2O_3 13·72, Fe_2O_3 3·64,   MgO 0·22,   CaO 1·54,
K_2O 4·98, Na_2O 5·39,   =99·23
    Weight per cub. ft. 176 lbs.
    Crushing strain per sqr. ft. 1645 tons (J. S. Butler, Col.)

610. NORTH AMERICAN GRANITE. *BIOTITE-GNEISS*
Light grey, medium-grained
    Holmesburg Quarries, near Philadelphia,
                    Pennsylvania, U.S.A.
Presented by The Holmesburg Granite Co., Philadelphia
    Weight per cub. ft. 167 lbs.
    Crushing strain per sqr. ft. 1415 tons

611. NORTH AMERICAN GRANITE. *OLIVINE-DOLERITE*
Dark grey, fine-grained
    The Quarries, Gettysburg, **Pennsylvania, U.S.A.**
    Weight per cub. ft. 170 lbs.

612. NORTH AMERICAN GRANITE. *BIOTITE-HORNBLENDE-GRANITE*
Bluish grey, very fine-grained
    Newell Quarries, Westerly, **Rhode Island, U.S.A.**
Presented by Messrs H. and I. Newell, Dalbeattie, N.B.
    Chemical Composition (F. W. Love):
SiO_2 71·64, Al_2O_3 15·66, Fe_2O_3 & FeO 2·34,   CaO 2·70,
K_2O 5·60, Na_2O 1·578, H_2O 0·482,   =100·000
Weight per cub. ft. 165 lbs.   Spec. grav. 2·64.   Crushing strain
    per sqr. ft. 2050 tons (U.S. Arsenal, Waterton, Mass.)

613. NORTH AMERICAN GRANITE. *Biotite-Granite*
Bluish grey, fine-grained
New England Quarries, Westerly,
Rhode Island, U.S.A.
Presented by The New England Granite Co., Westerly
Chemical Composition (Howard) :
SiO_2 67·90, Al_2O_3 18·80, Fe_2O_3 3·51,   CaO 2·15, K_2O 2·95,
Na_2O 4·51,   =99·82
Weight per cub. ft. 165 lbs.
Crushing strain per sqr. ft. 2055 tons

614. NORTH AMERICAN GRANITE. *Biotite-Granite*
Reddish brown, medium-grained
Redstone Quarries, Westerly, **Rhode Island, U.S.A.**
Presented by The New England Granite Co., Westerly
Chemical Composition (Howard):
SiO_2 67·80, Al_2O_3 18·70, Fe_2O_3 3·40,   CaO 2·18, K_2O 2·91,
Na_2O 4·55,   =99·54
Weight per cub. ft. 171 lbs.
Crushing strain per sqr. ft. 1834 tons

615. NORTH AMERICAN GRANITE. *Biotite-Granite*
Pink and black, coarse-grained
Granite Mountain Quarries, Burnett Co.,
Texas, U.S.A.
*Presented by Messrs Darragh Bros., Granite Co., Granite
Mountain*
Weight per cub. ft. 163·64 lbs.   Spec. grav. 2·62
Crushing strain per sqr. ft. 765 tons

616. NORTH AMERICAN GRANITE. *Muscovite-
Biotite-Granite*
Grey, fine-grained
Granite Quarries, Barre, **Vermont, U.S.A.**
Chemical Composition (*U.S.A. Geol. Survey*) :
SiO_2 69·56, Al_2O_3 15·38,   FeO 2·65,   CaO 1·76, K_2O 4·31,
Na_2O 5·38, H_2O & Loss 1·02,   =100·06
Weight per cub. ft. 176 lbs.

617. NORTH AMERICAN GRANITE. *Quartz-
Monzonite*
Very light grey, nearly white, medium-grained
Granite Quarries, Bethel, **Vermont, U.S.A.**
Weight per cub. ft. 176 lbs.

618. NORTH AMERICAN GRANITE. *HORNBLENDE-GRANITE*

Dark brownish red, fine-grained
    Montello Quarries, **Wisconsin, U.S.A.**
Presented by The Montello Granite Co., Wisconsin
    Chemical Composition (F. G. Weichmann) :
SiO_2 75·40, Al_2O_3 11·34, Fe_2O_3 4·16,   CaO 0·90, K_2O 6·44,
Na_2O 1·76,   =100·00
    Weight per cub. ft. 164 lbs.   Spec. grav· 2·640
    Crushing strain per sqr. ft. 2454 tons (Buckley)

619. NORTH AMERICAN GRANITE. *GRANITE*

Dark reddish brown, medium-grained
    Montello Quarries, **Wisconsin, U.S.A.**
Presented by The Geological and Natural History Survey, Madison, Wisconsin
Weight per cub. ft. 164·5 lbs.   Spec. grav. 2·64.   Absorp. 0·069⁰/₀
    Crushing strain per sqr. ft. 2877 tons

620. NORTH AMERICAN GRANITE. *GRANITE (some Biotite and Hornblende)*

Bright red, medium-grained
    Wasau Quarries, **Wisconsin, U.S.A.**
Presented by The Geological and Natural History Survey, Madison, Wisconsin
    Chemical Composition (W. W. Daniels) :
SiO_2 76·54, Al_2O_3 13·82, Fe_2O_3 1·62,   MgO 0·01,   CaO 0·85,
K_2O 2·31, Na_2O 4·32, H_2O 0·20,   =99·67
Weight per cub. ft. 163·3 lbs.   Spec. grav. 2·6.   Absorp. 0·46 °/₀
    Crushing strain per sqr. ft. 1704 tons

621. NORTH AMERICAN GRANITE. *GRANITE (some Biotite and Hornblende)*

Red, medium-grained
    Granite Heights Quarries, Wausau,
                 Wisconsin, U.S.A.
Presented by The Geological and Natural History Survey, Madison, Wisconsin
    Chemical Composition (W. W. Daniels) :
SiO_2 76·54, Al_2O_3 13·82, Fe_2O_3 1·62,   MgO 0·01,   CaO 0·85,
K_2O 2·31, Na_2O 4·32, H_2O 0·20,   =99·67
Weight per cub. ft. 163·3 lbs.   Spec. grav. 2·6.   Absorp. 0·46 °/₀
    Crushing strain per sqr. ft. 1704·5 tons

622. NORTH AMERICAN GRANITE. *Hornblende-*
Granite
Green and black, coarse-grained
  Parcher Quarries. Wausau, **Wisconsin, U.S.A.**
Presented by The Geological and Natural History Survey, Madison,
   Wisconsin
  Weight per cub. ft. 163 lbs.

623. NORTH AMERICAN GRANITE. *Biotite-Granite*
Grey, fine-grained
  Pike River Quarries, near Amberg, **Wisconsin, U.S.A.**
Presented by The Geological and Natural History Survey, Madison,
   Wisconsin
Weight per cub. ft. 167 lbs. Spec. grav. 2·684. Absorp. 0·094 %
  Crushing strain per sqr. ft. 1792 tons

624. NORTH AMERICAN RHYOLITE. *Rhyolite*
Gneiss
Black with light pink felspar crystals
  Berlin Quarries, **Wisconsin, U.S.A.**
Presented by The Geological and Natural History Survey, Madison,
   Wisconsin
   Chemical Composition (S. Weidman) :
SiO_2 73·65, Al_2O_3 11·19, Fe_2O_3 1·31, FeO 3·25, MgO 0·51,
CaO 2·78, K_2O 1·86, Na_2O 3·74, H_2O 0·44, =98·73
Weight per cub. ft. 164 lbs. Spec. grav. 2·641. Absorp. 0·075 %
  Crushing strain per sqr. ft. 3064 tons

625. NORTH AMERICAN GRANITE. *Crushed*
Granite
Pink and black, medium-grained
  Lohrville Quarries, Waushara, **Wisconsin, U.S.A.**
Presented by The Geological and Natural History Survey, Madison,
   Wisconsin
   Chemical Composition (S. Weidman) :
SiO_2 74·62, Al_2O_3 10·01, Fe_2O_3 3·85, FeO 1·72, MgO 0·33,
CaO 2·43, K_2O 3·38, Na_2O 3·33, H_2O 0·24, =99·91
Weight per cub. ft. 164 lbs. Spec. grav. 2·642. Absorp. 0·14 %
  Crushing strain per sqr. ft. 2315 tons

626. NORTH AMERICAN GRANITE. *Biotite-Granite*
Dark reddish brown and black, coarse-grained
  Waupaca Quarries, **Wisconsin, U.S.A.**
Presented by The Geological and Natural History Survey, Madison,
   Wisconsin
  Weight per cub. ft. 163 lbs.

627. CANADIAN GRANITE. *Muscovite-Biotite-Granite*
Light grey, medium-grained
    Standstead Quarries, Standstead, **Prov. Quebec**
Presented by Messrs Drummond, McCall & Co., Montreal
    Weight per cub. ft. 166 lbs.

628. CANADIAN GRANITE. *Hornblende-Granite*
Dark brownish red, medium-grained
    Staynerville Quarries, **Prov. Quebec**
Presented by The Laurentian Granite Co., Montreal
    Weight per cub. ft. 168 lbs. Spec. grav. 2·70
    Crushing strain per sqr. ft. 1989 tons (McGill University)

629. CANADIAN GRANITE. *Hornblende-Granite*
Warm red, coarse-grained
    Leek Island Quarries, **Thousand Islands**
Presented by The Forsyth Granite Co., Montreal
    Weight per cub. ft. 170·5 lbs.

630. CANADIAN GRANITE. *Hornblende-Granite*
Warm red, fine-grained
    Forsyth Island Quarries, **Thousand Islands**
Presented by The Forsyth Granite Co., Montreal
    Weight per cub. ft. 170·5 lbs.

631. CANADIAN GRANITE. *Hornblende-Granite*
Bright red, medium-grained
    Bay of Fundy Quarries, St George, **New Brunswick**
Presented by Messrs Milne Coutts & Co., St George
    Weight per cub. ft. 168 lbs. Crushing strain per sqr. ft. 1007·5
    tons (J. B. and J. M. Cornell, New York)

631A. SOUTH AMERICAN GRANITE. *Muscovite-Biotite-Granite*
Light pink, medium-grained
    La Falda Quarries, Córdoba, **Argentine**
Presented by Donald Fraser, Esq., Rosario
    Weight per cub. ft. 175 lbs.

631B. SOUTH AMERICAN GRANITE. *Muscovite-Biotite-Granite*
Light pinkish grey, medium-grained
    Capilla del Monte Quarries, Córdoba, **Argentine**
Presented by Donald Fraser, Esq., Rosario
    Weight per cub. ft. 173 lbs.

IGNEOUS ROCKS (VOLCANIC).

632. FRENCH GRANITE. *Kersantite*
Dark blue-grey, fine-grained
  Kersanton Quarries, near Brest, **Finistère**
Presented by L. Omnes, Esq., Brest
  Weight per cub. ft. about 170 lbs.

633. GERMAN LAVA. *Trachyte-Tuff* ("*Trass*")
Dark grey, cellular, basaltic
  Brohl Quarries, Eifel, **Rhenish Prussia**
Presented by Michel Lickes, Esq., Cologne
  Weight per cub. ft. 85·5 lbs.

633A. GERMAN BASALT
Dark grey, cellular
  Eastern Quarries, Eifel, **Rhenish Prussia**
Presented by Michel Lickes, Esq., Cologne
  Weight per cub. ft. 140 lbs.

634. RHENISH BASALT. *Olivine-Basalt*
Dark grey, compact
  Weilberg Quarries, Heisterbach, near **Königswinter**
Presented by Messrs Johnston Bros., London and Rotterdam
  Specimens of columns and blocks in Court-yard of Museum.

635. ITALIAN VOLCANIC TUFF. (*Tufa Litoide*)
Trachyte-Tuff
Dark straw-coloured, fragmental
  Tuff Quarries, near **Rome**
  Weight per cub. ft. 123·5 lbs.

636. ITALIAN PEPERINO. (*Lapis-Albanus*) *Leucite-Tuff*
Light blue and grey mottled agglomerate
    Marino Quarries, near **Rome**
Weight per cub. ft. 123·5 lbs.

637. ITALIAN LAVA. (*Piperno*) *Trachyte-Tuff*
Grey lava, containing black fragments
    Arecco Quarries, Pianura, near **Naples**
           Chemical Composition :
SiO_2 61·74, Al_2O_3 19·24, Fe_2O_3 4·12,   MgO 0·39,   CaO 1·14,
K_2O 5·50, Na_2O 6·68, H_2O 1·12,   NaCl 0·19,   =100·12
    Weight per cub. ft. 162·5 lbs.   Spec. grav. 2·596
    Crushing strain per sqr. ft. 520 tons

638. ITALIAN LAVA. *Trachyte*
Light grey trachyte with augite and hornblende
    Monte Olibano Quarries, Pozzuoli, near **Naples**
           Chemical Composition :
SiO_2 66·85 (other constituents not determined)
    Weight per cub. ft. 123·5 lbs.

639. ITALIAN LAVA. *Leucite-Tephrite*
Dark grey
    S. Sebastiano Quarries (West Slope of Vesuvius), **Naples**
Weight per cub. ft. 173 lbs.   Spec. grav. 2·78
    Crushing strain per sqr. ft. 374·5 tons

640. ITALIAN LAVA. *Leucite-Tephrite*
Light grey
    Camaldoli Quarries, Torre del Greco,
          (South-West Slope of Vesuvius), **Naples**
Weight per cub. ft. 173 lbs.   Spec. grav. 2·78
    Crushing strain per sqr. ft. 406·5 tons

641. ITALIAN TUFF. *Trachyte-Tuff*
Yellow, fragmental
    Fontanelli Quarries, near **Naples**
           Chemical Composition :
SiO_2 52·80, Al_2O_3 15·83, Fe_2O_3 & FeO 7·57,   MnO 0·54,
MgO 0·84,   CaO 3·13, K_2O 7·66, Na_2O 2·30, H_2O 3·26,
NaCl 0·15,   =94·08
    Weight per cub. ft. 153 lbs.   Spec. grav. 2·45
    Crushing strain per sqr. ft. 50 tons

642. ITALIAN TUFF. *TRACHYTE-TUFF*
Yellow, fragmental
    Fontanelli Quarries, near **Naples**
    Weight per cub. ft. 76 lbs.

643. ITALIAN TUFF. *TRACHYTE-TUFF*
Grey with black inclusions, fragmental
    Fiano Quarries, Nocera, Caserta, **Campania**
            Chemical Composition :
SiO_2 51·65, Al_2O_3 15·08, Fe_2O_3 6·21,   MgO 1·18,   CaO 5·43,
K_2O 6·19, Na_2O 1·01, H_2O & NaCl 11·40,   =98·15
    Weight per cub. ft. 95·5 lbs.
    Crushing strain per sqr. ft. 57 tons

644. SOUTH AFRICAN PORPHYRY. *QUARTZ-*
PORPHYRY
Dark grey, medium-grained
    Mills Quarries, Saldanha Bay, **Cape of Good Hope**
    Weight per cub. ft. 157 lbs.

645. TURKISH LAVA. *DACITE*
Reddish pink, slightly porphyritic
    Dikili Quarries, Mytilene Channel, **Asia Minor**
Presented by E. Whittall, Esq., Smyrna
    Weight per cub. ft. 147 lbs.

646. TURKISH LAVA. *HORNBLENDE-ANDESITE*
Light grey with small black angular spots
    Dikili Quarries, Mytilene Channel, **Asia Minor**
Presented by E. Whittall, Esq., Smyrna
    Weight per cub. ft. 142·5 lbs.

647. TURKISH LAVA. *HORNBLENDE-ANDESITE*
Greenish grey, compact, slightly porphyritic
    Foot of Mount Pagus, Smyrna, **Asia Minor**
Presented by E. Whittall, Esq., Smyrna
    Weight per cub. ft. 152 lbs.

648. TURKISH LAVA. ? *MICA-ANDESITE*
Light purple, tufaceous
    Bariakili Quarries, opposite Smyrna, **Asia Minor**
Presented by E. Whittall, Esq., Smyrna
    Weight per cub. ft. 138 lbs.

649. TURKISH LAVA. *ANDESITE-TUFF*
Light purple, slightly speckled with white
  Quarries, Island of Mytilene, **Asia Minor**
Presented by E. Whittall, Esq., Smyrna
  Weight per cub. ft. 133 lbs.

650. TURKISH LAVA. *HORNBLENDE-ANDESITE*
Purple-grey, compact
  Lithri Quarries, Karabouran Peninsula, **Asia Minor**
Presented by E. Whittall, Esq., Smyrna
  Weight per cub. ft. 142·5 lbs.

651. TURKISH LAVA. *MICA-ANDESITE-TUFF*
Pink, tufaceous
  Awaly Quarries, 10 miles N. of Smyrna, **Asia Minor**
Presented by E. Whittall, Esq., Smyrna
  Weight per cub. ft. 114 lbs.

652. TURKISH LAVA. *RHYOLITIC TUFF*
Light yellow, tufaceous
  Alatazata, Karabouran Peninsula, **Asia Minor**
Presented by E. Whittall, Esq., Smyrna
  Weight per cub. ft. 104·5 lbs.

653. TURKISH LAVA. *RHYOLITIC TUFF*
Light cream, tufaceous
  Fokies Quarries (or Focha), **Asia Minor**
Presented by E. Whittall, Esq., Smyrna
  Weight per cub. ft. 104·5 lbs.

654. TURKISH LAVA. *RHYOLITIC TUFF*
Light cream and yellow, tufaceous
  Fokies Quarries (or Focha), **Asia Minor**
Presented by E. Whittall, Esq., Smyrna
  Weight per cub. ft. 114 lbs.

655. SYRIAN BASALT. *OLIVINE-BASALT*
Dark grey with ferruginous specks
  Houran Quarries, near Lake Gennesaret, **Palestine**
Presented by E. Uengar, Esq., Jerusalem
  Weight per cub. ft. 170 lbs.

656. KURLA BLUE BASALT (Deccan Trap). *Quartz-*
Dolerite
Greenish grey, compact
    Kurla Quarries, near **Bombay**
Presented by P. G. Messent, Esq., M.Inst.C.E., Bombay
    Weight per cub. ft. 171 lbs.
    Crushing strain per sqr. ft. 771 tons

657. KURLA YELLOW BASALT (Deccan Trap).
    Quartz-Dolerite
Brownish yellow, compact
    Kurla Quarries, near **Bombay**
Presented by P. G. Messent, Esq., M.Inst.C.E., Bombay
    Weight per cub. ft. 142·5 lbs.
    Crushing strain per sqr. ft. 771 tons

658. BOMBAY BASALT (Deccan Trap). *Basalt*
Blue-grey, compact
    Basalt Quarries, **Bombay**
Presented by P. G. Messent, Esq., M.Inst.C.E., Bombay
    Weight per cub. ft. 165 lbs.

659. INDIAN FELSITE. *Porphyry*
Pale green, medium-grained
    Belagola Quarries, near **Mysore**
Presented by Dr W. F. Smeeth, Bangalore
    Weight per cub. ft. 153 lbs.

660. INDIAN PORPHYRY. *Bostonite-Porphyry*
Pink and dark brown mottled, medium-grained
    Hosur Quarries, district of **Seringapatam**
Presented by Dr W. F. Smeeth, Bangalore
    Weight per cub. ft. 153 lbs.

661. INDIAN PORPHYRY. *Porphyry*
Pink with light green tint, mottled
    Kesalgiri Quarries, district of **Seringapatam**
Presented by Dr W. F. Smeeth, Bangalore
    Weight per cub. ft. 149 lbs.

662. INDIAN PORPHYRY. *Quartz-Porphyry*
Pink and brown mottled, slightly veined
    Arathe Koppal Quarries, district of **Seringapatam**
Presented by Dr W. F. Smeeth, Bangalore
    Weight per cub. ft. 158 lbs.

663. INDIAN PORPHYRY. *Bostonite-Porphyry*
Dark brown with large phenocrysts of felspar
    6th Mile Stone Quarries, Mandya Bunnur Rd., **Mandya**
Presented by Dr W. F. Smeeth, Bangalore
    Weight per cub. ft. 158 lbs.

664. INDIAN PORPHYRY. *Porphyry (Orthophyre)*
with Melanite and Augite
Dark red with large white crystals of felspar
    Tadagwade Quarries, district of **Seringapatam**
Presented by Dr W. F. Smeeth, Bangalore
    Weight per cub. ft. 153 lbs.

665. INDIAN PORPHYRY. *Porphyry (Orthophyre)*
with Hornblende
Pink with dark prisms of felspar
    Kyatanhalli Quarries, district of **Seringapatam**
Presented by Dr W. F. Smeeth, Bangalore
    Weight per cub. ft. 158 lbs.

666. INDIAN PORPHYRY. *Orthoclase-Porphyry*
Greenish brown with white crystals of felspar
    Jakkenhalli Quarries, district of **Seringapatam**
Presented by Dr W. F. Smeeth, Bangalore
    Weight per cub. ft. 158 lbs.

667. INDIAN PORPHYRY. *Quartz-Porphyry with Hornblende*
Greenish grey, medium-grained
    Sidlingpur Quarries, **Mysore**
Presented by Dr W. F. Smeeth, Bangalore
    Weight per cub. ft. 158 lbs.

668. INDIAN PORPHYRY. *Orthoclase-Porphyry*
Pink and brown mottled, medium-grained
    Kengal Koppal and Rampin Quarries, **Seringapatam**
Presented by Dr W. F. Smeeth, Bangalore
    Weight per cub. ft. 158 lbs.

669. INDIAN PORPHYRY. *Quartz-Porphyry*
Dark brown with pink crystals of felspar
    Kempegandan Koppal Quarries, district of **Mandya**
Presented by Dr W. F. Smeeth, Bangalore
    Weight per cub. ft. 149 lbs.

670. SIAM PORPHYRY. *Tuff*
Light green and brown mottled, close-grained
Porphyry Quarries, near **Bangkok**
Presented by E. Bock, Esq., Bangkok
Weight per cub. ft. 175 lbs.

671. SIAM TUFF. *Andesite-Tuff*
Pinkish brown and yellow mottled, close-grained
Tuff Quarries, near **Bangkok**
Presented by E. Bock, Esq., Bangkok
Weight per cub. ft. 175 lbs.

672. CHINESE PORPHYRY. *Porphyrite*
Dark blue-grey, medium-grained
Kowloon Quarries, near **Hong Kong**
Presented by Messrs Leigh and Orange, Hong Kong
Weight per cub. ft. 179 lbs.

673. CHINESE GREEN STONE. *Tuff*
Light green, close-grained
Green Stone Quarries, near **Ning-po**
Presented by A. E. Algar, Esq., Shanghai
Weight per cub. ft. 153 lbs.

674. CHINESE RED STONE. *Tuff*
Light pink, medium-grained
Red Stone Quarries, near **Ning-po**
Presented by A. E. Algar, Esq., Shanghai
Weight per cub. ft. 155 lbs.

675. CHINESE WHITE STONE. *Tuff*
Light grey, coarse-grained
White Stone Quarries, near **Ning-po**
Presented by A. E. Algar, Esq., Shanghai
Weight per cub. ft. 155 lbs.

676. AUSTRALIAN TUFF.
Light pink with dark specks of scoria
Porphyry Quarries, near **Brisbane**
Presented by The Government of Queensland
Weight per cub. ft. 138 lbs.

677. AUSTRALIAN TUFF.
Light purplish pink with dark scoria specks
  Porphyry Quarries, near **Brisbane**
Presented by The Government of Queensland
  Weight per cub. ft. 138 lbs.

678. NEW ZEALAND BASALT. *OLIVINE-BASALT*
Dark grey, compact
  Paerata Quarries, near **Auckland**
Presented by Messrs Wm Parkinson & Co., Auckland
  Weight per cub. ft. 170 lbs.

679. NEW ZEALAND BASALT. *BASALT*
Dark grey, very dense
  Central Otago Plains (Boulders), **Otago**
  Weight per cub. ft. 178 lbs.

680. NEW ZEALAND BASALT. *BASALT*
Purplish grey, dense
  McBride's Quarries, **Timaru**
Presented by S. M. McBride, Esq., Timaru
  Weight per cub. ft. 140 lbs.

681. NEW ZEALAND TUFF. *VOLCANIC AGGLOMERATE*
Greenish grey, fragmental
  Port Chalmers Quarries, near **Dunedin**
  Weight per cub. ft. 161 lbs.

682. ST KITTS FIRE STONE. *HYPERSTHENE-ANDESITE*
Light purple, tufaceous
  Stone Quarries, near Basseterre,
     St Christopher, West Indies
  Weight per cub. ft. 123·5 lbs.

683. ST KITTS HARD STONE. *HYPERSTHENE-ANDESITE*
Light grey, compact, basaltic
  Stone Quarries, near Basseterre,
     St Christopher, West Indies
  Weight per cub. ft. 161·5 lbs.

684. GRENADA BASALT. *BASALT-TUFF*
Dark brown mottled, cellular
  Basalt Quarries, St George, **Grenada, West Indies**
  Weight per cub. ft. 150 lbs.

685. ST LUCIA BLUE STONE. *HYPERSTHENE-ANDESITE*
Blue-grey, compact
> Boulders on Surface, **St Lucia, West Indies**
> Weight per cub. ft. 170·5 lbs.

686. ST LUCIA CUT STONE. *TUFF*
Greenish grey, tufaceous
> Stone Quarries, **St Lucia, West Indies**
> Weight per cub. ft. 142·5 lbs.

687. ST LUCIA CUT STONE. *HORNBLENDE-ANDESITE*
Light fawn colour with black specks, tufaceous
> Stone Quarries, **St Lucia, West Indies**
> Weight per cub. ft. 138 lbs.

688. ST LUCIA STONE. *HORNBLENDE-DACITE*
Greenish grey, semi-compact
> Stone Quarries, **St Lucia, West Indies**
> Weight per cub. ft. 170·5 lbs.

689. MADEIRA LAVA. *BASALT*
Light purple, semi-compact
> Small quarries all over the **Island of Madeira**
> Weight per cub. ft. 171 lbs.   Spec. grav. 2·9

690. MADEIRA LAVA. *BASALT*
Light pinkish purple, cellular
> Camara de Lobos Quarries, **Caxäo**
> Weight per cub. ft. 121 lbs.   Spec. grav. 1·94

691. MADEIRA LAVA. *BASALT-TUFF*
Brownish red, coarse, cellular
> Quinta Granda Quarries, **Cabo Giraõ**
> Weight per cub. ft. 113·5 lbs.   Spec. grav. 1·82

692. MADEIRA LAVA. *BASALT-TUFF*
Dark brown, cellular
> Quinta Granda Quarries, **Cabo Giraõ**
> Weight per cub. ft. 162 lbs.   Spec. grav. 2·6

693. MADEIRA LAVA. *BASALT-TUFF*
Light yellowish brown, cellular
> Quinta Granda Quarries, **Cabo Giraõ**
> Weight per cub. ft. 109 lbs.   Spec. grav. 1·75

694. MADEIRA LAVA. *TRACHYTE*
White with buff specks, semi-compact
    Porto Santo Quarries, **Island of Porto Santo**
Weight per cub. ft.

695. CANARY ISLAND LAVA. *TRACHYTE-TUFF*
        (agglomerate)
Dark grey with black scoriaceous fragments
    Arnaeas Quarries, **Grand Canary**
Presented by Messrs Hamilton & Co., Teneriffe
    Weight per cub. ft. 147 lbs.

696. CANARY ISLAND LAVA. *TRACHYTE*
Dark purplish brown
        San Lorenzo en Niletso Quarries, **Grand Canary**
Presented by Messrs Hamilton & Co., Teneriffe
    Weight per cub. ft. 138 lbs.

697. CANARY ISLAND LAVA. *TRACHYTE*
Dark grey
        Iguesta de Andrés Quarries, **Teneriffe**
Presented by Messrs Hamilton & Co., Teneriffe
    Weight per cub. ft. 161·5 lbs.

698. CANARY ISLAND LAVA. *TRACHYTE*
Pink, slightly cellular
        Tajao Quarries, **South Teneriffe**
Presented by Messrs Hamilton & Co., Teneriffe
    Weight per cub. ft. 138 lbs.

699. CANARY ISLAND LAVA. *TRACHYTE-TUFF*
Light buff with brown specks
        Valley of Ajagua Quarries, **South Teneriffe**
Presented by Messrs Hamilton & Co., Teneriffe
    Weight per cub. ft. 129 lbs.

700. CANARY ISLAND TUFF. *TRACHYTE-TUFF*
Light yellow with brown fragmentary spots, cellular
        Tuff Quarries, Las Palmas, **Grand Canary**
    Weight per cub. ft. 85½ lbs.

701. ST HELENA TUFF. *TUFF*
Bright red, cellular
        Tuff Quarries, **St Helena**
Collected by Messrs Solomon & Co., St Helena
    Weight per cub. ft. 95 lbs.

702.   ST HELENA TUFF. *TUFF*
Pinkish brown, fairly compact
        Tuff Quarries, **St Helena**
Collected by Messrs Solomon & Co., St Helena
        Weight per cub. ft. 111 lbs.

703.   MAURITIUS BASALT. *OLIVINE-BASALT*
Grey, hard, somewhat vesicular
        Cassis Quarries, **Port Louis**
Presented by Messrs Ireland, Fraser & Co., Mauritius
        Weight per cub. ft. 160 lbs.

704.   SOUTH AMERICAN TUFF. *VOLCANIC TUFF*
Pink, cellular
        Huaris Quarries, Arequipa, **Peru**
Presented by G. Stafford, Esq., Arequipa
        Weight per cub. ft. 85 lbs.

705.   SOUTH AMERICAN TUFF. *VOLCANIC TUFF*
Light cream, cellular
        Anashuaico Quarries, Arequipa, **Peru**
Presented by G. Stafford, Esq., Arequipa
        Weight per cub. ft. 84 lbs.

706.   SOUTH AMERICAN TUFF. *VOLCANIC TUFF*
Light grey, close-grained
        Chilina Quarries, Arequipa, **Peru**
Presented by G. Stafford, Esq., Arequipa
        Weight per cub. ft. 90 lbs.

707.   SOUTH AMERICAN TUFF. *VOLCANIC TUFF*
Light pinkish yellow, close-grained
        Chilina Quarries, Arequipa, **Peru**
Presented by G. Stafford, Esq., Arequipa
        Weight per cub. ft. 90 lbs.

708.   SOUTH AMERICAN TUFF. *VOLCANIC TUFF*
Greyish white, cellular
        Chilina Quarries, Arequipa, **Peru**
Presented by G. Stafford, Esq., Arequipa
        Weight per cub. ft. 84 lbs.

METAMORPHIC ROCKS.

709. CENTRAL AFRICAN SOAP STONE
Light greenish yellow, soapy to the touch
Limbue Quarries, shores of **Lake Nyasa**
Presented by The Universities' Mission, Likoma
Weight per cub. ft. 176 lbs.

710. WEST AFRICAN QUARTZITE. *QUARTZITE*
Dull whitish red, fine-grained
Dodowa Quarries, **Gold Coast**
Presented by The Gold Coast Colony, Accra
Weight per cub. ft. 147 lbs.

711. WEST AFRICAN QUARTZITE. *QUARTZITE*
Grey with white bands, fine-grained
Dodowa Quarries, **Gold Coast**
Presented by The Gold Coast Colony, Accra
Weight per cub. ft. 152 lbs.

712. WEST AFRICAN QUARTZITE. *QUARTZITE*
Greenish grey, fine-grained
Dodowa Quarries, **Gold Coast**
Presented by The Gold Coast Colony, Accra
Weight per cub. ft. 151 lbs.

713. WEST AFRICAN QUARTZITE. *QUARTZITE*
Light grey, nearly white, fine-grained
Aburi Quarries, **Gold Coast**
Presented by The Gold Coast Colony, Accra
Weight per cub. ft. 152 lbs.

714. SOUTH AFRICAN DIORITE
Dark brownish grey, compact, foliated
   Stone Quarries, near Pretoria, **Transvaal**
Presented by J. J. Kirkness, Esq., Pretoria
   Weight per cub. ft. 171 lbs.
   Used for foundations of Law Courts, Pretoria

715. INDIAN POTSTONE. *MICA-TALC-ROCK*
Dark grey, close-grained
   Manhalli Quarries, Heggadadevankote, **Mysore**
Presented by Dr W. F. Smeeth, Bangalore
   Weight per cub. ft. 170 lbs.

716. INDIAN GREEN QUARTZITE. *QUARTZITE with*
            pale green Mica
Bright green, semi-transparent, fine-grained
   Belavadi Quarries, Kadur, **Mysore**
Presented by Dr W. F. Smeeth, Bangalore
   Weight per cub. ft. 153 lbs.

717. INDIAN LIMESTONE. *CRYSTALLINE LIMESTONE*
Dark grey with white veins
   Akkikalgudda Quarries, near Huliyar,
                                    Tumkur, Mysore
Presented by Dr W. F. Smeeth, Bangalore
   Weight per cub. ft. 161·5 lbs.

718. CEYLON DOLOMITE. *COARSELY CRYSTALLINE*
            DOLOMITE
White with few black specks, highly crystalline
   Dolomite Quarries, near Kandy, **Central Province**
Presented by The Mineral Survey, Colombo
   Weight per cub. ft. 171 lbs.

719. PERSIAN SCHIST.
Pinkish grey, compact
   Stone Quarries, Kuh-i-Sufi Mountain, near **Ispahan**
Presented by Dr D. W. Carr, Ispahan
   Weight per cub. ft. 178 lbs.

CAMBRIAN AND SILURIAN.

720. DANISH SANDSTONE. *CAMBRIAN*
Light brown, medium-grained, banded
    Frederiks Quarries, near Nexö, **Isle of Bornholm**
Presented by Messrs H. and J. Larsen, Copenhagen
    Weight per cub. ft. 155 lbs. Spec. grav. 2·50

721. RUSSIAN LIMESTONE. *SILURIAN*
Light buff, very fine-grained
    Limestone Quarries, near **Reval**
Presented by G. A. List, Esq., Moscow
    Weight per cub. ft. 171 lbs.

722. SWEDISH LIMESTONE. *LOWER SILURIAN*
Greenish grey, hard, compact
    Lanna Quarries, near **Nerike**
    Weight per cub. ft. 171 lbs.
    Crushing strain per sqr. ft. 1181 tons

723. SWEDISH LIMESTONE. *LOWER SILURIAN*
Light pink, hard, compact
    Borgholm Quarries, **Isle of Öland**
    Weight per cub. ft. 176 lbs.

724. TABLE MOUNTAIN SANDSTONE. *UPPER*
    SILURIAN (Cape System)
Yellow, crystalline, coarse-grained
    The Quarries, Table Mountain, **Cape of Good Hope**
    Weight per cub. ft. 163 lbs.

725. TABLE MOUNTAIN SANDSTONE. *UPPER*
SILURIAN (Cape System)
Light greenish grey, medium-grained
  Simon's Bay Quarries, **Cape of Good Hope**
Presented by Sir John Jackson, LL.D., London
  Weight per cub. ft. 147 lbs.

726. TABLE MOUNTAIN SANDSTONE. *UPPER*
SILURIAN (Cape System)
Light purple, close-grained
  Simon's Bay Quarries, **Cape of Good Hope**
Presented by Sir John Jackson, LL.D., London
  Weight per cub. ft. 166 lbs.

727. TABLE MOUNTAIN SANDSTONE. *UPPER*
SILURIAN (Cape System)
Light yellow and orange-coloured, close-grained
  Simon's Bay Quarries, **Cape of Good Hope**
Presented by Sir John Jackson, LL.D., London
  Weight per cub. ft. 157 lbs.

728. SOUTH AFRICAN RED FREESTONE. *SILURIAN*
Dark terra-cotta, fine-grained sandstone
  Buiskop Quarries, Waterberg, **Transvaal**
Presented by F. Watson, Esq., Game Pass, Natal
  Weight per cub. ft. 142 lbs.

729. INDIAN SANDSTONE. *SILURIAN*
Light yellow, fine-grained
  Rowali Quarries, near **Allahábád**
Presented by The Bengal Stone Co., Calcutta
  Weight per cub. ft. 142·5 lbs.

730. INDIAN SANDSTONE. *SILURIAN*
Light yellow, fine-grained
  Mirzapore Quarries, **Mirzapore**
Presented by The Bengal Stone Co., Calcutta
  Weight per cub. ft. 142·5 lbs.

731. INDIAN SANDSTONE. *SILURIAN*
Pinkish yellow, fine-grained
  Mirzapore Quarries, **Mirzapore**
Presented by The Bengal Stone Co., Calcutta
  Weight per cub. ft. 152 lbs.

732. INDIAN LIMESTONE. *Silurian*
Purple, fine-grained, compact stone
Fibyingyi Quarries, **Upper Burma**
Presented by Messrs R. Bagchi & Co., Rangoon
Weight per cub. ft. 180 lbs.

733. SIAM MARBLE. *Pre-Cambrian*
Grey and white streaked, compact
Limestone Quarries, near **Bangkok**
Presented by E. Bock, Esq., Bangkok
Weight per cub. ft. 193 lbs.

734. AUSTRALIAN SANDSTONE. *Lower Cambrian*
Light yellow and brown mottled, coarse-grained
Mitcham Quarries, near Adelaide, **South Australia**
Weight per cub. ft. 152 lbs.

735. AUSTRALIAN MUDSTONE. *Cambrian*
Drab and brown, banded, compact, fine-grained
Tapley Hill Quarries, near Adelaide, **South Australia**
Weight per cub. ft. 147 lbs.

736. AUSTRALIAN SANDSTONE. *Cambrian*
Cream with orange blotches, medium-grained
Shoeak Hill Quarries, near Belair, **South Australia**
Weight per cub. ft. 142·5 lbs.

737. NORTH AMERICAN SANDSTONE. *Upper*
Cambrian (Potsdam Series)
Bright red, medium-grained stone
Potsdam Quarries, **New York, U.S.A.**
Presented by The Potsdam Red Sand Stone Co., Potsdam
Weight per cub. ft. 162 lbs.
Spec. grav. 2·604. Absorp. 2·08 %

738. NORTH AMERICAN SANDSTONE. *Upper*
Cambrian (Potsdam Series)
Light red, slightly banded, medium-grained stone
Potsdam Quarries, **New York, U.S.A.**
Weight per cub. ft. 162 lbs.
Spec. grav. 2·604. Absorp. 2·08 %

739. NORTH AMERICAN SANDSTONE.   *Upper*
CAMBRIAN (Potsdam Series)
Light salmon-coloured, even, fine-grained
    Banning Quarries, Kettle River, **Minnesota, U.S.A.**
Presented by The Barber Asphalt Paving Co., Minneapolis
    Chemical Composition (Booth, Garrett & Blair, Philadelphia):
SiO_2 98·92, Al_2O_3 0·61, Fe_2O_3 0·17, H_2O 0·30, = 100·00
    Weight per cub. ft. 154·3 lbs. Spec. grav. 2·47. Absorp. 4·88 %
    Crushing strain per sqr. ft. 913 tons

740. NORTH AMERICAN SANDSTONE.   *Upper*
CAMBRIAN (Potsdam Series)
Light salmon-coloured, even, medium-grained
    Sandstone Quarries, Kettle River, **Minnesota, U.S.A.**
Presented by The Kettle River Quarries Co., Minneapolis
    Chemical Composition :
SiO_2 97·10, Al_2O_3 2·20, MgO 0·10, CaO 0·60, = 100·00
Weight per cub. ft. 139 lbs. Crushing strain per sqr. ft. 806·5 tons

741. NORTH AMERICAN SANDSTONE.   *Upper*
CAMBRIAN (Potsdam Series)
Light fawn-coloured, fine-grained
    Cofax Quarries, Cofax, **Wisconsin, U.S.A.**
Presented by The Geological and Natural History Survey, Madison,
    Wisconsin
Weight per cub. ft. 162·5 lbs. Spec. grav. 2·6. Absorp. 10 %
    Crushing strain per sqr. ft. 257 tons

742. NORTH AMERICAN SANDSTONE.   *Lower*
SILURIAN
Cream-coloured, fine-grained
    Fountain City Quarries, **Wisconsin, U.S.A.**
Presented by The Geological and Natural History Survey, Madison,
    Wisconsin

743. SOUTH AMERICAN QUARTZITE. *PRE-CAMBRIAN*
Very light fawn-coloured, coarse, crystalline
    Estancia Chapademalal Quarries, Mar del Plata,
                                 Buenos Ayres
Presented by Señor Don M. A. Martinez de Hoz, London
    Weight per cub. ft. 161·5 lbs.

744. SOUTH AMERICAN QUARTZITE. *PRE-CAMBRIAN*
Light cream, fine-grained
    Estancia Chapademalal Quarries, Mar del Plata,
                                 Buenos Ayres
Presented by Señor Don M. A. Martinez de Hoz, London
    Weight per cub. ft. 165 lbs.

DEVONIAN AND OLD RED SANDSTONE.

745. BELGIAN LIMESTONE. *DEVONIAN*
Dark blue-grey, fine-grained
    Meuse Quarries, **Hainault**
    Weight per cub. ft. 176 lbs.

746. BELGIAN LIMESTONE. *DEVONIAN*
Dark grey and white mottled, fine-grained
    Feluy-Arquennes Quarries, **Hainault**
    Weight per cub. ft. 176 lbs.

747. BELGIAN LIMESTONE. *DEVONIAN*
Light blue-grey, fine-grained
    Gawday à Samson Quarries, **Hainault**
    Weight per cub. ft. 176 lbs.

748. BELGIAN LIMESTONE. *DEVONIAN*
Blue-grey, highly crystalline, medium-grained, lenticular
    Frasnes-les-Couvin Quarries, **Hainault**
    Weight per cub. ft. 180·5 lbs.

749. BELGIAN LIMESTONE. *DEVONIAN*
Grey and white mottled, highly crystalline, lenticular
    Frasnes Quarries, **Hainault**
    Weight per cub. ft. 176 lbs.

749A. DANISH SANDSTONE. *DEVONIAN*
Light brown, medium-grained, laminated
    Frederiks Quarries, near Nexö, **Isle of Bornholm**
Presented by Messrs H. and J. Larsen, Copenhagen

750. GERMAN SANDSTONE. *Devonian*
Light purplish red, medium-grained
    Daufenback Quarries, Eifel, **Rhine**
Presented by Michel Lickes, Esq., Cologne
    Weight per cub. ft. 142·5 lbs.

751. GERMAN SANDSTONE. *Devonian*
Light drab, medium-grained, slightly micaceous
    Cordel Quarries, Eifel, **Rhine**
Presented by Michel Lickes, Esq., Cologne
    Weight per cub. ft. 133 lbs.

752. GERMAN SANDSTONE. *Devonian*
Greenish drab, fine-grained
    Borner Quarries, Eifel, **Rhine**
Presented by Michel Lickes, Esq., Cologne
    Weight per cub. ft. 123 lbs.

753. GERMAN SANDSTONE. *Devonian*
Pale greenish drab, fine-grained
    Hofweller Quarries, near Trier, Eifel, **Rhine**
Presented by Michel Lickes, Esq., Cologne
    Weight per cub. ft. 142·5 lbs.

754. NORTH AMERICAN SANDSTONE (*Warsaw Bluestone*). *Devonian*
Pale green, fine-grained
    Rock Glen Quarries, **New York, U.S.A.**
Presented by The Warsaw Blue Stone Co., New York
        Chemical Composition (Harris):
SiO_2 76·50, Al_2O_3 14·75, Fe_2O_3 6·35, H_2O 2·00,   =99·60
    Weight per cub. ft. 158 lbs.   Crushing strain per sqr. ft.
    1284 tons (D. W. Fayle, Lt.-Col.)

755. CANADIAN LIMESTONE. *Devonian*
Light cream, fine-grained
    Beamsville Quarries, Lincoln County, **Ontario**
Presented by Hon. W. Gibson, Ottawa
        Chemical Composition:
SiO_2 0·33, Al_2O_3 2·74,   FeO 0·49,   MnO 0·33,   MgO 13·41,
CaO 37·72, H_2O 0·27, CO_2 44·68, P_2O_5 0·02, SO_3 0·02,
=100·01
    Weight per cub. ft. 150 lbs.

CARBONIFEROUS.

756. BELGIAN LIMESTONE (*Petit Granit*). *CARBON-
IFEROUS LIMESTONE*
*Black compact stone, spotted with debris of small shells, corals and
crinoids*
  Perlonjour Quarries, Soignies, **Hainault**
Presented by E. de Baatard, Esq., Soignies
  Weight per cub. ft. 180 lbs.

757. BELGIAN LIMESTONE (*Petit Granit*). *CARBON-
IFEROUS LIMESTONE*
Black, compact, fine-grained, crinoidal, admits of good polish
  The Quarries, **Anthiseus**
  Weight per cub. ft. 180 lbs.

758. BELGIAN LIMESTONE. *CARBONIFEROUS LIMESTONE*
Blue-grey, compact, slightly crystalline stone
  Méhaigne Quarries, near **Namur**
  Weight per cub. ft. 180 lbs.

759. BELGIAN LIMESTONE. *CARBONIFEROUS LIMESTONE*
Dark blue-grey, very fine, compact
  Méhaigne Quarries, near **Namur**
  Weight per cub. ft. 180 lbs.

760. BELGIAN LIMESTONE. *CARBONIFEROUS LIMESTONE*
Light blue-grey, fine, crystalline
  Méhaigne Quarries, near **Namur**
  Weight per cub. ft. 180 lbs.

761. RUSSIAN LIMESTONE. *CARBONIFEROUS*
Yellow, fine-grained, slightly banded
    Quarries near Kreistadt, Govt. **Podolosk**
Presented by G. A. List, Esq., Moscow
    Weight per cub. ft. 161·5 lbs.

762. RUSSIAN LIMESTONE. *CARBONIFEROUS*
Light yellow, medium-grained, chalky
    Kreise Wenew Quarries, **Tula**
Presented by G. A. List, Esq., Moscow
    Weight per cub. ft. 133 lbs.

763. RUSSIAN LIMESTONE. *CARBONIFEROUS*
Light buff, fine-grained
    Barybine Quarries, near **Tula**
Presented by G. A. List, Esq., Moscow
    Weight per cub. ft. 171 lbs.

764. RUSSIAN SANDSTONE. *CARBONIFEROUS*
Pinkish yellow, fine-grained, crystalline
    Quarries near Murom, Govt. **Nizhni Novgorod**
Presented by G. A. List, Esq., Moscow
    Weight per cub. ft. 180 lbs.

765. SWISS GRIT STONE. *CARBONIFEROUS*
Dark grey, slightly micaceous, close-grained
    Finhaut Quarries, Trient Valley, **Valais**
    Weight per cub. ft. 171 lbs.

766. TURKISH SANDSTONE. *CARBONIFEROUS*
Yellowish grey, fine-grained
    Lefke Quarries, **Asia Minor**

767. AUSTRALIAN MARBLE. *CARBONIFEROUS*
Yellow and pink veined, compact
    Glanmore Quarries, Rockhampton, **Queensland**
Presented by The Government of Queensland
    Weight per cub. ft. 171 lbs.

768. AUSTRALIAN MARBLE. *CARBONIFEROUS*
Blue-grey streaked, compact
    Marmor Quarries, near Rockhampton, **Queensland**
Presented by The Government of Queensland
    Weight per cub. ft. 171 lbs.

769. AUSTRALIAN SANDSTONE. *CARBONIFEROUS*
Light straw-coloured, fine-grained
Warrigal Creek Quarries, 160 miles W. of Townsville,
Queensland
Presented by The Government of Queensland
Weight per cub. ft. 133 lbs.

770. AUSTRALIAN SANDSTONE. *CARBONIFEROUS*
Light drab, slightly banded, coarse
Murphy's Creek Quarries, 70 miles W. of **Brisbane**
Presented by The Government of Queensland
Weight per cub. ft. 152 lbs.

771. AUSTRALIAN SANDSTONE. *CARBONIFEROUS*
Light brown, medium-grained
Stanwell Quarries, Rockhampton, **Queensland**
Presented by The Government of Queensland
Weight per cub. ft. 142·5 lbs.

772. AUSTRALIAN FREESTONE. *CARBONIFEROUS*
Light dove-coloured, fine-grained sandstone
Stawell Quarries, **Victoria**
Weight per cub. ft. 152 lbs.

773. NORTH AMERICAN LIMESTONE. *SUB-CARBON-IFEROUS*
Light grey, fine-grained
Limestone Quarries, Bedford, **Indiana, U.S.A.**
Chemical Composition (*Geol. Survey*, Indiana):
SiO_2 0·50, Fe_2O_3 0·98, K_2O 0·31, Na_2O 0·40, H_2O 0·96,
$CaCO_3$ 96·60, $MgCO_3$ 0·13, =99·88
Weight per cub. ft. 152·39 lbs. Spec. grav. 2·44. Absorp. 3·16 %
Crushing strain per sqr. ft. 655 tons

774. NORTH AMERICAN SANDSTONE. *SUB-CARBON-IFEROUS*
Greenish grey, medium-grained, slightly banded
Sandstone Quarries, Berea, **Ohio, U.S.A.**
Chemical Composition (G. A. Gilmore):
SiO_2 84·40, Al_2O_3 7·49, Fe_2O_3 3·87, MgO 2·11, CaO 0·74,
K_2O 0·24, Na_2O 0·56, =99·41
Weight per cub. ft. 134 lbs.
Spec. grav. 2·14. Absorp. 1·21 %

775. NORTH AMERICAN SANDSTONE. *Sub-Carbon-iferous* (*Waverley*)

Greenish grey, fine-grained
> Sandstone Quarries, Euclid, **Ohio, U.S.A.**
> Chemical Composition :

SiO_2 91·00, Al_2O_3 5·20, Fe_2O_3 1·47, MgO 0·28, H_2O &c. 1·80, =99·75
> Weight per cub. ft. 142·5 lbs.

776. NORTH AMERICAN SANDSTONE. *Sub-Carbon-iferous*

Light grey, fine-grained
> Great Canyon Quarries, Cleveland, **Ohio, U.S.A.**
Presented by The Cleveland Stone Co., Cleveland
> Crushing strain per sqr. ft. 643 tons

777. CANADIAN LIMESTONE. *Carboniferous*

Dark grey, fine-grained, crystalline
> Limestone Quarries, near **Montreal**
Presented by Messrs Drummond, McCall & Co., Montreal
> Weight per cub. ft. 171 lbs.

778. CANADIAN SANDSTONE. *Millstone Grit*

Light greenish grey, slightly micaceous, fine-grained
> Chaleur Bay Quarries, Gloucester Co., **New Brunswick**
Presented by The Read Stone Co., Stonehaven, N.B.
> Weight per cub. ft. 142·5 lbs.

779. CANADIAN SANDSTONE. *Millstone Grit*

Light greenish grey, very fine-grained
> Chaleur Bay Quarries, Gloucester Co., **New Brunswick**
Presented by The Read Stone Co., Stonehaven, N.B.
> Weight per cub. ft. 152 lbs.

780. CANADIAN SANDSTONE. *Millstone Grit*

Brownish grey, medium-grained
> Miramichi Quarry, Indian-Town, **New Brunswick**
Presented by The Miramichi Quarry Co., Ltd., Renous Bridge, N.B.

781. CANADIAN SANDSTONE. *Millstone Grit*

Dark greenish brown, medium-grained
> Miramichi Quarries, Indian-Town, **New Brunswick**
Presented by Messrs Drummond, McCall & Co., Montreal

782. CANADIAN SANDSTONE. *MILLSTONE GRIT*
Light dove-coloured, medium-grained
Rockport Quarries, near Sackville, **New Brunswick**
Presented by The Read Stone Co., Stonehaven, N.B.
Weight per cub. ft. 152 lbs.

783. CANADIAN SANDSTONE. *MILLSTONE GRIT*
Light brown, medium-grained
Rockport Quarries, near Sackville, **New Brunswick**
Presented by The Read Stone Co., Stonehaven, N.B.
Weight per cub. ft. 152 lbs.

784. CANADIAN SANDSTONE. *MILLSTONE GRIT*
Brownish purple, medium-grained
Woodpoint Quarries, near Sackville, **New Brunswick**
Presented by The Read Stone Co., Stonehaven, N.B.
Weight per cub. ft. 152 lbs.

785. CANADIAN SANDSTONE. *COAL FORMATION*
Light greenish grey, slightly micaceous, medium-grained
Pictou Quarries, **Nova Scotia**
Presented by The Pictou Quarries Co., Pictou, Nova Scotia
Chemical Composition (C. E. Bogardus):
SiO_2 74·97, Al_2O_3 10·27, FeO 7·02, MnO 0·31, CaO 2·42,
H_2O 2·31, =97·30
Weight per cub. ft. 150 lbs. Spec. grav. 2·63

786. CANADIAN SANDSTONE. *MILLSTONE GRIT*
Light blue-grey, slightly banded, fine-grained
Wallace Quarries, Cumberland Co., **Nova Scotia**
Weight per cub. ft. 147·3 lbs.
Crushing strain per sqr. ft. 566 tons (McGill University)

787. CANADIAN SANDSTONE. *MILLSTONE GRIT*
Light brown, medium-grained
Wallace Quarries, Cumberland Co., **Nova Scotia**
Weight per cub. ft. 146 lbs.
Crushing strain per sqr. ft. 760 tons (McGill University)

PERMIAN.

788. GERMAN SANDSTONE. *Lower Permian*
Yellow and brown banded, fine-grained, slightly micaceous
Medarder Quarries, near **Kreuznach**
Presented by Michel Lickes, Esq., Cologne
Weight per cub. ft. 142 lbs.

789. GERMAN SANDSTONE. *Lower Permian*
Yellow and brown banded, fine-grained
Oderheim Quarries, **Hesse**
Presented by Michel Lickes, Esq., Cologne
Weight per cub. ft. 133 lbs.

790. GERMAN SANDSTONE. *Lower Permian*
Drab with brown ferruginous spots, medium-grained
Oderheim Quarries, **Hesse**
Presented by Michel Lickes, Esq., Cologne
Weight per cub. ft. 133 lbs.

791. RUSSIAN LIMESTONE. *Permian*
Light yellow, slightly mottled, compact
St Koloma Quarries, Govt. **Moscow**
Presented by G. A. List, Esq., Moscow
Weight per cub. ft. 171 lbs.

792. SOUTH AFRICAN SANDSTONE (*Greytown Stone*).
Ecca Beds, Lower Karroo
Cream with small ferruginous spots, medium-grained
Waterfeet Quarries, near Greytown, **Natal**
Presented by Messrs Cornelius and Hollis, Durban
Weight per cub. ft. 171 lbs.

793. SOUTH AFRICAN SANDSTONE (*Rosetta Stone*).
Ecca Beds, Lower Karroo
Pale greenish grey, fine-grained
Downing Farm Quarries, Rosetta, **Natal**
Presented by Messrs Cornelius and Hollis, Durban
Weight per cub. ft. 142 lbs.

794. SOUTH AFRICAN SANDSTONE (*Rosetta Stone*).
ECCA BEDS, LOWER KARROO
Pale greenish grey, fine-grained
  Hunwick Quarries, Rosetta, **Natal**
Presented by Messrs Cornelius and Hollis, Durban
Weight per cub. ft. 142 lbs.

795. SOUTH AFRICAN SANDSTONE (*Rosetta Stone*).
ECCA BEDS, LOWER KARROO
Greenish grey, medium-grained
  Downing Farm Quarries, Rosetta, **Natal**
Presented by Messrs Jesse Smith & Son, Pietermaritzburg
Weight per cub. ft. 142 lbs.

796. SOUTH AFRICAN SANDSTONE (*Town Bush Stone*). *ECCA BEDS, LOWER KARROO*
Dark drab, coarse-grained
  Town Bush Quarries, Pietermaritzburg, **Natal**
Presented by Messrs Jesse Smith & Son, Pietermaritzburg
Weight per cub. ft. 163 lbs.

797. SOUTH AFRICAN SANDSTONE (*Avoca Stone*).
ECCA BEDS, LOWER KARROO
Light brown, slightly banded, medium-grained
  Effingham Quarries, near Avoca, **Natal**
Presented by Messrs Adlam Reid & Co., Durban
Weight per cub. ft. 166 lbs.

798. SOUTH AFRICAN FREESTONE. *ECCA BEDS, LOWER KARROO*
Light yellowish brown, fine-grained sandstone
  Kockemoed Quarries, Klerksdorp, **Transvaal**
Presented by Frank Watson, Esq., Game Pass, Natal
Weight per cub. ft. 133 lbs.

799. SOUTH AFRICAN FREESTONE. *ECCA BEDS, LOWER KARROO*
Light drab, fine-grained sandstone
  Waring and Gillow's Quarries, East Rand, **Transvaal**
Presented by Frank Watson, Esq., Game Pass, Natal
Weight per cub. ft. 152 lbs.

800. SOUTH AFRICAN FREESTONE. *ECCA BEDS, LOWER KARROO*
Light yellowish grey, medium-grained sandstone
  Machavie Quarries, Potchefstroom, **Transvaal**
Presented by Frank Watson, Esq., Game Pass, Natal
Weight per cub. ft. 142 lbs.

801. SOUTH AFRICAN FREESTONE. *Ecca Beds,*
Lower Karroo
Light yellowish drab, medium-grained sandstone
    Heidelberg Quarries, South Rand, **Transvaal**
Presented by Frank Watson, Esq., Game Pass, Natal
Weight per cub. ft. 163 lbs.

802. SOUTH AFRICAN SANDSTONE. *Ecca Beds,*
Lower Karroo
Light dove-coloured, very fine-grained
    Orange River Quarries, near Johannesburg, **Transvaal**
Presented by Messrs Weightman and Avery, Johannesburg
Weight per cub. ft. 147 lbs.

803. SOUTH AFRICAN SANDSTONE. *Ecca Beds,*
Lower Karroo
Light delicate pink, fine-grained
    The Quarries, Pretoria, **Transvaal**
Presented by J. J. Kirkness, Esq., Pretoria
Weight per cub. ft. 138 lbs.

804. SOUTH AFRICAN SANDSTONE. *Beaufort*
Beds, Middle Karroo
Brown and pink mottled, medium-grained
    Gombio Quarries, **Cape of Good Hope**
Weight per cub. ft. 142·5 lbs.

805. SOUTH AFRICAN FREESTONE. *Beaufort*
Beds, Middle Karroo
Greenish grey, medium-grained sandstone
    Queenstown Quarries, **Cape of Good Hope**
Weight per cub. ft. 133 lbs.

806. SOUTH AFRICAN SANDSTONE. *Beaufort*
Beds, Middle Karroo
Light purplish brown, fine-grained
    Kivelagher Quarries, near East London, **Cape of**
                                 Good Hope
Weight per cub. ft. 142 lbs.

807. TASMANIAN FREESTONE. *Permo-Carboniferous*
Light drab, fine-grained, calcareous sandstone
    Spring Bay Quarries, **Swan Port**
Weight per cub. ft. 163 lbs.

TRIAS.

808. GERMAN SANDSTONE. *Lower Bunter*
Dull red, pebbly, with rounded inclusions of clay, medium-grained
    Freudenstadt Quarries, **Würtemberg**
    Weight per cub. ft. 151 lbs.

809. GERMAN SANDSTONE. *Middle Bunter*
Light red, medium-grained
    Freudenstadt Quarries, **Würtemberg**
    Weight per cub. ft. 147 lbs.

810. GERMAN SANDSTONE. *Middle Bunter*
Light chocolate-coloured, medium-grained
    Freudenstadt Quarries, **Würtemberg**
    Weight per cub. ft. 151 lbs.

811. GERMAN SANDSTONE. *Middle Bunter*
Pink and drab, banded, medium-grained
    Freudenstadt Quarries, **Würtemberg**
    Weight per cub. ft. 147 lbs.

812. GERMAN SANDSTONE. *Upper Bunter*
Dull red, fine-grained
    Hausen Quarries, near Leonberg, **Würtemberg**
    Weight per cub. ft. 151 lbs.

813. GERMAN LIMESTONE. *Upper Muschelkalk*
Drab, hard, shelly
    Crailsheim Quarries, **Würtemberg**
    Weight per cub. ft. 142·5 lbs.

814. GERMAN LIMESTONE. *Upper Muschelkalk*
Light brown, sub-crystalline
    Crailsheim Quarries, **Würtemberg**
    Weight per cub. ft. 151 lbs.

815. GERMAN LIMESTONE. *Upper Muschelkalk*
Dark grey, compact
    Rottenburg Quarries, **Würtemberg**
    Weight per cub. ft. 171 lbs.

816. GERMAN SANDSTONE. *Lower Keuper*
Light grey, medium-grained
    Altingen Quarries, Herrenberg, **Würtemberg**
    Weight per cub. ft. 129 lbs.

817. GERMAN SANDSTONE. *Lower Keuper*
Light drab, fine-grained
    Wendelsheim Quarries, Rottenburg, **Würtemberg**
    Weight per cub. ft. 147 lbs.

818. GERMAN SANDSTONE. *Lower Keuper*
Light yellow, fine-grained
    Hochdarf Quarries, near Harb, **Würtemberg**
    Weight per cub. ft. 133 lbs.

819. GERMAN SANDSTONE. *Lower Keuper*
Drab with yellow spots, fine-grained
    Crailsheim Quarries, **Würtemberg**
    Weight per cub. ft. 138 lbs.

820. GERMAN SANDSTONE. *Lower Keuper*
Light brown, fine-grained
    Maulbronn Quarries, **Würtemberg**
    Weight per cub. ft. 152 lbs.

821. GERMAN SANDSTONE. *Lower Keuper*
Light brown, fine-grained
    Schwab Hall Quarries, **Würtemberg**
    Weight per cub. ft. 133 lbs.

822. GERMAN SANDSTONE. *Lower Keuper*
Warm brown, banded, medium-grained
    Maulbronn Quarries, **Würtemberg**
    Weight per cub. ft. 138 lbs.

823. GERMAN SANDSTONE. *MIDDLE KEUPER (SCHILF-SANDSTEIN)*
Brown with purple streaks, fine-grained
    Maulbronn Quarries, **Würtemberg**
Weight per cub. ft. 152 lbs.

824. GERMAN SANDSTONE. *MIDDLE KEUPER (SCHILF-SANDSTEIN)*
Brown, fine-grained
    Heilbronn Quarries, **Würtemberg**
Weight per cub. ft. 133 lbs.

825. GERMAN SANDSTONE. *MIDDLE KEUPER (SCHILF-SANDSTEIN)*
Red and brown mottled, fine-grained
    Maulbronn Quarries, **Würtemberg**
Weight per cub. ft. 147 lbs.

826. GERMAN SANDSTONE. *MIDDLE KEUPER (SCHILF-SANDSTEIN)*
Light yellow with brown spots, coarse-grained
    Rattenburg Quarries, **Würtemberg**
Weight per cub. ft. 142 lbs.

827. GERMAN SANDSTONE. *MIDDLE KEUPER (SCHILF-SANDSTEIN)*
Light grey (nearly white), medium-grained
    Maulbronn Quarries, **Würtemberg**
Weight per cub. ft. 142 lbs.

828. GERMAN SANDSTONE. *MIDDLE KEUPER (SCHILF-SANDSTEIN)*
Light drab, coarse-grained
    Dettenhausen Quarries, Tübingen, **Würtemberg**
Weight per cub. ft. 142 lbs.

829. GERMAN SANDSTONE. *MIDDLE KEUPER (SCHILF-SANDSTEIN)*
Light yellow, fine-grained
    Renningen Quarries, Magstadt, **Würtemberg**
Weight per cub. ft. 119 lbs.

830. GERMAN SANDSTONE. *MIDDLE KEUPER (STUBEN-SANDSTEIN)*
Light drab, medium-grained
    Hirshau Quarries, Rottenburg, **Würtemberg**
Weight per cub. ft. 133 lbs.

831.   GERMAN SANDSTONE. *Middle Keuper (Stuben-sandstein)*
Light drab, coarse-grained
   Nürtingen Quarries, **Würtemberg**
   Weight per cub. ft. 147 lbs.

832.   GERMAN SANDSTONE. *Middle Keuper (Stuben-sandstein)*
Light drab, hard, medium-grained
   Nürtingen Quarries, **Würtemberg**
   Weight per cub. ft. 147 lbs.

833.   GERMAN SANDSTONE. *Middle Keuper (Stuben-sandstein)*
Light drab, medium-grained
   Nürtingen Quarries, **Würtemberg**
   Weight per cub. ft. 147 lbs.

834.   GERMAN SANDSTONE. *Middle Keuper (Stuben-sandstein)*
Light drab, yellow specks, coarse-grained
   Nürtingen Quarries, **Würtemberg**
   Weight per cub. ft. 142 lbs.

835.   GERMAN SANDSTONE. *Middle Keuper (Stuben-sandstein)*
Light drab with yellow spots, coarse-grained
   Nürtingen Quarries, **Würtemberg**
   Weight per cub. ft. 142 lbs.

836.   GERMAN SANDSTONE. *Middle Keuper (Stuben-sandstein)*
Light drab, coarse-grained
   Dettenhausen Quarries, Tübingen, **Würtemberg**
   Weight per cub. ft. 142 lbs.

837.   GERMAN SANDSTONE. *Rhætic*
Light brown, fine-grained
   Pfrondarf Quarries, near Tübingen, **Würtemberg**
   Weight per cub. ft. 152 lbs.

838.   GERMAN SANDSTONE. *Trias*
Purplish brown, coarse-grained
   Olsbrucken Quarries, **Pfalz**
Presented by Michel Lickes, Esq., Cologne
   Weight per cub. ft. 142 lbs.

839. GERMAN SANDSTONE. *TRIAS*
Yellow and brown banded, medium-grained
    Klingenmunster Quarries, **Pfalz**
    Weight per cub. ft. 172·5 lbs.

840. RUSSIAN SANDSTONE. *TRIAS*
Dark red, fine-grained
    Stone Quarries, Govt. Radom, **Poland**
Presented by G. A. List, Esq., Moscow
    Weight per cub. ft. 142·5 lbs.

841. RUSSIAN SANDSTONE.. *TRIAS*
Bright yellow, fine-grained
    Stone Quarries, Govt. Radom, **Poland**
Presented by G. A. List, Esq., Moscow
    Weight per cub. ft. 142·5 lbs.

842. RUSSIAN SANDSTONE. *TRIAS*
Light buff, fine-grained
    Stone Quarries, Govt. Radom, **Poland**
Presented by G. A. List, Esq., Moscow
    Weight per cub. ft. 142·5 lbs.

843. WEST AFRICAN SANDSTONE. *TRIAS*
Yellow banded with red, medium-grained
    Church Quarries, Accra, **Gold Coast**
Presented by The Gold Coast Colony, Accra
    Weight per cub. ft. 128 lbs.

844. WEST AFRICAN SANDSTONE. *TRIAS*
Yellow, slightly banded with pink, fine-grained
    Horse Road Quarries, Accra, **Gold Coast**
Presented by The Gold Coast Colony, Accra
    Weight per cub. ft. 128 lbs.

845. SOUTH AFRICAN SANDSTONE. *STORMBERG*
    SERIES, UPPER KARROO (Rhœtic)
Light yellow, fine-grained
    Platberg Quarries, near Ladybrand,
               Orange River Province
Presented by The Town Council of Ladybrand
    Weight per cub. ft. 133 lbs.

846. SOUTH AFRICAN SANDSTONE. *Stormberg*
Series, Upper Karroo (Rhœtic)
Light yellow, fine-grained
Platberg Quarries, near Ladybrand,
Orange River Province
Presented by The Town Council of Ladybrand
Weight per cub. ft. 133 lbs.

847. SOUTH AFRICAN SANDSTONE. *Stormberg*
Series, Upper Karroo (Rhœtic)
Light yellow, fine-grained
Platberg Quarries, near Ladybrand,
Orange River Province
Presented by The Town Council of Ladybrand
Weight per cub. ft. 133 lbs.

848. SOUTH AFRICAN SANDSTONE. *Stormberg*
Series, Upper Karroo (Rhœtic)
Light yellowish brown, medium-grained
Steinpan Quarries, Kroonstadt,
Orange River Province
Presented by Frank Watson, Esq., Game Pass, Natal
Weight per cub. ft. 133 lbs.

849. SOUTH AFRICAN SANDSTONE. *Pink Forest*
Sandstone. Stormberg Series, Upper Karroo
(Rhœtic)
Light pink, medium-grained
Taba-z-Induna Quarries, near Buluwayo, **Rhodesia**
Presented by The British South Africa Co., London
Weight per cub. ft. 133 lbs.

850. SOUTH AFRICAN SANDSTONE. *White Forest*
Sandstone. Stormberg Series, Upper Karroo
(Rhœtic)
White with delicate pink tinge, fine-grained
Kopje Quarries, Forest Vale, near Buluwayo, **Rhodesia**
Presented by The British South Africa Co., London
Weight per cub. ft. 133 lbs.

851. SOUTH AFRICAN SANDSTONE. *Red Forest*
Sandstone. Stormberg Series, Upper Karroo
(Rhœtic)
Rich salmon-coloured, medium-grained
Pasipas Quarries, near Buluwayo, **Rhodesia**
Presented by The British South Africa Co., London
Weight per cub. ft. 133 lbs.

852. AUSTRALIAN FREESTONE. *HAWKESBURY*
          SANDSTONE. TRIAS
Yellowish brown, medium-grained
    Pyrmont Quarries, Sydney, **New South Wales**
    Weight per cub. ft. 142 lbs.

853. AUSTRALIAN FREESTONE. *HAWKESBURY*
          SANDSTONE. TRIAS
Light yellow, medium-grained
    Pyrmont Quarries, Sydney, **New South Wales**
Presented by The Technological Museum, Sydney
    Weight per cub. ft. 142 lbs.

854. AUSTRALIAN FREESTONE. *HAWKESBURY*
          SANDSTONE. TRIAS
Light drab, medium-grained
    Hunter's Hill Quarries, Sydney, **New South Wales**
    Weight per cub. ft. 142 lbs.

855. AUSTRALIAN FREESTONE. *HAWKESBURY*
          SANDSTONE. TRIAS
Light grey, fine-grained
    Bundanoon Quarries, 60 miles S. of Sydney,
                       New South Wales
Presented by The Technological Museum, Sydney
    Weight per cub. ft. 147 lbs.

856. AUSTRALIAN SANDSTONE. *TRIAS*
Light brown, slightly banded, medium-grained
    Yangan Quarries, near Warwick, **Queensland**
Presented by The Government of Queensland
    Weight per cub. ft. 142·5 lbs.

857. AUSTRALIAN SANDSTONE. *TRIAS*
Light brown, coarse-grained
    Helidon Quarries, 70 miles W. of Brisbane, **Queensland**
Presented by The Government of Queensland
    Weight per cub. ft. 147 lbs.

858. AUSTRALIAN SANDSTONE. *TRIAS*
Light drab, medium-grained
    Helidon Quarries, 70 miles W. of Brisbane, **Queensland**
Presented by The Government of Queensland
    Weight per cub. ft. 147 lbs.

859. NEW ZEALAND SANDSTONE. *TRIAS*
Light greenish grey, close-grained
        Waikola Quarries, **South Island**
    Weight per cub. ft. 178 lbs.

860. NORTH AMERICAN SANDSTONE. *TRIAS*
Chocolate-coloured, fine-grained
        Seneca Creek Quarries, **Maryland, U.S.A.**
    Weight per cub. ft. 167 lbs. Spec. grav. 2·67
    Crushing strain per sqr. ft. 1157 tons

861. NORTH AMERICAN SANDSTONE. *TRIAS*
Purple-brown, medium-grained
        Brown Stone Quarries, Waltonville,
                                Pennsylvania, U.S.A.
Presented by The Hummelston Brown Stone Co., Waltonville
    Weight per cub. ft. 166 lbs. Spec. grav. 2·66. Absorp. 1·37%
        Crushing strain per sqr. ft. 948 tons (Buckley)

862. SOUTH AMERICAN SANDSTONE. *TRIAS*
Pink and grey mottled, fine-grained
        Basilica Quarries, near Colon, **Entre Rios**
Presented by Cura V. M. Davani, Injár
    Weight per cub. ft. 147 lbs.

JURASSIC.

863. SWISS LIMESTONE. *Lias*
Blue-black with white veins, compact, takes a polish
Vionnaz Quarries, near Muraz, **Valais**
Weight per cub. ft. 176 lbs.

864. SWISS LIMESTONE. *Lias*
Dark grey, fine-grained, compact, admits of a polish
Stone Quarries, Colombay, **Valais**
Weight per cub. ft. 166 lbs.

865. SWISS LIMESTONE. *Lias*
Grey, close-grained, compact, admits of a good polish
Stone Quarries, Villeneuve, **Vaud**
Weight per cub. ft. 180 lbs.

866. SWISS LIMESTONE. *Lias*
Dark purplish grey, veined, compact, takes a good polish
Stone Quarries, Villeneuve, **Vaud**
Weight per cub. ft. 171 lbs.

867. AUSTRIAN LIMESTONE. *Lower Jurassic*
Light cream, medium-grained, oolitic
Sóskút Quarries, near Budapest, **Hungary**
Presented by Seenyer Béla, Esq., Budapest
Weight per cub. ft. 123½ lbs.

868. FRENCH LIMESTONE. *Lower Jurassic*
Light yellowish grey, fine-grained
Villette Quarries, **Calvados**
Weight per cub. ft. 171 lbs.

869. FRENCH LIMESTONE. *Lower Jurassic*
Greyish white, fine-grained
Chauvigny Quarries, **Vienne**
Presented by Messrs Civit Pommier & Co., Poitiers
Weight per cub. ft. 168 lbs.
Crushing strain per sqr. ft. 432 tons

870. FRENCH LIMESTONE. *Lower Jurassic*
Greyish white, fine-grained, soft
    Tercé Normandoux Quarries, **Vienne**
    Weight per cub. ft. 154 lbs.
    Crushing strain per sqr. ft. 262 tons

871. FRENCH LIMESTONE *(Caen Stone).* *Middle Jurassic*
Pale cream, fine-grained limestone
    The Quarries, Caen, **Calvados**
Presented by Messrs W. and J. R. Freeman, London
    Weight per cub. ft. 168 lbs.
    Crushing strain per sqr. ft. 198 tons (Rivington)

872. FRENCH LIMESTONE. *Middle Jurassic*
Silvery white, very fine, crystalline
    Euville Quarries, **Meuse**
Presented by Messrs Fevre & Co., Paris
    Weight per cub. ft. 150 lbs.
    Crushing strain per sqr. ft. 288 tons

873. FRENCH LIMESTONE. *Middle Jurassic*
White, fine-grained, calcareous
    Chassignelles Quarries, **Yonne**
Presented by Messrs Fevre & Co., Paris
    Chemical Composition (Stanger):
SiO_2 trace, Al_2O_3 & Fe_2O_3 0·06, MgO 0·32, CaO 55·58, H_2O &c. 0·52, CO_2 43·52, = 100·00
    Weight per cub. ft. 150 lbs.
    Crushing strain per sqr. ft. 410 tons

874. FRENCH LIMESTONE. *Middle Jurassic*
Light brown, fine-grained
    Massangis Quarries, **Yonne**
Presented by Messrs Fevre & Co., Paris
    Weight per cub. ft. 150 lbs.
    Crushing strain per sqr. ft. 578 tons

875. FRENCH LIMESTONE. *Middle Jurassic*
Cream, fine-grained, granular
    Massangis Quarries, **Yonne**
Presented by Messrs Fevre & Co., Paris
    Weight per cub. ft. 156 lbs.
    Crushing strain per sqr. ft. 612 tons

876. FRENCH LIMESTONE. *MIDDLE JURASSIC*
Light cream, fine-grained, granular
    Méreuil Quarries, **Yonne**
Presented by Messrs Fevre & Co., Paris
    Weight per cub. ft. 156 lbs.
        Crushing strain per sqr. ft. 792 tons

877. FRENCH LIMESTONE. *MIDDLE JURASSIC*
White, fine-grained, soft chalky
    Palotte Quarries, **Yonne**
Presented by Messrs Fevre & Co., Paris
    Weight per cub. ft. 137 lbs.
        Crushing strain per sqr. ft. 176 tons

878. FRENCH LIMESTONE. *MIDDLE JURASSIC*
Whitish grey, fine-grained
    Forêt de Brousses Quarries, **Yonne**
Presented by Messrs Fevre & Co., Paris
    Weight per cub. ft. 123½ lbs.

879. FRENCH LIMESTONE. *MIDDLE JURASSIC*
Greyish white, fine-grained, crystalline
    Villars Quarries, **Côte d'Or**
Presented by Messrs Fevre & Co., Paris
    Weight per cub. ft. 175 lbs.
        Crushing strain per sqr. ft. 965 tons

880. FRENCH LIMESTONE. *MIDDLE JURASSIC*
Milk white, fine-grained, crystalline
    Lignerolles Quarries, **Côte d'Or**
Presented by Messrs Fevre & Co., Paris
    Weight per cub. ft. 162 lbs.
        Crushing strain per sqr. ft. 352 tons

881. FRENCH LIMESTONE. *MIDDLE JURASSIC*
White, fine-grained, even texture
    Chateau-Gaillard Quarries, **Vienne**
Presented by Messrs Civit Pommier & Co., Poitiers
    Weight per cub. ft. 140 lbs.
        Crushing strain per sqr. ft. 184 tons

882. FRENCH LIMESTONE. *MIDDLE JURASSIC*
White, fine-grained, rather chalky
    Migne Quarries, **Vienne**
    Weight per cub. ft. 137 lbs.
        Crushing strain per sqr. ft. 176 tons

883.   FRENCH LIMESTONE. *MIDDLE JURASSIC*
Cream, fine-grained, slightly chalky
    Fremigère Quarries, Lavoux, **Vienne**
    Weight per cub. ft. 149 lbs.
    Crushing strain per sqr. ft. 220 tons

884.   FRENCH LIMESTONE. *MIDDLE JURASSIC*
White, hard, shelly stone
    Vilhonneur Raillats Quarries, **Charente**
    Weight per cub. ft. 167 lbs.
    Crushing strain per sqr. ft. 571 tons

885.   FRENCH LIMESTONE. *MIDDLE JURASSIC*
Cream, fine-grained, compact
    Lens Quarries, **Drôme**
    Weight per cub. ft. 152 lbs.

886.   FRENCH LIMESTONE. *MIDDLE JURASSIC*
Light buff-coloured, fine-grained, compact, admits of polish
    Napoleon's Plateau Quarries, Grasse, **Alpes-Maritimes**
    Weight per cub. ft. 171 lbs.

887.   FRENCH LIMESTONE. *MIDDLE JURASSIC*
Light grey, fine-grained, compact, admits of good polish
    Napoleon's Plateau Quarries, Grasse, **Alpes-Maritimes**
    Weight per cub. ft. 171 lbs.

888.   FRENCH LIMESTONE. *MIDDLE JURASSIC*
Light cream with pink streak, compact
    Napoleon's Plateau Quarries, Grasse, **Alpes-Maritimes**
    Weight per cub. ft. 171 lbs.

889.   FRENCH LIMESTONE. *MIDDLE JURASSIC*
Dark buff-coloured, fine-grained, compact, takes a good polish
    La Turbie Quarries, near Nice, **Alpes-Maritimes**
Presented by Messrs Fontana and Gamba, Monaco
    Weight per cub. ft. 171 lbs.

890.   FRENCH LIMESTONE. *MIDDLE JURASSIC*
Dark buff with brown streaks, compact, takes polish
    La Turbie Quarries, near Nice, **Alpes-Maritimes**
Presented by Messrs Fontana and Gamba, Monaco
    Weight per cub. ft. 176 lbs.

891. FRENCH LIMESTONE. *Upper Jurassic*
Light grey, fine-grained, compact
 Stone Quarries, Mont Salève, **Haute Savoie**
 Weight per cub. ft. 171 lbs.

892. FRENCH LIMESTONE. *Upper Jurassic*
Dark blue-grey, fine-grained
 Stone Quarries, Mont Salève, **Haute Savoie**
 Weight per cub. ft. 133 lbs.

893. GERMAN LIMESTONE. *Upper Jurassic*
Light cream, fine-grained
 Schneistheim Quarries, **Würtemberg**
 Weight per cub. ft. 165 lbs.

894. ITALIAN LIMESTONE. *Boticino. Jurassic*
Light dove-coloured, fine-grained, compact
 Brescia Quarries, **Lombardy**
 Weight per cub. ft. 170 lbs.

895. ITALIAN LIMESTONE. *Dark Mazzana. Jurassic*
Light brown, compact, takes a good polish
 Mazzano Quarries, near Brescia, **Lombardy**
Presented by Messrs A. Guttridge & Co., London
 Weight per cub. ft. 163 lbs.

896. ITALIAN LIMESTONE. *White Boticino. Jurassic*
Cream, fine-grained, compact
 Rezzato Quarries, near Brescia, **Lombardy**
Presented by Messrs A. Guttridge & Co., London
 Weight per cub. ft. 171 lbs.

897. PORTUGUESE LIMESTONE. *Vidraco. Upper*
  Jurassic
Light cream, fine-grained, compact
 Maceira Quarries, Pero Pinheiro, **Cintra**
 Weight per cub. ft. 165 lbs.

898. PORTUGUESE LIMESTONE. *Almiscado Escuro.*
  Upper Jurassic
Light yellow and blue veined, compact
 Maceira Quarries, Pero Pinheiro, **Cintra**
 Weight per cub. ft. 165 lbs.

899. PORTUGUESE LIMESTONE. *Almiscado Amerello.*
Upper Jurassic
Light blue and yellow mottled, compact
 Maceira Quarries, Pero Pinheiro, **Cintra**
Weight per cub. ft. 165 lbs.

900. PORTUGUESE LIMESTONE. *Abancado.*
Upper Jurassic
Light pink and white mottled, compact
 Lameiras Quarries, Pero Pinheiro, **Cintra**
Weight per cub. ft. 165 lbs.

901. PORTUGUESE LIMESTONE. *Lioz.*   *Upper*
 Jurassic
Light cream, fine-grained, crystalline stone
 Lameiras Quarries, Pero Pinheiro, **Cintra**
Weight per cub. ft. 165 lbs.

902. PORTUGUESE LIMESTONE. *Upper Jurassic*
Blue-black, slightly veined, compact
 Limestone Quarries, **Cintra**
Presented by J. Coverley, Esq., Lisbon
Weight per cub. ft. 152 lbs.

903. PORTUGUESE LIMESTONE. *Canataria.*
Upper Jurassic
Light buff-coloured, fine-grained, compact
 Paco d'Arcos Quarries, near **Leiria**
Weight per cub. ft. 161·5 lbs.

904. PORTUGUESE LIMESTONE. *Emperor's Red.*
Upper Jurassic
Pink veined, fine-grained, compact, takes a good polish
 Pedra Furada Quarries, Pero Pinheiro, **Cintra**
Weight per cub. ft. 165 lbs.

905. SWISS LIMESTONE. *Upper Jurassic*
Dark blue-grey, fine-grained, compact
 Trofon Quarries, **Vaud**
Weight per cub. ft. 170 lbs.

906 SWISS LIMESTONE. *Upper Jurassic*
Light drab, fine-grained, compact
 Chaumont Quarries, **Neuchâtel**
Weight per cub. ft. 170 lbs.

906A. SWISS LIMESTONE. *UPPER JURASSIC*
Blue-grey, fine-grained, compact
Stone Quarries, **Soleure** (Solothurn)
Weight per cub. ft. 168 lbs. Spec. grav. 2·69. Absorp. 0·30 %
Crushing strain per sqr. ft. 943 tons

907. SWISS LIMESTONE. *UPPER JURASSIC*
Light cream, fine-grained, compact
Stone Quarries, **Soleure** (Solothurn)
Weight per cub. ft. 171 lbs.
Crushing strain per sqr. ft. 943 tons

908. SWISS LIMESTONE. *UPPER JURASSIC*
Light yellow, fine-grained, compact, takes a polish
Stone Quarries, **Soleure** (Solothurn)
Weight per cub. ft. 167 lbs. Spec. grav. 2·67. Absorp. 0·30 %
Crushing strain per sqr. ft. 943 tons

909. SWISS LIMESTONE. *UPPER JURASSIC*
Cream, slightly banded, compact
Salz Quarries, Muttenz, near **Basle**
Weight per cub. ft. 165 lbs.

910. SWISS LIMESTONE. *UPPER JURASSIC*
Cream, close-grained
Homburg Quarries, near Arlesheim, **Basle**
Presented by The Société Baloise de Construction, Basle
Weight per cub. ft. 170 lbs.

911. SWISS LIMESTONE. *JURASSIC*
Blue-drab, fine-grained, compact
Merligen Quarries, **Lake Thun**
Weight per cub. ft. 170 lbs.

912. MOROCCO LIMESTONE. *JURASSIC*
Light blue-grey, fine-grained, compact
Boulders in the Bay, 2 miles S.W. of **Tangier**
Presented by H. E. White, Esq., M.A. Cantab., Consul-General, Tangier
Weight per cub. ft. 170·5 lbs.

913. MOROCCO LIMESTONE. *JURASSIC*
Light yellow, fine-grained
Boulders in the Bay, 2 miles S.W. of **Tangier**
Presented by H. E. White, Esq., M.A. Cantab., Consul-General, Tangier
Weight per cub. ft. 142·5 lbs.

914.　AUSTRALIAN FREESTONE.　*JURASSIC*
Dark drab, fine-grained, calcareous sandstone
　　　Burrabool Hills Quarries, **Victoria**
　　　Weight per cub. ft. 142 lbs.

915.　AUSTRALIAN SANDSTONE.　*JURASSIC*
Light buff, fine-grained
　　　Donnybrook Quarries, **Western Australia**
Presented by The Government of Western Australia
　　　Weight per cub. ft. 133 lbs.

916.　AUSTRALIAN SANDSTONE.　*JURASSIC*
Yellow and brown banded, medium-grained
　　　Donnybrook Quarries, **Western Australia**
Presented by The Government of Western Australia
　　　Weight per cub. ft. 138 lbs.

917.　AUSTRALIAN SANDSTONE.　*JURASSIC*
Light yellowish brown, medium-grained
　　　Donnybrook Quarries, **Western Australia**
Presented by The Government of Western Australia
　　　Chemical Composition (S. S. Dougall, Perth):
SiO_2 96·88,　Al_2O_3 0·24,　Fe_2O_3 1·26,　K_2O 0·03,　Na_2O 0·06,
H_2O 1·10,　$CaCO_3$ 0·13,　$MgCO_3$ 0·06,　= 99·76
　　　Weight per cub. ft. 133 lbs.　Spec. grav. 2·48

918.　AUSTRALIAN SANDSTONE.　*JURASSIC*
Very bright buff, fine-grained
　　　Donnybrook Quarries, **Western Australia**
Presented by the Government of Western Australia
　　　Chemical Composition (S. S. Dougall, Perth):
SiO_2 98·28,　Al_2O_3 0·10,　Fe_2O_3 0·56,　K_2O 0·01,　Na_2O 0·03,
H_2O 0·78,　$CaCO_3$ 0·13,　$MgCO_3$ 0·24,　= 100·13
　　　Weight per cub. ft. 133 lbs.　Spec. grav. 2·41

919.　NORTH AMERICAN LIMESTONE.　*JURASSIC*
Light drab, fine-grained, calcareous
　　　White Stone Quarries, Memphis Junction,
　　　　　　　　　　　　　　　　Kentucky, U.S.A.
Presented by The Bowling Green White Stone Co., Memphis Junction
　　　Chemical Composition:
SiO_2 1·42,　H_2O & loss 1·76,　$CaCO_3$ 95·31,　$MgCO_3$ 1·12,
= 99·61
　　　Weight per cub. ft. 165 lbs.
　　　Crushing strain per sqr. ft. 432 tons

CRETACEOUS.

920. AUSTRIAN LIMESTONE. *Upper Cretaceous*
White, fine-grained, soft, chalky
  Melera Quarries, near Pola, **Istria**
Weight per cub. ft. 131 lbs. Spec. grav. 2·09. Absorp. 10·74 %
  Crushing strain per sqr. ft. 227 tons

921. AUSTRIAN LIMESTONE. *Upper Cretaceous*
Light cream, fine-grained
  Marzana Quarries, near Pola, **Istria**
Weight per cub. ft. 146 lbs. Spec. grav. 2·35. Absorp. 5·28 %
  Crushing strain per sqr. ft. 525 tons

922. AUSTRIAN LIMESTONE. *Upper Cretaceous*
Light cream, fine-grained
  Vincurial Quarries, near Pola, **Istria**
Weight per cub. ft. 148 lbs. Spec. grav. 2·38. Absorp. 6·07 %
  Crushing strain per sqr. ft. 444 tons

923. AUSTRIAN LIMESTONE. *Upper Cretaceous*
Light grey with dark streaks, compact, takes polish
  Reppen Quarries, **Istria**
Weight per cub. ft. 180 lbs. Spec. grav. 2·60. Absorp. 1·23 %
  Crushing strain per sqr. ft. 1246·5 tons

924. AUSTRIAN LIMESTONE. *Upper Cretaceous*
Light fawn-coloured, fine-grained, compact, takes a good polish
  San Stefano Quarries, **Istria**
Weight per cub. ft. 162 lbs. Spec. grav. 2·51. Absorp. 3·22 %
  Crushing strain per sqr. ft. 1021·5 tons

925. AUSTRIAN LIMESTONE. *Upper Cretaceous*
Light greyish brown mottled, compact, takes a good polish
  Nabresina Quarries, near **Trieste**
Weight per cub. ft. 172 lbs. Spec. grav. 2·57. Absorp. 1·34 %
  Crushing strain per sqr. ft. 1393 tons

926. DANISH LIMESTONE. *Upper Cretaceous*
Cream, coarse-grained, cellular
  Faxe Quarries, **Zealand**
Presented by Messrs H. and J. Larsen, Copenhagen
  Weight per cub. ft. 114 lbs.
  Crushing strain per sqr. ft. 345 tons

927. DANISH CHALK. *Upper Cretaceous*
Light cream, medium-grained, chalky
  The Cliff of Stevns Quarries, **Zealand**
Presented by Messrs H. and J. Larsen, Copenhagen
  Weight per cub. ft. 125 lbs. Spec. grav. 2·00
  Crushing strain per sqr. ft. 90 tons

928. FRENCH LIMESTONE. *Upper Cretaceous*
Cream, hard, medium-grained, crystalline
  Arros Quarries, near Nay, **Basses-Pyrénées**
  Weight per cub. ft. 171 lbs.

929. FRENCH LIMESTONE. *Upper Cretaceous*
White, fine-grained, soft, chalky
  Arros Quarries, near Nay, **Basses-Pyrénées**
  Weight per cub. ft. 166 lbs.

930. FRENCH LIMESTONE. *Lower Cretaceous*
Light grey, hard, medium-grained, crystalline
  Louvie Quarries, near Nay, **Basses-Pyrénées**
  Weight per cub. ft. 171 lbs.

931. FRENCH LIMESTONE. *Lower Cretaceous*
Grey, fine-grained, hard, compact
  Rebenacq Quarries, near Nay, **Basses-Pyrénées**
  Weight per cub. ft. 166 lbs.

932. FRENCH LIMESTONE. *Upper Cretaceous*
White, very cellular, soft, chalky stone
  Stone Quarries, Angoulême, **Charente**
  Weight per cub. ft. 133 lbs.
  Crushing strain per sqr. ft. 60 tons

933. FRENCH LIMESTONE. *UPPER CRETACEOUS*
Greyish white, medium-grained
    Sireuil Quarries, **Charente**
    Weight per cub. ft. 127 lbs.
    Crushing strain per sqr. ft. 95·7 tons

934. FRENCH LIMESTONE. *UPPER CRETACEOUS*
Light cream, medium-grained
    Nersac Quarries, near Couronne, **Charente**
    Weight per cub. ft. 125 lbs.
    Crushing strain per sqr. ft. 66 tons

935. FRENCH LIMESTONE. *UPPER CRETACEOUS*
Cream, coarse-grained, crystalline
    Bagni Quarries, **Vaucluse**
    Weight per cub. ft. 114 lbs.

936. FRENCH LIMESTONE. *UPPER CRETACEOUS*
Light cream, medium-grained, crystalline
    Estaillades Quarries, **Vaucluse**
    Weight per cub. ft. 123·5 lbs.

937. FRENCH LIMESTONE. *UPPER CRETACEOUS*
Light cream, fine-grained
    Estaillades Quarries, **Vaucluse**
Presented by The Société Général des Carrières du Midi, Lyon
    Weight per cub. ft. 144 lbs.
    Crushing strain per sqr. ft. 153·7 tons

938. FRENCH LIMESTONE. *UPPER CRETACEOUS*
Brownish grey, compact, shelly, admits of a polish
    Cassis Quarries, **Bouches-du-Rhône**
    Weight per cub. ft. 176 lbs.

939. FRENCH LIMESTONE. *UPPER CRETACEOUS*
Cream, coarse-grained, arenaceous
    Arles Quarries, **Bouches-du-Rhône**
    Weight per cub. ft. 114 lbs.

940. FRENCH LIMESTONE. *UPPER CRETACEOUS*
Light cream, fine-grained
    Fontveille Quarries, **Bouches-du-Rhône**
Presented by The Société Général des Carrières du Midi, Lyon
    Weight per cub. ft. 124 lbs.
    Crushing strain per sqr. ft. 65·8 tons

941. FRENCH LIMESTONE. *Upper Cretaceous*
Cream, fine-grained, compact
  Tarascon Quarries, **Bouches-du-Rhône**
Weight per cub. ft. 123·5 lbs.

942. FRENCH LIMESTONE. *Upper Cretaceous*
Cream, fine-grained, compact
  Brouzet Quarries, near Alais, **Gard**
Weight per cub. ft. 174 lbs.

943. FRENCH LIMESTONE. *Upper Cretaceous*
Light grey, fine-grained, compact
  Aix-les-Bains Quarries, **Savoie**
Presented by Paul Blache, Esq., Aix-les-Bains
  Weight per cub. ft. 163 lbs.

944. GERMAN LIMESTONE. *Lower Cretaceous*
Light yellow, fine-grained, argillaceous
  Jaumont Quarries, near Metz, **Lorraine**
Weight per cub. ft. 142 lbs.

945. GERMAN SANDSTONE. *Upper Cretaceous*
Light drab and brown mottled, fine-grained, calcareous
  Cotta Quarries, near Pirna, **Saxony**
Weight per cub. ft. 142 lbs.

946. GERMAN SANDSTONE. *Upper Cretaceous*
Yellow and drab mottled, medium-grained, calcareous
  Cotta Quarries, near Pirna, **Saxony**
Weight per cub. ft. 142 lbs.

947. GERMAN SANDSTONE. *Upper Cretaceous*
Light yellow and drab, medium-grained, calcareous
  Cotta Quarries, near Pirna, **Saxony**
Weight per cub. ft. 142 lbs.

948. GERMAN SANDSTONE. *Upper Cretaceous*
Light drab, coarse-grained, grit
  Posta Quarries, near Pirna, **Saxony**
Weight per cub. ft. 112 lbs.

949. GERMAN SANDSTONE. *Upper Cretaceous*
Light yellow, coarse-grained, grit
  Postelwitz Quarries, near Schöna, **Saxony**
Weight per cub. ft. 142 lbs.

950. GERMAN SANDSTONE. *UPPER CRETACEOUS*
Light drab, very coarse-grained, grit
  Postelwitz Quarries, near Schöna, **Saxony**
  Weight per cub. ft. 152 lbs.

951. GERMAN SANDSTONE. *UPPER CRETACEOUS*
Very light grey, fine-grained
  Herzogwalden Quarries, Breslau, **Silesia**
  Weight per cub. ft. 156 lbs.

952. GERMAN SANDSTONE. *UPPER CRETACEOUS*
Very light grey, fine-grained
  Herzogwalden Quarries, near Breslau, **Silesia**
  Weight per cub. ft. 156 lbs.

953. GREEK LIMESTONE. *UPPER CRETACEOUS*
Light blue-grey mottled, compact
  Mount Lycabettus Quarries, near **Athens**
Collected by H. I. W. Tillyard, Esq., Cambridge
  Weight per cub. ft. 142 lbs.

954. GREEK LIMESTONE. *UPPER CRETACEOUS*
Light chocolate-coloured, compact
  Psychiko Quarries, near **Athens**
Collected by H. I. W. Tillyard, Esq., Cambridge
  Weight per cub. ft. 166 lbs.

955. ITALIAN LIMESTONE. *CRETACEOUS*
Light grey, fine-grained, compact
  Monte Gerusalemme Quarries, Bellona, Caserta,
                          Campania
  Weight per cub. ft. 176 lbs. Spec. grav. 2·68
  Crushing strain per sqr. ft. 725 tons

956. ITALIAN LIMESTONE. *CRETACEOUS*
Dark grey, fine-grained, compact
  Castellammare Quarries, near **Naples**
  Weight per cub. ft. 171 lbs. Spec. grav. 2·65

957. PORTUGUESE LIMESTONE. *LOWER CRETACEOUS*
Light blue-grey, fine-grained, compact
  Collares Quarries, near **Cintra**
  Weight per cub. ft. 171 lbs.

958. RUSSIAN LIMESTONE. *Upper Cretaceous*
Pink, fine-grained, rather arenaceous
  Elkair Quarries, Kutais, **Caucasus**
    Chemical Composition:
SiO_2 10·21, Al_2O_3 0·33, FeO 3·32, H_2O 0·49, $CaCO_3$ 83·55,
$MgCO_3$ 1·50, =99·40
  Weight per cub. ft. 142 lbs.

959. RUSSIAN LIMESTONE. *Upper Cretaceous*
Pink and brown, fine-grained, compact, takes polish
  Elkair Quarries, Kutais, **Caucasus**
  Weight per cub. ft. 161 lbs.

960. RUSSIAN LIMESTONE. *Upper Cretaceous*
Light fawn colour, fine-grained, compact
  Elkair Quarries, Kutais, **Caucasus**
  Weight per cub. ft. 161 lbs.

961. RUSSIAN LIMESTONE. *Upper Cretaceous*
Very light drab, fine-grained, compact
  Elkair Quarries, Kutais, **Caucasus**
  Weight per cub. ft. 128 lbs.

962. RUSSIAN LIMESTONE. *Upper Cretaceous*
Light cream, fine-grained, chalky
  Elkair Quarries, Kutais, **Caucasus**
    Chemical Composition:
SiO_2 5·10, Al_2O_3 trace, FeO 1·75, H_2O 0·63, $CaCO_3$ 90·60,
$MgCO_3$ 1·17, =99·25
  Weight per cub. ft. 128 lbs.

963. SWISS LIMESTONE. *Lower Cretaceous*
Yellow, medium-grained
  Favarage Quarries, **Neuchâtel**
  Weight per cub. ft. 157 lbs.

964. TUNISIAN LIMESTONE. *Cretaceous*
Light grey, fine-grained, compact
  Djebal Djelond Quarries, near **Tunis**
  Weight per cub. ft. 163 lbs.

965. TUNISIAN LIMESTONE. *Cretaceous*
Light pink, fine-grained, compact
  Djebal Karrouba Quarries, near **Tunis**
  Weight per cub. ft. 163 lbs.

966. TUNISIAN LIMESTONE. *CRETACEOUS*
Light yellowish grey, fine-grained, compact
 Djebal Karrouba Quarries, near **Tunis**
 Weight per cub. ft. 163 lbs.

967. EGYPTIAN SANDSTONE. *NUBIAN SANDSTONE.*
    CRETACEOUS
Orange-brown banded, medium-grained
 Gebel Selsella Quarries, near **Aswan**
Presented by The Egyptian Government
 Weight per cub. ft. 119 lbs. Spec. grav. 1·90
 Crushing strain per sqr. ft. 175·14 tons

968. EGYPTIAN SANDSTONE. *NUBIAN SANDSTONE.*
    CRETACEOUS
Red and white streaked, medium-grained
 Gebel Selsella Quarries, near **Aswan**
Presented by The Egyptian Government
 Weight per cub. ft. 119 lbs. Spec. grav. 1·90
 Crushing strain per sqr. ft. 175·14 tons

969. EGYPTIAN SANDSTONE. *NUBIAN SANDSTONE.*
    CRETACEOUS
Straw-coloured, medium-grained
 Gebel Selsella Quarries, near **Aswan**
Presented by The Egyptian Government
 Weight per cub. ft. 119 lbs. Spec. grav. 1·90
 Crushing strain per sqr. ft. 175·14 tons

970. EGYPTIAN SANDSTONE. *NUBIAN SANDSTONE.*
    CRETACEOUS
Cream, medium-grained
 Gebel Selsella Quarries, **Aswan**
Presented by The Egyptian Government
 Weight per cub. ft. 119 lbs. Spec. grav. 1·90
 Crushing strain per sqr. ft. 175·14 tons

971. EGYPTIAN SANDSTONE. *NUBIAN SANDSTONE.*
    CRETACEOUS
Light pink, slightly banded, coarse-grained
 Gebel Auli Quarries, White Nile, 25 miles S. of
            Khartoum
Presented by Major P. R. Phipps, Khartoum
 Weight per cub. ft. 123·5 lbs.

972. EGYPTIAN SANDSTONE. *Nubian Sandstone.*
Cretaceous
Light yellowish grey, coarse-grained
    Gebel Auli Quarries, White Nile, 25 miles S. of
                              Khartoum
Presented by Major P. R. Phipps, Khartoum
    Weight per cub. ft. 123·5 lbs.

973. WEST AFRICAN LIMESTONE. *Cretaceous*
Light yellow, compact
    Cacuaco Quarries, **Angola**
Presented by H. G. Mackie, Esq., Loanda
    Weight per cub. ft. 161·5 lbs.

974. SOUTH AFRICAN SANDSTONE. *Hard Cœga*
              Stone. Lower Cretaceous
Light greenish grey, close-grained
    Coega Quarries, Uitenhage, Port Elizabeth, **Cape**
                        of Good Hope
Presented by Messrs A. and S. Gough, Port Elizabeth
    Weight per cub. ft. 163 lbs.

975. SOUTH AFRICAN SANDSTONE. *Soft Cœga*
              Stone. Lower Cretaceous
Light brown, medium-grained
    Coega Quarries, Uitenhage, Port Elizabeth, **Cape**
                        of Good Hope
Presented by Messrs A. and S. Gough, Port Elizabeth
    Weight per cub. ft. 171 lbs.
    Chiefly used for internal work

976. SOUTH AFRICAN SANDSTONE. *Cathcart*
            Freestone. Lower Cretaceous
Dark drab, fine-grained
    Cathcart Quarries, **Cape of Good Hope**
    Weight per cub. ft. 147 lbs.

977. CHIOS LIMESTONE. *Cretaceous*
Grey with red veins, saccharoidal
    Limestone Quarries, **Chios Island (Asia Minor)**
Presented by P. Graticola, Esq., Chios
    Weight per cub. ft. 171 lbs.

978. SYRIAN LIMESTONE (*Mizzeh Yahudi*). *Upper*
    Cretaceous
Pink and yellow mottled, fine-grained, hard
    Catamon Quarries, near **Jerusalem**
    Weight per cub. ft. 144 lbs.

979. SYRIAN LIMESTONE (*MIZZEH YAHUDI*). *UPPER CRETACEOUS*
Cream and brown mottled, fine-grained, hard
    Catamon Quarries, near **Jerusalem**
    Weight per cub. ft. 144 lbs.

980. SYRIAN LIMESTONE (*MIZZEH YAHUDI*). *UPPER CRETACEOUS*
Pink with red veins, compact, hard
    Ta-la-bi-eli Quarries, ½ mile W. of **Jerusalem**
Collected by E. Unga, Esq., Jerusalem
    Weight per cub. ft. 149 lbs.

981. SYRIAN LIMESTONE (*MIZZEH AZRAQ*). *UPPER CRETACEOUS*
Light drab with thin white quartz veins, fine-grained, hard
    Schafat Quarries, 5 miles N. of **Jerusalem**
    Weight per cub. ft. 136 lbs.

983. SYRIAN LIMESTONE (*MIZZEH ASFAR*). *UPPER CRETACEOUS*
Bright pink and yellow veined, fine-grained, hard
    Safafa Quarries, 3 miles S. of **Jerusalem**
    Weight per cub. ft. 136 lbs.

984. SYRIAN LIMESTONE (*MIZZEH ASFAR*). *UPPER CRETACEOUS*
Dull pink and yellow veined, fine-grained, hard
    Safafa Quarries, 3 miles S. of **Jerusalem**
    Weight per cub. ft. 136 lbs.

985. SYRIAN LIMESTONE (*MIZZEH ASFAR*). *UPPER CRETACEOUS*
Light pink and yellow veined, fine-grained, hard
    Safafa Quarries, 3 miles S. of **Jerusalem**
    Weight per cub. ft. 136 lbs.

986. SYRIAN LIMESTONE (*MIZZEH ASFAR*). *UPPER CRETACEOUS*
Straw-coloured with pink veins, fine-grained, hard
    Safafa Quarries, 3 miles S. of **Jerusalem**
    Weight per cub. ft. 136 lbs.

987. SYRIAN LIMESTONE (*Mizzeh Akhdar*). *Upper Cretaceous*
Light pink and green mottled, fine-grained, hard
Ibn-son-Abed Quarries, 5 miles S. of **Jerusalem**
Weight per cub. ft. 144 lbs.

988. SYRIAN LIMESTONE (*Mizzeh Abyad*). *Upper Cretaceous*
Light straw-coloured, fine-grained, fairly hard
Jasin Quarries, 4 miles N. of **Jerusalem**
Weight per cub. ft. 144 lbs.

989. SYRIAN LIMESTONE (*Mizzeh Ahmar*). *Upper Cretaceous*
Pink, slightly crystalline, fairly hard
Chirbet Quarries, 14 miles N. of **Jerusalem**
Weight per cub. ft. 136 lbs.

990. SYRIAN LIMESTONE (*Mizzeh Ahmar*). *Upper Cretaceous*
Pink and yellow banded, fine-grained, hard
Sleyib Quarries, 8 miles S. of **Jerusalem**
Weight per cub. ft. 136 lbs.

991. SYRIAN LIMESTONE (*Mizzeh Ahmar*). *Upper Cretaceous*
Pink mottled, fine-grained, hard
Jasin Quarries, 4 miles W. of **Jerusalem**
Weight per cub. ft. 136 lbs.

992. SYRIAN LIMESTONE (*Mizzeh Helu*). *Upper Cretaceous*
Light cream mottled, fairly hard
Beit Lahem Quarries, near **Jerusalem**
Weight per cub. ft. 136 lbs.

993. SYRIAN LIMESTONE (*Mizzeh Helu*). *Upper Cretaceous*
Light cream, slightly crystalline, fairly hard
Beit Hanina Quarries, 6 miles N.W. of **Jerusalem**
Weight per cub. ft. 136 lbs.

994. SYRIAN LIMESTONE (*Melekeh*). *Upper Cretaceous*
Light cream with yellow spots, rather soft
Hanina Quarries, 6 miles N.W. of **Jerusalem**
Weight per cub. ft. 119 lbs.

995. SYRIAN LIMESTONE (*MELEKEH*). *UPPER CRETACEOUS*
Pink and white mottled, rather soft
> Hanina Quarries, 6 miles N.W. of **Jerusalem**
> Weight per cub. ft. 119 lbs.

996. SYRIAN LIMESTONE (*NARI*). *UPPER CRETACEOUS*
Fawn-coloured and white mottled, rather chalky
> Sur Bahir Quarries, 4 miles S.E. of **Jerusalem**
> Weight per cub. ft. 93 lbs.

997. SYRIAN LIMESTONE (*KAKULI*). *UPPER CRETACEOUS*
Light straw-coloured, fine-grained, soft, chalky
> Beit Safafa Quarries, 4 miles S. of **Jerusalem**
> Weight per cub. ft. 102 lbs.

998. SYRIAN LIMESTONE (*KAKULI*). *UPPER CRETACEOUS*
Light cream, fine-grained, soft, chalky
> Beit Hanina Quarries, 6 miles N.W. of **Jerusalem**
> Weight per cub. ft. 102 lbs.

999. SYRIAN LIMESTONE (*KAKULI*). *UPPER CRETACEOUS*
Cream, fine-grained, soft, chalky
> A-Na-Ta Quarries, near **Jerusalem**
Collected by E. Ungar, Esq., Jerusalem
> Weight per cub. ft. 133 lbs.

1000. PETRA SANDSTONE. *UPPER CRETACEOUS*
Mauve and pink banded, fine-grained
> Cliffs of Wády Músá, **Syria**
Presented by E. Ungar, Esq., Jerusalem
> Weight per cub. ft. 152 lbs.

1001. SYRIAN SANDSTONE (*NEBI MUSA*). *UPPER CRETACEOUS*
Black, fine-grained, compact, hard
> Hagar Quarries, near **Jericho**
> Weight per cub. ft. 119 lbs.

1002. SYRIAN LIMESTONE. *UPPER CRETACEOUS*
Cream, medium-grained, chalky
> Chalasa Quarries, near **Beersheba**
Presented by Rev. R. Sterling, Gaza
> Weight per cub. ft. 142·5 lbs.

1003. NORTH AMERICAN SANDSTONE. *UPPER CRETACEOUS*

Greenish grey, fine-grained

Sites Quarries, Colusa Co., **California, U.S.A.**

Presented by The Colusa Sandstone Co., San Francisco

Weight per cub. ft. 161 lbs.

1004. CANADIAN SANDSTONE. *CRETACEOUS*

Light blue-grey, medium-grained

Saturna Island Quarries, **British Columbia**

Presented by G. G. Taylor, Esq., Saturna Island

Weight per cub. ft. 144 lbs.

1005. TRINIDAD LIMESTONE. *CRETACEOUS*

Blue-grey, compact, with calcite veins

Laventille Quarries, near **Port of Spain**

Presented by Professor Cadman, Birmingham

Weight per cub. ft. 171 lbs.

EOCENE.

1006. AUSTRIAN LIMESTONE. *EOCENE*
Light fawn-coloured, fine-grained, calcareous
Beszterezebánya Quarries, **Hungary**
Presented by Seenger Béla, Esq., Budapest
Weight per cub. ft. 152 lbs.

1007. AUSTRIAN LIMESTONE. *EOCENE*
Orange yellow, fine-grained
Gobanka Quarries, **Hungary**
Presented by Seenger Béla, Esq., Budapest
Weight per cub. ft. 147 lbs.

1008. AUSTRIAN LIMESTONE. *EOCENE*
Cream-coloured, coarse-grained
Zogelsdorf Quarries, near **Eggenburg**
Weight per cub. ft. 116 lbs. Spec. grav. 1·88. Absorp. 15·79 %
Crushing strain per sqr. ft. 144 tons

1009. AUSTRIAN SANDSTONE. *EOCENE*
Light greenish yellow with brown bands, medium-grained
Rekawinkel Quarries, near **Vienna**
Weight per cub. ft. 143 lbs. Spec. grav. 2·29. Absorp. 6·02 %
Crushing strain per sqr. ft. 602·5 tons

1010. AUSTRIAN LIMESTONE. *EOCENE*
Dark cream, compact, nummulitic, takes good polish
Orsera Quarries, **Istria**
Weight per cub. ft. 114 lbs.

1011. FRENCH LIMESTONE (*MEULIÈRE ROUGE*).
OLIGOCENE
Bright orange red, highly cellular
Palaiseau Quarries, **Seine-et-Oise**
Presented by G. Collet, Esq., Palaiseau
Weight per cub. ft. 76 lbs.

1012. FRENCH LIMESTONE (*MEULIÈRE BLANC*).
OLIGOCENE
Light cream and yellow, highly cellular
    Palaiseau Quarries, **Seine-et-Oise**
Presented by G. Collet, Esq., Palaiseau
    Weight per cub. ft. 89 lbs.

1013. FRENCH SANDSTONE (*GRÈS DE FONTAINEBLEAU*).
OLIGOCENE
Light dove-coloured, medium-grained
    St Cherons Quarries, **Seine-et-Oise**
Presented by M. L. Reiquard, Esq., Paris
    Weight per cub. ft. 171 lbs.

1014. ITALIAN LIMESTONE (*RED VERONA*).   *EOCENE*
Pink and red streaked, compact, takes a good polish
    Limestone Quarries, near **Verona**
    Weight per cub. ft. 163 lbs.

1015. ITALIAN LIMESTONE (*ALMOND VERONA*). *EOCENE*
Pink and white mottled, fine-grained, admits of a polish
    Limestone Quarries, near **Verona**
    Weight per cub. ft. 163 lbs.

1016. ITALIAN LIMESTONE (*WHITE VERONA*).   *EOCENE*
Light cream-coloured, fine-grained, takes a polish
    Limestone Quarries, near **Verona**
    Weight per cub. ft. 163 lbs.

1016A. ITALIAN SANDSTONE. *EOCENE*
Light grey, fine-grained, slightly micaceous
    Signa Quarries, **Tuscany**
    Weight per cub. ft. 171 lbs.

1017. ITALIAN LIMESTONE. *EOCENE*
Light cream, fine-grained
    Aquila Quarries, **Abruzzi**
    Weight per cub. ft. 157 lbs.

1018. ITALIAN SANDSTONE. *EOCENE*
Light brown, medium-grained, arenaceous
    Sandstone Quarries, near **Bordighera**
Presented by E. Berry, Esq., Bordighera
    Weight per cub. ft. 152 lbs.

1019. ITALIAN SANDSTONE. *EOCENE*
Light brown, coarse-grained, arenaceous
  Sandstone Quarries, near **Bordighera**
Presented by E. Berry, Esq., Bordighera
  Weight per cub. ft. 156 lbs.

1020. ITALIAN SANDSTONE. *EOCENE*
Light blue-grey, medium-grained
  Sandstone Quarries, near **Bordighera**
Presented by E. Berry, Esq., Bordighera
  Weight per cub. ft. 161·5 lbs.

1021. TURKISH LIMESTONE. *EOCENE*
White and grey banded, fine-grained, takes a good polish
  Limestone Quarries, **Island of Marmora**
  Weight per cub. ft. 161·5 lbs.

1022. MALTESE LIMESTONE. *UPPER EOCENE*
  (*OLIGOCENE*)
Light cream, fine-grained stone
  Limestone Quarries, **Island of Malta**
    Chemical Composition (J. Murray):
Al_2O_3 & Fe_2O_3 3·87, $CaCO_3$ 16·75, $MgCO_3$ 0·44, $CaPO_3$ 2·22,
Residue insoluble in dilute acid, 76·42, = 99·70
  Weight per cub. ft. 140 lbs. Spec. grav. 2·24
  Crushing strain per sqr. ft. 116 tons

1023. EGYPTIAN LIMESTONE. *EOCENE*
Light cream, compact, slightly shelly
  Tura Quarries, near **Ghizeh**
Presented by H. E. C. E. Coles Pasha, Cairo
    Chemical Composition (A. Lucas, F.C.S.):
SiO_2 7·92, Al_2O_3 & Fe_2O_3 1·34, MgO trace, CaO 82·90,
H_2O &c. 4·62, = 96·78
  Weight per cub. ft. 128·3 lbs. Spec. grav. 2·09

1024. EGYPTIAN LIMESTONE. *EOCENE*
Light yellow, fine, compact, slightly shelly
  Tura Prisons Quarry, near **Ghizeh**
Presented by Dr W. F. Hume, Geological Survey, Cairo
    Chemical Composition (A. Lucas, F.C.S.):
SiO_2 7·92, Al_2O_3 & Fe_2O_3 1·34, MgO trace, CaO 82·90,
H_2O &c. 4·62, = 96·78
  Weight per cub. ft. 138·3 lbs. Spec. grav. 2·09

1025. EGYPTIAN LIMESTONE. *Ducka Stone.*
Eocene
Light cream, fine-grained, slightly chalky
    Palomba Quarries, White Mountain, **Abbassieh**
Presented by Dr W. F. Hume, Geological Survey, Cairo
    Weight per cub. ft. 142·5 lbs.

1026. EGYPTIAN LIMESTONE *Batur Stone*
Eocene
Cream-coloured, close-grained
    Atar el Nabi Quarries, near **Old Cairo**
Presented by Dr W. F. Hume, Geological Survey, Cairo
    Chemical Composition (A. Lucas, F.C.S.):
SiO_2 4·05, Al_2O_3 & Fe_2O_3 0·90, CaO 92·98, H_2O &c. 0·66,
=98·59
    Weight per cub. ft. 151 lbs. Spec. grav. 2·42
    Crushing strain per sqr. ft. 106 tons

1027. EGYPTIAN LIMESTONE. *Chemari Stone*
Eocene
Dark cream, medium-grained, chalky
    Choukri Bey's Quarries, **El Meaddasa Citadel**
Presented by Dr W. F. Hume, Geological Survey, Cairo
    Chemical Composition (A. Lucas, F.C.S.):
SiO_2 2·54, Al_2O_3 & Fe_2O_3 0·61, MgO 2·25, CaO 94·85,
H_2O 0·85, =101·10
    Weight per cub. ft. 142·5 lbs.

1028. EGYPTIAN LIMESTONE. *Chemari Stone.*
Eocene
Dark cream, fine-grained, chalky, slightly shelly
    Gussman Quarries, Giuschi, **El Meaddasa Citadel**
Presented by Dr W. F. Hume, Geological Survey, Cairo
    Chemical Composition (A. Lucas, F.C.S.):
SiO_2 1·35, Al_2O_3 & Fe_2O_3 0·35, MgO 1·27, CaO 97·84,
H_2O 0·40, =101·21
    Weight per cub. ft. 142·5 lbs.

1029. EGYPTIAN LIMESTONE. *Dar Samanond Stone.*
Eocene
Light yellow, fine-grained, chalky
    Dentamaro Quarries, **Dar Samanond Citadel**
Presented by Dr W. F. Hume, Geological Survey, Cairo
    Weight per cub. ft. 142·5 lbs.

1030. MOROCCO LIMESTONE. *Eocene*
Light buff, slightly banded, fine-grained
    Jews River Quarries, 1½ miles E. of **Tangier**
Presented by H. E. White, Esq., M.A. Cantab., Consul-General,
        Tangier
    Weight per cub. ft. 157 lbs.

1031. MOROCCO LIMESTONE. *Eocene*
Dark buff, slightly banded, fine-grained
    Three Saints Quarries, 2 miles S.E. of **Tangier**
Presented by H. E. White, Esq., M.A. Cantab., Consul-General,
        Tangier
    Weight per cub. ft. 147 lbs.

1032. TURKISH CONGLOMERATE. *Eocene*
Pinkish grey mottled, coarse-grained
    Héréké Quarries, near Ismid, **Asia Minor**
    Weight per cub. ft. 161·5 lbs.

1033. INDIAN LIMESTONE. *Eocene*
Light yellow, medium-grained
    Myaungmyau Quarries, **Lower Burma**
Presented by Messrs R. Bagchi & Co., Rangoon
    Weight per cub. ft. 85·5 lbs.

1034. AUSTRALIAN LIMESTONE. *Eocene*
Yellow, medium-grained, crystalline
    Waurn Ponds Quarries, **Victoria**
    Weight per cub. ft. 142·5 lbs.

1035. AUSTRALIAN SANDSTONE. *Eocene*
Light brown, medium-grained
    Murray Bridge Quarries, Co. Stuart,
                    South Australia
    Weight per cub. ft. 95 lbs.

1036. AUSTRALIAN FREESTONE (*Rottnest Free-*
        stone, Fine). *Eocene*
Light yellow, slightly shelly, medium-grained, calcareous
    Rottnest Island Quarries, **Western Australia**
Presented by The Government of Western Australia
    Weight per cub. ft. 142·5 lbs.

1037. AUSTRALIAN FREESTONE (ROTTNEST FREE-
STONE, COARSE). *EOCENE*
Light yellow, slightly shelly, coarse-grained, calcareous
Rottnest Island Quarries, **Western Australia**
Presented by The Government of Western Australia
Weight per cub. ft. 129 lbs.

1038. NEW ZEALAND LIMESTONE (*O. K. STONE*).
OLIGOCENE
Light cream, chalky, medium-grained
Limestone Quarries, **Oamaru**
Presented by The Oamaru T. T. and O. K. Stone Co., Oamaru
Weight per cub. ft. 95 lbs.

1039. NEW ZEALAND LIMESTONE. *OLIGOCENE*
Light cream, chalky, medium-grained
K. Quarries, Tetchmakies, near **Oamaru**
Presented by Messrs Rogers and Clarkson, Dunedin
Weight per cub. ft. 95 lbs.

1040. NEW ZEALAND LIMESTONE. *OLIGOCENE*
Light cream, fine-grained
Mount Somers Quarries, near **Ashburton**
Presented by Messrs Rogers and Clarkson, Dunedin
Weight per cub. ft. 128 lbs.

1041. JAMAICA SANDSTONE. *EOCENE*
Reddish brown, medium-grained, calcareous
Kellett Quarries, **Clarendon Parish**
Presented by F. Cundall, Esq., Kingston
Weight per cub. ft. 152 lbs.

1042. JAMAICA SANDSTONE. *EOCENE*
Reddish brown, medium-grained, calcareous
Kellett Quarries, **Clarendon Parish**
Presented by F. Cundall, Esq., Kingston
Weight per cub. ft. 152 lbs.

1043. JAMAICA SANDSTONE. *EOCENE*
Light yellow, medium-grained, calcareous
Serge Island Quarries,
St Thomas in the East Parish
Presented by F. Cundall, Esq., Kingston
Weight per cub. ft. 133 lbs.

1044. JAMAICA SANDSTONE. *Eocene*
Light yellow, medium-grained, calcareous
    Serge Island Quarries,
              St Thomas in the East Parish
Presented by F. Cundall, Esq., Kingston
    Weight per cub. ft. 133 lbs.

1045. ANTIGUA LIMESTONE. *Eocene*
Cream, fine-grained, with occasional ferruginous spots
    Gilbert's Quarries, **St John**
Presented by Messrs W. G. Bennett, Bryson & Co., St John
    Weight per cub. ft. 142 lbs.

MIOCENE.

1046. AUSTRIAN LIMESTONE. *MIOCENE*
Light grey, fine-grained, fossiliferous
Mannerdorf Quarries, **Leithargebirger**
Weight per cub. ft. 151 lbs. Spec. grav. 2·44. Absorp. 3·39 %
Crushing strain per sqr. ft. 841 tons

1047. AUSTRIAN LIMESTONE. *MIOCENE*
Light grey, medium-grained
Kaiserstein Quarries, **Leithargebirger**
Weight per cub. ft. 155 lbs. Spec. grav. 2·49. Absorp. 2·75 %
Crushing strain per sqr. ft. 868 tons

1048. AUSTRIAN LIMESTONE. *MIOCENE*
Light brown and grey, fine-grained, conglomerate
Baden Quarries, near **Vienna**
Weight per cub. ft. 156 lbs. Spec. grav. 2·50. Absorp. 2·76 %
Crushing strain per sqr. ft. 510 tons

1049. AUSTRIAN LIMESTONE. *MIOCENE*
Light brown, fossiliferous, admits of a good polish
Duna Almas Quarries, near Budapest, **Hungary**
Weight per cub. ft. 146 lbs. Spec. grav. 2·34. Absorp. 5·11 %
Crushing strain per sqr. ft. 885 tons

1050. AUSTRIAN LIMESTONE. *MIOCENE*
Cream, medium-grained
St Margit Quarries, near Budapest, **Hungary**
Weight per cub. ft. 112 lbs. Spec. grav. 1·77. Absorp. 16·29 %
Crushing strain per sqr. ft. 141 tons

1051. AUSTRIAN LIMESTONE. *Miocene*
White, chalky, medium-grained
    Breitenbrunner Quarries, near Budapest, **Hungary**
Weight per cub. ft. 112 lbs. Spec. grav. 1·77. Absorp. 21·01 %
    Crushing strain per sqr. ft. 113·5 tons

1052. AUSTRIAN LIMESTONE. *Miocene*
Cream, fine-grained, siliceous
    Bascolorgh Quarries, near Budapest, **Hungary**
Presented by Seenger Béla, Esq., Budapest
    Weight per cub. ft. 142 lbs.

1053. AUSTRIAN LIMESTONE. *Miocene*
Dark cream, compact, slightly shelly
    Suttö Quarries, near Budapest, **Hungary**
Presented by Seenger Béla, Esq., Budapest
    Weight per cub. ft. 180 lbs.

1054. AUSTRIAN LIMESTONE. *Miocene*
White, medium-grained, calcareous
    Lorrettorn Quarries, near Budapest, **Hungary**
Weight per cub. ft. 133 lbs. Spec. grav. 2·15. Absorp. 13·55 %
    Crushing strain per sqr. ft. 295 tons

1055. AUSTRIAN LIMESTONE. *Miocene*
Light cream, medium-grained, chalky
    Aflenz Quarries, near **Leibnitz**
Weight per cub. ft. 109 lbs. Spec. grav. 1·75. Absorp. 18·65 %
    Crushing strain per sqr. ft. 98·5 tons

1055A. FRENCH SANDSTONE (*Gres Molasse*). *Miocene*
Greenish grey, fine-grained
    Bouville Quarries, **Haute Savoie**
    Weight per cub. ft. 151 lbs.

1056. GREEK LIMESTONE. *Miocene*
Light cream, fine-grained, chalky
    Kalamaki Quarries, near **Corinth**
Collected by H. I. W. Tillyard, Esq., Cambridge
    Weight per cub. ft. 104·5 lbs.

1057. GREEK LIMESTONE. *Miocene*
Light cream, medium-grained, chalky
    Ægina Quarries, **Island of Ægina**
Collected by H. I. W. Tillyard, Esq., Cambridge
    Weight per cub. ft. 85·5 lbs.

1058. GREEK LIMESTONE. *MIOCENE*
Light pink, medium-grained, cellular
    Ægina Quarries, **Island of Ægina**
Collected by H. I. W. Tillyard, Esq., Cambridge
Weight per cub. ft. 114 lbs.

1059. GREEK LIMESTONE. *MIOCENE*
Light cream, crystalline
    Piræus Quarries, near **Athens**
Collected by H. I. W. Tillyard, Esq., Cambridge
Weight per cub. ft. 147 lbs.

1060. SPANISH LIMESTONE. *MIOCENE*
Light buff, fine-grained
    Cornicabra Quarries, **Madrid**
Weight per cub. ft. 152 lbs.

1061. SPANISH LIMESTONE. *MIOCENE*
Light pinkish cream, fine-grained
    Guadalix Quarries, **Madrid**
Weight per cub. ft. 141 lbs.

1062. SPANISH LIMESTONE. *MIOCENE*
Light pinkish cream, fine, compact
    Colmenar de Oreja Quarries, **Madrid**
Weight per cub. ft. 152 lbs.

1063. SPANISH LIMESTONE. *MIOCENE*
Light pinkish cream, fine, compact
    Morata de Tajuna Quarries, **Madrid**
Weight per cub. ft. 152 lbs.

1064. SPANISH LIMESTONE. *MIOCENE*
Light cream, fine, chalky
    Otero Quarries, **Segovia**
Weight per cub. ft. 141 lbs.

1065. SPANISH LIMESTONE. *MIOCENE*
Light grey, medium-grained
    Cuenca Quarries, **Cuenca**
Weight per cub. ft. 152 lbs.

1066. SPANISH LIMESTONE. *MIOCENE*
Light cream, fine-grained
 Balseta Quarries, **Alicante**
Weight per cub. ft. 141 lbs.

1067. SPANISH LIMESTONE. *MIOCENE*
Light straw-coloured, fine-grained
 Monovar Quarries, **Alicante**
Weight per cub. ft. 152 lbs.

1068. SPANISH LIMESTONE. *MIOCENE*
Light cream, medium-grained
 Almorgni Quarries, **Alicante**
Weight per cub. ft. 152 lbs.

1069. SPANISH LIMESTONE. *MIOCENE*
Light grey, medium-grained
 Almorgni Quarries, **Alicante**
Weight per cub. ft. 152 lbs.

1070. SPANISH LIMESTONE. *MIOCENE*
Light cream, fine-grained
 Petrel Quarries, **Alicante**
Weight per cub. ft. 141 lbs.

1071. SPANISH LIMESTONE. *MIOCENE*
Light cream, fine-grained
 Villena Quarries, **Alicante**
Weight per cub. ft. 141 lbs.

1072. SPANISH SANDSTONE. *MIOCENE*
Light straw-coloured, very fine-grained
 Novelda Quarries, **Alicante**
Weight per cub. ft. 141 lbs.

1073. SWISS SANDSTONE (*MOLASSE DE BERN*). *MIOCENE*
Greenish grey, medium-grained
 Stockern Quarries, near **Berne**
Weight per cub. ft. 133 lbs.

1074. SWISS SANDSTONE (*MOLASSE DE BERN*). *MIOCENE*
Greenish yellow, fine-grained
 Ostimundigen Quarries, near **Berne**
Weight per cub. ft. 152 lbs.

1075. SWISS SANDSTONE (*Molasse Marine*). *Miocene*
Greenish grey, fine-grained
    Savigny Quarries, near **Lausanne, Vaud**
    Weight per cub. ft. 180 lbs.

1076. SWISS SANDSTONE (*Molasse Marine*). *Miocene*
Greenish grey, fine-grained
    The Quarries, **Lucerne**
    Weight per cub. ft. 176 lbs.

1077. SWISS SANDSTONE (*Grès de la Molière*).
Miocene
Light grey and yellow banded, cellular
    Estavayer Quarries, **Fribourg**
    Weight per cub. ft. 163 lbs.

1078. INDIAN LIMESTONE. *Miocene*
Straw-coloured granular, medium-grained
    Hands Hill Quarries, near **Karáchi, Sind**
Presented by M. J. H. Somaki, Esq., Karáchi
    Chemical Composition :
SiO_2 9·20, Al_2O_3 5·60, Fe_2O_3 2·00, H_2O & loss 1·20,
$CaCO_3$ 80·00, $MgCO_3$ 1·80, =99·80
    Weight per cub. ft. 142 lbs.

1079. INDIAN LIMESTONE. *Miocene*
Yellow, tough, medium-grained
    Hands Hill Quarries, near **Karáchi, Sind**
Presented by M. J. H. Somaki, Esq., Karáchi
    Weight per cub. ft. 133 lbs.

1080. INDIAN LIMESTONE. *Miocene*
Light straw-coloured, compact, calcareous
    Chizree Quarries, near **Karáchi, Sind**
Presented by M. J. H. Somaki, Esq., Karáchi
    Weight per cub. ft. 133 lbs.

1081. INDIAN SANDSTONE. *Miocene*
Yellow, slightly banded, fine-grained
    Pakhanjee Quarries, Lower Chindwin, **Upper Burma**
Presented by Messrs R. Bagchi & Co , Rangoon
    Weight per cub. ft. 133 lbs.

1082. INDIAN SANDSTONE. *MIOCENE*
Dull red, fine-grained
    Hsipau Quarries, **Northern Shan States**
Presented by Messrs R. Bagchi & Co., Rangoon
    Weight per cub. ft. 152 lbs.

1083. CUBA LIMESTONE. *MIOCENE*
White, medium-grained
    Bermeja Quarries, near **Havana**
Presented by Carlos Booth, Esq., Havana
    Weight per cub. ft. 95 lbs.

1084. CUBA LIMESTONE. *MIOCENE*
White, fine-grained, chalky
    Aguacate Quarries, near **Havana**
Presented by Carlos Booth, Esq., Havana
    Weight per cub. ft. 85·5 lbs.

1085. CUBA LIMESTONE. *MIOCENE*
Cream, coarse, shelly
    Jaimanito Quarries, near **Havana**
Presented by Carlos Booth, Esq., Havana
    Weight per cub. ft. 85·5 lbs.

1086. JAMAICA LIMESTONE. *MIOCENE*
White and pink mottled and streaked, chalky
    Portland Chalk Stone Quarries, **Portland Parish**
Presented by F. Cundall, Esq., Kingston
    Weight per cub. ft. 157 lbs.

PLIOCENE.

1087. GREEK CONGLOMERATE. *Pliocene*
Light pink and brown, cellular limestone
      Amarousi Quarries, near Kephisia, **Attica**
Collected by H. I. W. Tillyard, Esq., Cambridge
    Weight per cub. ft. 171 lbs.

1088. GREEK CONGLOMERATE. *Pliocene*
Red and black mottled, coarse-grained
      Peristeri Quarries, near Kato Liosia, **Attica**
Collected by H. I. W. Tillyard, Esq., Cambridge
    Weight per cub. ft. 171 lbs.

1089. CYPRUS LIMESTONE. *Pliocene*
White, coarse-grained, cellular
      Helofaghton Quarries, near **Larnaca**
Presented by C. S. Cramby, Esq., Larnaca
    Weight per cub. ft. 123·5 lbs.

1090. CYPRUS LIMESTONE. *Pliocene*
Cream-coloured, coarse-grained, cellular
      Voroklini Quarries, near **Larnaca**
Presented by C. S. Cramby, Esq., Larnaca
    Weight per cub. ft. 123·5 lbs.

1091. ALGERIAN LIMESTONE. *Pliocene*
Cream, fine-grained, chalky
      Raz-el-Ain Quarries, **Oran**
Presented by T. A. Barber, Esq., Oran
    Weight per cub. ft. 161 lbs.

1092. ALGERIAN LIMESTONE. *PLIOCENE*
Cream-coloured, compact
    Karguentah Quarries, near Gambetta, **Oran**
Presented by T. A. Barber, Esq., Oran
    Weight per cub. ft. 161 lbs.

1093. TURKISH LIMESTONE. *PLIOCENE*
Light cream, chalky
    Sarakeni Quarries, Mœander Valley, **Asia Minor**
Presented by E. Whittall, Esq., Smyrna
    Weight per cub. ft. 133 lbs.

PLEISTOCENE AND RECENT.

1094. GERMAN LIMESTONE. *Pleistocene*
Dark cream, vesicular
Seeburg Quarries, near Urach, **Würtemberg**
Weight per cub. ft. 90 lbs.

1095. GERMAN LIMESTONE. *Pleistocene*
Light brown, very vesicular
Gröningen Quarries, near Tübingen, **Würtemberg**
Weight per cub. ft. 90 lbs.

1096. GERMAN LIMESTONE. *Pleistocene*
Dark brown, hard, cellular
Munster Quarries, **Würtemberg**
Weight per cub. ft. 156 lbs.

1097. ITALIAN LIMESTONE (*Travertine*). *Pleistocene*
Light cream, fine-grained, compact
Sabino Quarries, near **Rome**
Weight per cub. ft. 157 lbs.

1098. ITALIAN LIMESTONE (*Travertine*) *Pleistocene*
Light cream, medium-grained, cellular
Limestone Quarries, **Tivoli**
Weight per cub. ft. 157 lbs.

1099. SWISS LIMESTONE (*Tuffstein*). *Pleistocene*
Light straw-coloured, cellular, calcareous
Aellfluh Quarries, **Grindelwald**
Weight per cub. ft. 76 lbs.

1100. SWISS LIMESTONE (*Tuffstein*). *Pleistocene*
Light straw-coloured, cellular, calcareous
Tuffen Quarries, near **Berne**
Weight per cub. ft. 75 lbs.

1101. SOUTH NIGERIAN SANDSTONE. *Pleistocene*
Reddish brown, ferruginous, medium-grained
Stone Quarries, near **Onitsha**
Presented by Rev. H. Proctor, Onitsha
Weight per cub. ft. 170 lbs.

1102. SOUTH AFRICAN SANDSTONE. *Pleistocene*
Light drab, medium-grained, calcareous
Hoetjes Quarries, Saldanha Bay, **Cape of Good Hope**
Weight per cub. ft. 104½ lbs.

1103. ZANZIBAR LIMESTONE. *Pleistocene?*
Light cream, cellular, chalky, with brown ferruginous spots
Limestone Quarries, near **Zanzibar**
Presented by Messrs Cowasjee Dinshaw & Bros., Zanzibar
Weight per cub. ft. 142·5 lbs.

1104. ZANZIBAR LIMESTONE. *Pleistocene*
Light cream, hard, very vesicular, shelly
Limestone Quarries, near **Zanzibar**
Presented by Messrs Cowasjee Dinshaw & Bros., Zanzibar
Weight per cub. ft. 142·5 lbs.

1105. ZANZIBAR LIMESTONE. *Pleistocene?*
Light cream, hard, slightly vesicular
Limestone Quarries, near **Zanzibar**
Presented by Messrs Cowasjee Dinshaw & Bros., Zanzibar
Weight per cub. ft. 104·5 lbs.

1106. ZANZIBAR LIMESTONE. *Pleistocene?*
Light cream, porous, cellular, chalky
Limestone Quarries, near **Zanzibar**
Presented by Messrs Cowasjee Dinshaw & Bros., Zanzibar
Weight per cub. ft. 95·5 lbs.

1107. ZANZIBAR SANDSTONE. *Pleistocene*
Light grey, close-grained, siliceous
Sandstone Quarries, near **Zanzibar**
Presented by Messrs Cowasjee Dinshaw & Bros., Zanzibar
Weight per cub. ft. 147 lbs.

1108. SYRIAN LIMESTONE. *Pleistocene*
Light yellow, medium-grained, cellular
   Cave Quarries, near **Gaza**
Presented by Rev. R. Sterling, Gaza
   Weight per cub. ft. 123·5 lbs.

1109. SYRIAN LIMESTONE. *Pleistocene*
Light yellow, medium-grained, cellular
   Samson Hill Quarries, near **Gaza**
Presented by Rev. R. Sterling, Gaza
   Weight per cub. ft. 142·5 lbs.

1110. SYRIAN LIMESTONE. *Pleistocene*
Cream, consolidated mass of comminuted shells
   Caves on Sea Shore, near **Gaza**
Presented by Rev. R. Sterling, Gaza
   Weight per cub. ft. 100 lbs.

1111. INDIAN LATERITE. *Pleistocene*
Reddish, cellular, ferruginous
   Laterite Quarries, near Cochin, **Malabar Coast**
Presented by Messrs Aspinwall & Co., Cochin
   Weight per cub. ft. 95 lbs.

1112. INDIAN SANDSTONE. *Pleistocene*
Light yellowish drab with brown streaks, medium-grained
   Surat Quarries, **Surat, Bombay**
Presented by P. G. Messent, Esq., M.Inst.C.E., Bombay
   Weight per cub. ft. 133 lbs.

1113. INDIAN SANDSTONE. *Porbander Stone.*
              Pleistocene
Light yellowish drab, medium-grained, calcareous
   Porbander Quarries, near **Káthiáwár**
Presented by P. G. Messent, Esq., M.Inst.C.E., Bombay
   Weight per cub. ft. 95 lbs.
   Crushing strain per sqr. ft. 165 tons 14 cwts.

1114. CEYLON LATERITE (*Cabook*). *Pleistocene*
Reddish brown, brecciated, cellular
   Boralasgamawa Quarries, **Western Province**
Presented by The Mineral Survey, Colombo
   Weight per cub. ft. 133 lbs

1115. CEYLON LATERITE (*CABOOK*). *PLEISTOCENE*
Reddish brown, brecciated, cellular
    Laterite Quarries, near **Galle**
Presented by J. R. Black, Esq., Galle
    Weight per cub. ft. 133 lbs.

1116. CEYLON LATERITE (*CABOOK*). *PLEISTOCENE*
Reddish brown, brecciated, cellular
    Laterite Quarries, near **Colombo**
Presented by J. H. Bostock, Esq., Colombo
    Weight per cub. ft. 133 lbs.

1117. SINGAPORE CONGLOMERATE. *PLEISTOCENE*
Light buff and grey, coarse-grained
    Mount Faber Quarries, **Singapore**
Presented by Messrs Swan and Maclaren, Singapore
    Weight per cub. ft. 171 lbs.

1118. SINGAPORE CORAL LIMESTONE. *PLEISTOCENE*
Light cream with shells interbedded
    Limestone Quarries, **Singapore**
Presented by Messrs Swan and Maclaren, Singapore
    Weight per cub. ft. 90 lbs.

1118A. PERSIAN SANDSTONE. *PLEISTOCENE*
    Shushter Quarries, **Khazistan**
Presented by Dr D. N. Carr, Ispahan
    Weight per cub. ft.

1119. BARBADOES LIMESTONE. *PLEISTOCENE*
Light cream, slightly cellular coral rock
    The Mount Quarries, **St Georges**
Presented by J. R. Bovell, Esq., Barbadoes
    Weight per cub. ft. 85 lbs.

1120. BARBADOES LIMESTONE. *PLEISTOCENE*
Light cream, slightly cellular coral rock
    The Mount Quarries, **St Georges**
Presented by J. R. Bovell, Esq., Barbadoes
    Weight per cub. ft. 85·5 lbs.

1121. BARBADOES LIMESTONE. *PLEISTOCENE*
Light cream, very cellular coral rock
    The Mount Quarries, **St Georges**
Presented by J. R. Bovell, Esq., Barbadoes
    Weight per cub. ft. 76 lbs.

1122. BARBADOES LIMESTONE. _PLEISTOCENE_
Light cream, porous coral rock
    The Mount Quarries, **St Georges**
Presented by J. R. Bovell, Esq., Barbadoes
    Weight per cub. ft. 113 lbs.

1123. BAHAMA LIMESTONE. _PLEISTOCENE_
White, close-grained coral rock
    Nassau Quarries, **New Providence**
Presented by Messrs R. H. Sawyer & Co., Nassau
    Weight per cub. ft. 114 lbs.

1124. BERMUDA LIMESTONE. _PLEISTOCENE_
Light cream, granular, rather friable
    Limestone Quarries, near **Hamilton**
Presented by Messrs Thompson, Roberts & Co., Hamilton
    Weight per cub. ft. 81 lbs.

1125. BERMUDA LIMESTONE. _PLEISTOCENE_
Light cream, granular, slightly banded
    Limestone Quarries, near **Hamilton**
Presented by Messrs Thompson, Roberts & Co., Hamilton
    Weight per cub. ft. 81 lbs.

1126. BERMUDA LIMESTONE. _PLEISTOCENE_
Light cream, granular, slightly banded
    Limestone Quarries, near **Hamilton**
Presented by Messrs Thompson, Roberts & Co., Hamilton
            Chemical Composition (F. A. Manning):
SiO_2 0·05,   FeO 0·52,   MgO 1·69,   CaO 52·47, K_2O 0·06,
Na_2O 0·24, CO_2 42·87, P_2O_5 0·08,   NaCl 0·02, SO_3 0·20,
Organic matter 3·80,   = 102·00
    Weight per cub. ft. 81 lbs.

INDEX.

Abbassieh, limestone quarries near, 223, 414

Abbey of Barnwell, 169, 170, 199; Battle, 194; Bury St Edmunds, 169, 170; Croxdale, 154; Glastonbury, 167; Hartshill, 99; Kirkstall, 129; Malmesbury, 175; Ramsey, 169, 170; Roche, 147; Rushen, 122; Thorney, 169, 170, 171; Tintern, 134; Westminster, 146, 176, 183, 185, 196; Whitby, 175; Woburn, 200

Abbey Church of Bath, 175; Selby, 147; Shrewsbury, 142, 154

Abbontiakoon granite quarries, 46, 336

Abbot of St Albans, 185; St Edmund, 169; Malmesbury, 175; Peterborough, 169

Abenbury sandstone quarries, 291

Abeokuta granite quarries, 46, 336

Aberdalgie sandstone quarries, 113, 266

Aberdeen, granite quarries near, 25, 252

Aberdeenshire, granites of, 22, 252; Old Red Sandstone of, 113, 267; quarries, how worked, 24

Aberdour, sandstone quarries near, 124, 275

Abriachan granite quarries, 27, 254

Abruzzi, Eocene rocks of the, 219, 412

Aburi quartzite quarries, 96, 367

Abuski Court House, 46; Hospital, 46; Police Barracks, 46

Academy of Fine Arts, Venice, 219

Accra, granite quarries near, 46; sandstone quarries near, 160, 387

Ackworth stone, 294

Adam County, granites of, 70

Addison granite quarries, 65, 349

Adelaide, buildings of, 106, 225; granite quarries near, 345; sandstone quarries near, 371; Telegraph Offices, 225

Admiralty Docks, Portsmouth, 133; Harbour, Dover, 18

Ægina Island, temple of Athena, 229; limestone quarries, 228, 419, 420

Aellfluh limestone quarries, 426

Aflenz limestone quarries, 419

Agra Palace, 105

Aguacate limestone quarries, 423

Aislaby, sandstone quarries near, 174, 311

Aix-les-Bains, buildings of, 204; limestone quarries near, 204, 402

Ajaccio, granite quarries near, 33, 326

Akkikalgudda limestone quarries, 368

Akusi granite quarries, 46, 336

Alatazata, lava quarries near, 85, 359; restoration of, 85

Albert Memorial, Hyde Park, 29

Alderley Edge sandstone quarries, 155, 300

Alexander III Bridge, Paris, 33, 186

Alford, granite quarries near, 253

Algeria, Pliocene rocks of, 235, 424; Tell Zone of, 235

Algoa Bay, sandstone beds of, 103

Alicante, buildings of, 230; Province, Miocene rocks of, 229, 421

Allahábád, Muir Central College 105; Silurian rocks of, 105, 370

Alleghany, Carnegie Library at, 64

Allens granite quarries, 49, 338

Allinge, granite quarries near, 31, 324

Alloa dolorite quarries, 78, 259; "granite" quarries, 78, 259; sandstone quarries near, 136, 289

All Saints' Church, Bordighera, 219; Kettering, 172

Almorgni limestone quarries, 421

Alnmouth, Millstone Grit near, 131

Alnwick Castle, building of, 120

Alnwick, sandstone quarries near, 120, 271

Alpes-Maritimes, Middle Jurassic rocks of, 187, 394

Alpirsbach granite quarries, 34, 327

Altingen sandstone quarries, 384

Alzo granite quarries, 35, 327

Amarousi conglomerate quarries, 424

Amatongas granite quarries, 337

Amber, Palace of, 105

Amberg, granite quarries near, 74, 354

Ambleside, slate stone quarries near, 101, 262

Amenophis II, 222

American Bank, New York, 71

Amphibolite of Russia, 331; Sweden, 334

Analcime-dolerite of Russia, 333

Anashuaico tuff quarries, 366

A-Na-Ta limestone quarries, 409

Anatolian Railway, construction of, 138

Ancaster limestone quarries, 174, 309, 310

Anderson granite quarries, 69, 351

Andesites of Asia Minor, 84, 358; St Christopher, 89, 363; St Lucia, 90, 364

Anglesey, Carboniferous Limestone of, 121; Penmon marble of, 121, 272

Angola, Cretaceous rocks of, 208, 406; granites of, 47, 336

Angoulême, Cathedral, 187; limestone quarries near, 400; Post Office, 186

Annan, sandstone quarries near, 156, 301

Annanlea Stone of Dumfriesshire, 302

Anston Stone, 145, 293

Anthiseus limestone quarries, 375

Antigua, Cathedral, 226; Eocene rocks of, 226, 417

Antrea, granite quarries near, 332

Antrim, Lower Carboniferous rocks of, 127, 277; quartzporphyry rocks of, 259

Antwerp, East Railway Station, 137

Anuradhapura, ruins near, 52

Ape's Hill of Morocco, 223

Appleby, sandstone quarries near, 144, 292

Appley Road sandstone quarries, 288

Aqueduct of Jamaica, 226

Aqueducts of Vienna, 228

Aquila, buildings of, 219; limestone quarries, 412

Arathe Koppal porphyry quarries, 360

Arbroath, sandstone quarries near, 113, 267

Ardingly, sandstone quarries near, 315

Arecco, Piperno quarries near, 83, 357

Arequipa Cathedral, 93; Ladies' College, 93; tuff quarries, 93, 366

Argentine Republic, granites of, 76, 355; Triassic rocks of, 164

Argyllshire, granites of, 27, 254; Metamorphic rocks of, 95, 260

Arles limestone quarries, 401

Arlesheim, limestone quarries near, 190, 397

Arnaeas trachyte quarries, 365

Arros limestone quarries, 400

Arsenal Buildings, Woolwich, 133

Arthwith sandstone quarries, 133, 282

Arundel Castle, 193

Ashantee, granites of, 46, 336

Ashburton, limestone quarries near, 225, 416

Asia Minor, andesites of, 84, 358; Carboniferous rocks of, 138, 376; Cretaceous rocks of, 209, 406; Eocene rocks of, 224, 415; lavas of, 84, 358; Pliocene rocks of, 235, 425; rhyolitic tuffs of, 85, 359

Aspatria, sandstone quarries of, 153, 296

Aswan, Dam, 43, 208; granite quarries, 43; grey granite, 45, 335; red granite, 45, 335; sandstone quarries near, 207, 405

Atar el Nabi limestone quarries, 223, 414

Athena, Temple of, 229

Athens, ancient buildings of, 205; British School, 229; limestone quarries near, 403, 420

Atherstone, quartzite rocks of, 98, 260

Atlanta, granite quarries near, 62

Atmospheric influence on stone, 3

Attica conglomerate quarries, 234, 424; limestone quarries, 205

Aubrey, John, antiquary, 175

Auchincarroch sandstone quarries 113, 266

Auchinheath sandstone quarries, 266

Auchinlea sandstone quarries, 136, 289

Auckland, Bank of New Zealand, 60; basalt quarries near, 363; granite quarries near, 345; Post Office, 60

Augustus, Forum of, 82

Austerlitz Bridge, Paris, 185

Austin, Texas State Capitol Building at, 72

Australia, Bowral trachyte of, 57; Cambrian rocks of, 106, 371; Carboniferous rocks of, 138, 376; granites of, 56, 343; mudstones of, 106, 371; sandstones of, 106, 371; syenites of, 57, 344; Triassic rocks of, 161, 389; tuffs of, 88, 362

Austria, Cretaceous rocks of, 202, 399; Eocene rocks of, 217, 411; granites of, 30, 323; Jurassic rocks of, 184, 391; Miocene rocks of, 227, 418; quartz-diorites of, 325

Avoca, sandstone quarries near, 149, 381

Avonmouth New Dock, 134

Awaly lava quarries, 359

Ayrshire, Triassic rocks of, 157, 302; Upper Carboniferous rocks of, 136, 289

Baden conglomerate quarries, 228, 418; Triassic rocks of, 158

Bagni limestone quarries, 401

Bahamas, Pleistocene rocks of, 243, 430

Bakka, granite quarries of, 328

Bala series of Shropshire, 99

Balater Roman Catholic Church, New York, 116

Ballagh granite quarries, 257

Ballochmyle sandstone quarries, 157, 302

Ball's Green limestone quarries, 168, 306

Ballycastle, sandstone quarries near, 126, 277

Ballyknockan granite quarries, 29, 256

Ballymoney Parish Church, 127

Balseta limestone quarries, 421

Baltimore County granite quarries, 65

Baltimore Post Office, 67

Bamburgh Castle, Northumberland, 144

Banbury, limestone quarries near, 166, 305

Banchory, granite quarries near, 27, 254

Bangalore, granite quarries near, 339

Bangkok, buildings of, 87; marble quarries near, 106, 371; tuff quarries near, 87, 362

Bangor, New University College of, 134; slate stone quarries near, 101, 263

Bank, of Africa, Durban, 35; of Australia, Melbourne, 25; of Australia, Wellington, N.Z., 60; buildings, Pictou, 141; of England, London, 123; of France, Grasse, 187; of Japan, 56; of Liverpool, Keswick, 135; of New Zealand, Auckland, N.Z., 60; of New Zealand, Dunedin, N.Z., 225; of Shanghai, Hong Kong, 55

Bank Note Buildings, New York, 73

Bankend sandstone quarries, 295

Banks Peninsula of New Zealand, 89

Banning, Potsdam stone quarries at, 107, 372

Banqueting Hall, Whitehall, London, 180

Barbadoes, Pleistocene rocks of, 242, 429: "Sawstone" of, 243

Bargate Stone of Surrey, 196, 317

Barhut, stupa of, 105

Bariakili lava quarries, 84, 358

Barna granite quarries, 30, 257

Barnack, Church, 169; "Hills and Holes" of, 169; Stone, 168, 307

Barnard Castle, Bowes Museum, 130; sandstone quarries near, 130, 280

Barnton Park sandstone quarries, 123, 273

Barnwell Abbey, 170, 199

Barrabool Hills, Jurassic rocks of, 191

Barracks of Salisbury, 134; Vienna, 227

Barre granite quarries, 72, 352

Barren Jack granite quarries, 56, 343; Reservoir, 56

Barrow-in-Furness Docks, 153

Barskimnieg sandstone quarries, 303

Bartlett Building, Chicago, 74

Barybine limestone quarries, 137, 376

Basalts, of Bombay Presidency, 86, 360; British Isles, 77; the Eifel, 79, 356; Grenada, 89, 363; India, 86, 360; Madeira, 90, 364; Mauritius, 93, 366; New Zealand, 88, 363; Palestine, 85, 359; Rhenish Prussia, 79, 356; St Christopher, 89

Bascolorgh limestone quarries, 419

Basilica, sandstone quarries of the, 390

Basle, High School for Girls, 190; Railway Station, 190; limestone quarries near, 190, 397

Basses - Pyrénées, Cretaceous rocks of, 204, 400

Basseterre andesite quarries, 89, 363

Bass Rock Lighthouse, 124

Bass Straits, Australia, 138

Bastard Freestone of Shepton Mallet, 165, 304

Basutoland, Upper Karroo system of, 160

Bath, Abbey Church of, 175; limestone quarries near, 175, 311; Oolites of Somersetshire, 175, 312; Oolites of Wiltshire, 175, 311

Bath Road limestone quarries, 165, 304

Bath Stone quarries, how worked, 176

Batley, sandstone quarries near, 285

Batoum, Commercial Bank, 206; Cathedral, 206

Battle Abbey, 194

"Batur Stone" of Egypt, 223

Baveno granite quarries, 35, 327

Bawtry limestone quarries, 293

Bay of Fundy granite quarries, 76, 355

Bayonne Cathedral, 187

Beachy Head Lighthouse, 19

Beamsville limestone quarries, 116, 374

Bearl sandstone quarries, 131, 281

Beaufort series of Cape of Good Hope, 150, 382
Beaumaris limestone quarries, 121, 272
Beche, Sir H. de la, 146
Bécon granite quarries, 325
Bedding of stone, 9
Bedford, Earl of, 170
Bedfordshire, Lower Chalk of, 199, 317; Totternhoe Stone of, 199, 317
Bedford, U.S.A., limestone quarries, 139, 377
Beersheba, limestone quarries near, 213, 409
Beer Stone of Devonshire, 6, 200, 318
Beggars Well sandstone quarries, 297
Beira Prov., limestone quarries of, 188
Beira, E. Africa, granites of, 47, 336
Beit Hanina limestone quarries, 408
Beit Lahem limestone quarries, 408
Beit Safafa limestone quarries, 409
Belagola porphyry quarries, 360
Belair, sandstone quarries near, 371
Belfast Old Court House, 127
Belgium, Carboniferous Limestone of, 136, 375; Devonian rocks of, 115, 373; *Petit Granit* of, 136, 375
Bellavadi quartzite quarries, 96, 368
Bellona limestone quarries, 403
Benares, Queen's College, 105
Ben Cruachan granite quarries, 255
Bendigo, granite quarries near, 58
Benguela granite quarries, 47, 336
Ben Lomond, Tasmania, 151
Ben Rhydding Hydropathic Establishment, 129
"Berea Grit" quarries, 140
Berea, U.S.A., sandstone quarries, 140, 377

Berlin, Imperial Post Office, 34; Potsdam Railway Station, 34
Berlin, U.S.A., rhyolite quarries, 74, 354
Bermeja limestone quarries, 423
Bermuda, Pleistocene rocks of, 243, 430
Berne Canton, Miocene rocks of, 421
Berne, Federal Palace, 190; limestone quarries near, 421
Bernese Oberland, "tuffstein" of, 237, 426
Bernician rocks of Northumberland, 119, 270
Berwickshire, Lake Cheviot area of, 112; Old Red Sandstone of, 112, 266
Berwick-upon-Tweed, Barracks, 121; sandstone quarries near, 120, 271
Beszterezebánya limestone quarries, 217, 411
Bethel, U.S.A., granite quarries, 72, 352
"Bethel White Granite," 72
Biasca granite quarries, 42, 335
Bicknell Museum, Bordighera, 219
Bigrigg, sandstone quarries near, 295
Bilton, sandstone quarries near, 120
Binsted Limestone, 216
"Bird's Eye" granite of Norway, 37
Birkenhead, Central Library, 155; Town Hall, 155
Birling sandstone quarries, 281
Birmingham Gaol, 128
Birr Castle, 125
Birr, limestone quarries near, 125, 275
Birstall sandstone quarries, 282
Bjärnā granite quarries, 331
Bjergbakken granite quarries, 324
Bjoerkelund Hill granite quarries, 41, 334
Black Forest, granites of, 34, 327
Black granites, of Italy, 36, 327; Maine, U.S.A., 65; Russia, 38; Sweden, 40, 334
Black marble of Poolvash, 122

Black Pasture sandstone quarries, 119, 270

Black Prince Monument, Leeds, 36

Black Trap quarries, of Kadêhalli, 339; of Tumkur, 51, 339

Blackenstone granite quarries, 19, 249

Blackpool sea wall, 80

Blackstone Library, Chicago, 68

Blackwater Creek, Queensland, 162

Blatchcombe breccia quarries, 290

Blue Hailes stone quarries, 123, 273

Blue Hills granite quarries, 66

Blue Liver Rock quarries, 123, 273

Blue Liver Stone of Forfarshire, 267

Blue Nelt stone quarries, 282

Blue Stone, of China, 87; Nova Scotia, 141; Warsaw, U.S.A., 116, 374

Bluff Harbour, New Zealand, 61

Board Schools, Sheffield, 128

Bodmin, granite quarries near, 17, 248

Bole Hill sandstone quarries, 277

Bolsover Moor limestone quarries, 144, 293

Bolton Abbey, limestone quarries near, 270

Bolton Wood sandstone quarries, 133, 284

Bombay, granite quarries near, 50; New Docks, 51; Presidency, Deccan Trap, 86, 360; sandstone quarries near, 241, 428; Trust, Chairman of, 50

Bonney, Professor T. G., 189

Boralasgamawa granite quarries, 341; laterite quarries, 428

Bordeaux, Channel Islands, granite quarries, 257; France, St Louis Church, 204

Bordighera, All Saints' Church, 219; Bicknell Museum, 219; buildings of, 219; sandstone quarries near, 219, 412

Borgholm limestone quarries, 102, 369

Borjon, Grand Duke's residence, 206

Borner sandstone quarries, 374

Bornholm Island, Cambrian rocks of, 102, 369; Devonian rocks of, 115, 373; granites of, 31, 324; sandstones of, 102, 369

Borough Green limestone quarries, 316

Borrowdale, green slate quarries, 77, 259; volcanic ash, 77, 259

Boston, U.S.A., County Court House, 63; General Post Office, 67; South Union Railway Station, 62

Botanic Gardens of Copenhagen, 204

"Boticino" of Italy, 188

Bouches-du-Rhône, Cretaceous rocks of, 204, 401

Boulders of basalt, Bay of Tangier, 191, 397; Central Otago, 88, 363; of granite, Chelmsford, U.S.A., 65; Quincy Common, U.S.A., 65

Bouville sandstone quarries, 419

Bovey Tracey, granite quarries near, 18

Bowes Museum, Barnard Castle, 130

Bowral trachyte of Australia, 57

Box Ground limestone quarries, 175, 311

Bracken Hill sandstone quarries, 285

Brackenhill limestone quarries, 294

Brackenwell limestone quarries, 294

Bradford-on-Avon, 175

Bradford, sandstone quarries near, 132, 133, 282; Town Hall, 133; Waterworks, 128

Bradford Township, U.S.A., granites of, 62, 346

Braehead sandstone quarries, 131, 281

Bramley Fall stone quarries, 129, 280

Breadalbane, Earl of, 27

Breakwater, Plymouth, 19; Singapore, 54

Breitenbrunner limestone quarries, 419

Brescia, limestone quarries near, 188, 395

Breslau, sandstone quarries near, 403

Brest, granite quarries near, 79, 326

Bride Church, Isle of Man, 22

Bridge and Cureton sandstone quarries, 299

Bridge, of Alexander III, Paris, 33; Austerlitz, Paris, 185; of the Firth of Forth, 25, 113; of the Firth of Tay, 25; London, 18; Old London, 195; Putney, 19; Tower, London, 19; Vauxhall, London, 18; Waterloo, London, 18, 25; Westminster, 19, 28, 182

Bridgend sandstone quarries, 155, 300

Bridges of the Grand Trunk Railway, 116

Bridgeton, New Jersey, buildings of, 70

Bridgetown Railway, W. Australia, 191

Brisbane, Cathedral, 163; granite quarries near, 58, 344; porphyry quarries near, 88, 362; sandstone quarries near, 377, 389

Bristol, buildings of, 133; Cathedral, 165, 167; sandstone quarries near, 286; Waterworks, 18

British Columbia, Cretaceous rocks of, 214

British Museum, London, 123, 132, 182

British School at Athens, 229

Brittany, granite quarries of, 33, 326

Brohl lava quarries, 79, 356

Bromsgrove Stone of Worcestershire, 154, 298

Bronzite, Mysore quarries of, 340

Broomhouse sandstone quarries, 113, 266

Broughton sandstone quarries, 134, 287

Brouzet limestone quarries, 402

Brown Hill sandstone quarries, 279

Bruce, Dr J. Collingwood, 119

Bruni Island, Tasmania, 151

Brunswick Dock, London, 129

Brussels, Law Courts, 137; Palais de Justice, 186

Bryn-Teg sandstone quarries, 134, 287

Buar granite quarries, 334

Buckingham Palace, London, 123, 176, 185

Buckinghamshire, Heath Stone of, 316; Upper Greensand of, 195, 316

Buckley, E. R., 5

Budapest, limestone quarries near, 184, 228, 391, 419

Buenos Ayres, quartzite quarries near, 108, 372

Buiskop sandstone quarries, 370

Bukit Timah Hills granite quarries, 54, 342

Buluwayo, Government Buildings, 161; Market Hall, 161; sandstone quarries near, 161,

Bulwell limestone quarries, 292

Bundanoon sandstone quarries, 162, 389

Bungalow granite quarries, 342

Bunker Hill Monument, Charlestown, 66

Bunter beds, of Cumberland, 153, 295; Germany, 158, 383; Lancashire, 152, 295; Würtemberg, 158, 383

Burham, St John's Church, 201

Burlescombe limestone quarries, 269

Burma, Eocene rocks of, 224, 415; Miocene rocks of, 231, 422; Nummulitic limestones of, 224; Silurian rocks of, 106, 371

Burnett Co., granites of, 72, 352

Burnley, sandstone quarries near, 134, 288

Burns Monument, Kilmarnock, 157

Burrabool Hills limestone quarries, 191, 398

Burradon sandstone quarries, 135, 288

"Burradon Firestone," 135, 288

Burrum beds of Queensland, 162
Burton-on-Trent, Trinity Church, 154
Burtonport granite quarries, 257
Burwell rocks of Cambridgeshire, 199
Bury St Edmunds, 169
Butchard's Buildings, Durban, 49
Bute Docks, Cardiff, 153

Cabo Giraõ, basalts of, 91, 364
"Cabook" of Ceylon, 241
Cacuaco limestone quarries, 208, 406
Caen, Cathedral, 185; Stone of Normandy, 167, 184, 392
Cairngall granite quarries, 27, 254
Cairo, limestone quarries near, 223, 414; Lady Cromer Hospital, 222
Caithness flagstone, 114, 268
Caithness-shire, Old Red Sandstone of, 114, 268
Caius, Dr, 170
Calcaire grossier of the Eocene system, 218
Calciferous Sandstone, of Edinburgh, 122, 273; Fifeshire, 124, 274; Haddingtonshire, 123, 274
Calcutta, buildings of, 105; High Court, 105
Caldecote Hill, Cambrian rocks of, 98
Caldy Island, limestone quarries of, 121
California, Chico rocks of, 214; Chinese granite used in, 55; Upper Cretaceous rocks of, 214, 410
Calvados, granite quarries of, 33; Lower Jurassic rocks of, 391; Middle Jurassic rocks of, 184, 392
Camaldoli lava quarries, 357
Camara de Lobos basalt quarries, 364
Cambrian rocks, of Australia, 106, 371; Island of Bornholm, 102, 369; Denmark, 102, 369; North Wales, 101, 263; South Australia, 106, 371; United States,

106, 371; Warwickshire, 98, 261; Co. Wexford, 101, 263
Cambridge, buildings of, 7; College: Christ's, 172, 197, 198; Clare, 171, 172; Corpus Christi, 170, 172, 197; Downing, 172; Emmanuel, 200; Gonville and Caius, 170, 171, 172, 174, 197, Gate of Honour of, 172; Jesus, 171; King's, 129, 143, 182, Bridge of, 125, Chapel, 146, 147, 171, 173; Pembroke, 167, 173, 174; Peterhouse, 172, 197, 198; Queens' Chapel, 199; St John's, 199, Chapel, 26, 170, 174, 189; Sidney Sussex, 171, 198; Trinity, 172, 173, 199, Chapel, 171, Fountain, 173; Trinity Hall, 197; Cemetery Lodge, 202; Cornish granite, examples in, 19; Fitzwilliam Museum, 26, 182; Gas Company Offices, 182; Messrs Hallack and Bond's business premises, 156; Messrs Macintosh's business premises, 37, 40; Madingley Hall, 199; Medical Schools, 165, 174; Messrs Pryor's business premises, 19, 27; Roman Catholic Church, 176; St Mary's (Little) Church, 198; St Paul's Church, 202; Messrs Sayle's business premises, 20; Sedgwick Museum, 173; Senate House, 182; Shap granite, examples in, 20; University Library, 175, 182; University Offices, 20
Cambridgeshire, Burwell rock of, 199; clunch of, 197, 317; Lower Chalk of, 197, 317
Campania, limestone quarries of, 206, 403; trachyte tuff quarries of, 358
Campanile, Venice, 219
Canada, Devonian rocks of, 116, 374; granites of, 75, 355; limestones of, 116, 374; Lower Carboniferous rocks of, 140, 378; Millstone Grit of, 140, 378
Canary Islands, trachytes of, 92, 365
Canical basalt tuff quarries, 91

Canterbury Cathedral, 167, 185; Monks of, 185

Cape Ann granite quarries, 66, 349

Cape of Good Hope, Beaufort series of, 150, 382; Coega Stone of, 209, 406; granites of, 49, 338; Lower Cretaceous rocks of, 209, 406; Matsáp beds of, 104; Middle Karroo system of, 150, 382; Pleistocene rocks of, 238, 427; quartz porphyry of, 84, 358; Silurian rocks, 103, 369; Uitenhage series of, 209

Cape system of South Africa, 103, 369

Cape Town, Cathedral, 104; granite quarries near, 338; New Town Hall, 177; Museum, 238; Parliament Houses, 49; Post Office, 49, 238

Cape York, Queensland, 58

Capilla del Monte granite quarries, 76, 355

Caradoc series, Horderley Stone of, 100, 262; Hoar Edge Grit of, 99

Carboniferous rocks of Asia Minor, 138, 376; Australia, 138, 376; Belgium, 136, 375; British Isles, 117, 269; Canada, 140, 379; Russia, 137, 376; Switzerland, 138, 376; United States, 139, 377

Carboniferous Limestone of Anglesey, 121, 272; Belgium, 136, 375; Co. Antrim, 126, 277; Co. Carlow, 126, 276; Co. Cork, 126, 277; Co. Donegal, 126, 277; Co. Kildare, 125, 276; Co. Kilkenny, 126, 276; King's Co., 125, 275; Co. Limerick, 126, 276; Derbyshire, 117, 269; Isle of Man, 122, 272; Northumberland, 119, 270; North Wales, 121, 272; South Wales, 121, 272; Yorkshire, 118, 269

Cardiff, buildings of, 133; Bute Docks, 153

Carey Hill limestone quarries, 306

Carlinghoe sandstone quarries, 285

Carlisle, Citadel Railway Station, 136; St Aidan's Church, 143

Carlow, Carboniferous Limestone of, 126, 276

Carlton Club, London, 26; Hotel, Johannesburg, 150

Carmyllie sandstone quarries, 113, 267

Carnarvonshire, granites of, 21, 251

Carnegie Library, Alleghany, 64; Victoria, B.C., 214; Cockermouth, 135

Carnsew granite quarries, 19, 247

Carrick limestone quarries, 126, 276

Carstone of Norfolk, 194, 316

Carthagena, buildings of, 230

Cary Hill limestone quarries, 167, 306

Caserta limestone quarries, 403; trachyte-tuff quarries, 358

Cassis (France) limestone quarries, 401: (Mauritius) basalt quarries, 366

Castellammare limestone quarries, 403

Casterton limestone quarries, 173, 309

Castle, Bamburgh, Northumberland, 144; Dalgaty, 113; Morton Corbet, 154; Rochester, 194; Windsor, 153, 195, 201

Castle Hill flagstone quarries, 268

Castle Cary Church, 167

Castle sandstone quarries, 135, 288

Castletown, limestone beds of Isle of Man, 122; limestone quarries, 122, 272

Catamon limestone quarries, 406

Cathcart sandstone quarries, 209, 406

Cathedral of Angoulême, 187; Antigua, 226; Arequipa, 93; Batoum, 206; Bayonne, 187; Brisbane, 163; Bristol, 165, 167; Caen, 185; Canterbury, 167, 185; Cape Town, 104; Chester, 154; Chichester, 143, 185; Christ Church, Dublin, 126; Dunblane, 112; Dunkeld,

112; Durham, 119, 185; Ely, 143, 169, 170, 173, 199; Exeter, 6, 167, 173, 200; Gloucester, 168; Hereford, 154; Iona, 124; Khartoum, 208; Lakoma, Lake Nyasa, 46, 95; Lausanne, 230; Letterkenny, 126; Lichfield, 154; Lincoln, 174; Liverpool, 111, 143, 152, 153, 154; Llandaff, 155, 167, 168; Madeira, 91; Isle of Man, 111; Manchester, 128, 153; Melbourne, 224; Milan, 35; Narva, 39; Newark, U.S.A., 64; Norwich, 173; Peel, Isle of Man, 111; Peterborough, 168, 173; Port of Spain, Trinidad, 215; Ripon, 147; Rochester, 183; St Albans, 185, 201; St Giles', Edinburgh, 124; St John the Divine, New York, 63; St Mary's, Edinburgh, 21, 124; St Paul's, London, 12, 19, 122, 171, 177, 180, 181, 185, 194; Salisbury, 183; Tiflis, 206; Warsaw, 39; Wells, 167; Winchester, 167, 195; York, 7, 146, 147, 173

Caucasus, granites of, 40, 333; Upper Cretaceous rocks of, 206, 404

Cave limestone quarries, 428

"Cave Sandstone" of South Africa, 161

Caxäo, basalt of, 91, 364

"Cefn Stone" of Ruabon, 134, 287

Celtic crosses of Cornwall, 17

Cemeteries, of Genoa, 35; Milan, 35

Cemetery, Cambridge, 202; Lake View, Minneapolis, 75

Central Africa, granites of, 46, 336

Central Arcade, Newcastle-upon-Tyne, 130

Central Library, Birkenhead, 155

Central Otago, basalt boulders of, 88, 363

Central Park, New York, Obelisk of, 42, 43

Central Railway Station, Newcastle-upon-Tyne, 119; Vienna, 31

Cesterbeck, Russia, 39

Ceylon, "Cabook" of, 241; dolomite of, 97, 368; gneiss of, 52, 340; granites of, 52, 340; laterite of, 241, 428; Pleistocene rocks of, 241, 428

Chalasa limestone quarries, 213, 409

Chaleur Bay sandstone quarries, 140, 378

Chalk, Lower, of Bedfordshire, 199, 317; Cambridgeshire, 197, 317; Wiltshire, 200, 318

Chalk, Middle, of Devonshire, 200, 318

Chalk quarries, Cherryhinton, 197, 198, 317; near Devizes, 201, 318; near Reach, 197

Chalk stone of Wiltshire, 200

Chamonix-Martigny Railway, 138

Chamundi, granite quarries near, 51, 339

Chapel, of King's College, Cambridge, 146, 147, 171, 173; St John's College, Cambridge, 26; Trinity College, Cambridge, 171

Charente, Middle Jurassic rocks of, 186, 394; Upper Cretaceous rocks of, 204, 400

Charlotte Co., granite quarries of, 76, 355

Charmouth, St Andrew's Church, 200

Chassignelles limestone quarries, 186, 392

Chateau-Gaillard, limestone quarries of, 393

Chaumont limestone quarries, 189, 396

Chatelard granite quarries, 324

Chatham Docks, 30

Chathill, sandstone quarries near, 120, 271

Chatsworth, Derbyshire, 117

Chatwall group of rocks, 99

Chauvigny limestone quarries, 186, 391

Cheesewring granite quarries, 19, 248

"Cheeses" of Lundy Island, 22

Che-Kiang Prov., granites of, 55

Chelmsford, granite boulders of, 65

"Cheltenham Beds" of Gloucestershire, 168

Cheltenham, limestone quarries near, 168

"Chelynch Beds" of Somersetshire, 167

Chemari Stone of Egypt, 414

Chemical composition of stone, 3

Chemical National Bank, New York, 64

Cheops, builder of the Great Pyramid, 42, 221

Cherryhinton chalk quarries, 197, 198, 317

Cheshire "Flecked sandstone," 154; Keuper beds of, 154, 299; Storeton Stone of, 154, 300

Chester Cathedral, 154

Chicago, Bartlett Buildings, 74; Blackstone Library, 68; buildings of, 73; Majestic Theatre, 68; May Memorial Chapel, 62

Chichester Cathedral, 143, 185

Chickamauga, Tenn., State Soldiers Monument, 75

"Chico" rock of California, 214

Chilina, tuff quarries of, 366

Chilmark limestone quarries, 182, 314

China, "Blue Stone" of, 87; granites of, 55, 342; Great Wall of, 54; porphyry of, 87, 362; tuffs of, 87, 362

Chios, Cretaceous rocks of, 209, 406; "Porta Santa" of, 210

Chipping Campden limestone quarries, 168, 306

Chirbet limestone quarries, 408

Chivive Creek, East Africa, 47

Chizree, limestone quarries near, 231, 422

Choukri limestone quarries, 414

Christiania Fjord, granites of, 36, 329

Christ Church, Oxford, 168

Christ's College, Cambridge, 172, 197, 198

Church sandstone quarries, 160, 387

Church Stretton Hills, Lower Silurian rocks of, 99

Church Stretton Parish Church, 99

Cintra, limestone quarries of, 188, 206, 396, 403

Citadel Railway Station, Carlisle, 136

Citizen National Bank, Green Bay, Wis., U.S.A., 74

City Hall, Toronto, 140

Clackmannanshire, dolerite quarries of, 78, 259; Upper Carboniferous rocks of, 136, 289

Clacton-on-Sea, sea wall of, 80

Clare College, Cambridge, New Hall, 172

Clare Co., Millstone grit of, 131, 282

Clarendon Parish sandstone quarries, 226, 416

Clarendon Press buildings, Oxford, 178

Cleopatra's Needle, London, 42

Cleveland, Queensland, granite quarries, 344

Cleveland, Ohio, sandstone quarries, 378

Cliff of Stevns limestone quarries, 203, 400

Clifton, Westmorland, Parish Church, 135

Clinton township limestone quarries, 116

Clipsham limestone quarries, 173, 309

Clive Co. granite quarries, 57

Closeburn sandstone quarries, 156, 301

Clunch of Cambridgeshire, 197, 317

Coal measures of Ayrshire, 136, 289; Clackmannan, 136, 289; Cumberland, 135, 288; Derbyshire, 287; Durham, 134, 288; Gloucestershire, 133, 286; Kincardine, 136, 289; Lanarkshire, 136, 289; Lancashire, 134, 288; Monmouthshire, 133, 286; Northumberland, 134, 288; North Wales, 134, 287; South Wales, 133, 287; Yorkshire, 132, 282

Cobo granite quarries, 257

Cochin, granite quarries near, 52, 340; Laterite quarries near, 239, 428

Cockermouth, Carnegie Library, 135

Coega Stone of Cape of Good Hope, 209, 406

Cofax sandstone quarries, 108, 372

Colcerrow granite quarries, 19, 247

Collares limestone quarries, 206, 403

College, of Dundee, 125; Eton, 177, 185, 196, 201; of Industry, Vienna, 227; Bangor, 134; Trinity, Dublin, 29; Fettes, Edinburgh, 124

Colmenar de Oreja limestone quarries, 229, 420

Cologne, buildings of, 79

Colombay limestone quarries, 391

Colombo, granite quarries near, 53, 340; harbour of, 53; Laterite quarries near, 241, 429

Colon, sandstone quarries near, 164, 390

Colonial Offices, Pietermaritz-burg, 149

Colonial Trust Buildings, Pitts-burg, 71

Colosseum, Leeds, 132; Rome, 82, 83, 237

Colour of Stone, 12; of Old Red Sandstone, 114

Columbia, granites near, 70

Columbia University, New York, 62

Column of Liberty, Copenhagen, 102, 115

Colusa Co. sandstone quarries, 214, 410

Combe Down limestone quarries, 176, 312

Commercial Bank of Batoum, 206

Commercial name of specimens, 1, 2

"Common Rock" quarries, 123, 273

Compression, test of, 5

Concord granite quarries, 68, 350

Conglomerates of Greece, 234, 424

Congressional Library, Washing-ton, 68

Coniston green slate quarries, 100, 263

Connecticut, granites of, 61, 345; "White Granite," 61

Constantinople, buildings of, 220

Conway, U.S.A., granite quarries near, 68

Copenhagen, Botanic Gardens, 204; Column of Liberty, 102, 115; State Life Insurance Offices, 32; Town Hall, 32

Cordel sandstone quarries, 374

Córdoba Province, granites of, 76, 355

Corgrigg limestone quarries, 276

Corinth, limestone quarries near, 234, 419

Cork, Carboniferous Limestone of, 126, 277

Cork Exhibition, 131

Corncockle sandstone quarries, 156, 301

Cornicabra limestone quarries, 420

Cornish granites, colour of, 19

Cornwall, granites of, 16, 247; Metamorphic rocks of, 94, 260

Coromandel Co. granite quarries, 60, 345

Corpus Christi College, Cam-bridge, 172, 197

Correnie granite quarries, 25, 253

Corsehill sandstone quarries, 156, 301

Corsham Down limestone quar-ries, 176, 311

Corsica, granites of, 33, 326

Côte d'Or, Middle Jurassic rocks of, 186, 393

Cotta limestone quarries, 205, 402

Cottesloe limestone quarries, 225

Cotton Exchange, New York, 192

Country, buildings in the, 7

County Council Offices, Durham, 130

County Court House, Boston, U.S.A., 63

Couronne, limestone quarries near, 401

Court House, Abuski, 46 ; Galveston, Texas, 72 ; Harris, Texas, 72 ; St John's, Antigua, 226 ; Wise, Texas, 72

Covehithe Church, Suffolk, 202, 318

Coxbench, sandstone quarries near, 297

C. P. R. Telegraph Buildings, Montreal, 140

Craglee limestone quarries, 277

Craighall granite quarries, 48, 338

Craigleith sandstone quarries, 122, 273

Craig-y-Herg sandstone quarries, 287

Crailsheim sandstone quarries, 158, 383, 384

Crathie Church, Balmoral, 27

Crathie granite quarries, 253

Creetown granite quarries, 28, 255

Cretaceous rocks of Angola (West Africa), 208, 406 ; Austria, 202, 399 ; Bedfordshire, 199, 317 ; British Columbia, 214, 410 ; Buckinghamshire, 195, 316 ; California, 214, 410; Cambridgeshire, 197, 317 ; Canada, 214, 410 ; Cape of Good Hope 209, 406; Chios, 209, 406; Denmark, 203, 400 ; Devonshire, 200, 318 ; Egypt, 207, 405 ; France, 204, 400 ; Germany, 204, 402 ; Greece, 205, 403 ; Hampshire, 196, 317; Italy, 205, 403 ; Kent, 194, 201, 316, 318 ; Norfolk, 194, 196, 316, 317; Portugal, 206, 403 ; Russia, 206, 404 ; Saxony, 205, 402; South Africa, 209, 406; Suffolk, 202, 318; Surrey, 195, 316; Sussex, 193, 315; Switzerland, 207, 404 ; Syria, 210, 406 ; Trinidad, 215, 410 ; Tunis, 207, 404 ; United States, 214, 410 ; West Africa, 208, 406 ; Wiltshire, 200, 318

Criffel, granite blocks of, 22

Croft granite quarries, 21, 250

Cromford sandstone quarries, 128, 278

Crosby granite quarries, 22, 252

Crosland Hill sandstone quarries, 132, 284

Croxdale Abbey, 154

Crusaders' Mosque, Gaza, 214

Crushing strain of stone, 5

Cuba, Miocene rocks of, 232, 423

Cubic foot of stone, weight of, 4

Cuenca limestone quarries, 420

Cuenca Province, Miocene rocks of, 229, 420

Cullaloe sandstone quarries, 124, 275

Cumberland, Bunter beds of, 153, 295 ; Coal Measures of, 135, 288 ; granites of, 21, 249 ; Lower Silurian rocks of, 100, 262 ; "Penrith sandstone" of, 143

Cumberland Co., Nova Scotia, sandstone of, 141, 379

Cumberland County Court House, Bridgeton, N.J., 70

Cummock, sandstone quarries near, 136, 289

Curraes granite quarries, 330

Currook Billy Hills granite quarries, 57

Custom House, Liverpool, 155 ; Nashville, 192

Cut Stone of St Lucia, 90

Cuyahoga Co. sandstone quarries, 140

Cybele, Temple of, 82

Cyprus, Pliocene rocks of, 234, 424

Dacites, of Asia Minor, 84, 358 ; of St Lucia, 90, 364

Dāgobas of Ceylon, 52

Dalbeattie granite quarries, 28, 255

Dalgaty Castle, 113

Dalgaty sandstone quarries, 113, 267

Dalkey granite quarries, 28, 255

Dalston, sandstone quarries near, 153

Dam of Aswan, 43

Damascus Gate of Jerusalem, 212

Dampier Co., granite quarries of, 57

Dancing Cairns granite quarries, 25, 253

Darka Hills of Morocco, 223
Darley Dale Millstone grit, 127, 278
Darlington Co., U.S.A., 70
Darling Downs of Queensland, 162
Dar Samanond Citadel limestone quarries, 414
Dar Samanond Stone of Egypt, 414
Dartmoor, granites of, 16, 17, 248
Dartmouth Naval College, 110
Daufenbach sandstone quarries, 374
Dauphin Co. sandstone quarries, 164
"Deccan Trap" of Bombay, 86, 360
Deccan, granites of, 51, 339
Deer Park granite quarries, 256
De Lank granite quarries, 17, 248
Delhi, Palaces of, 105
Denbigh, sandstone quarries near, 287
Denmark, Cambrian rocks of, 102, 369; Cretaceous rocks of, 203, 400; Devonian rocks of, 115, 373; Faxe chalk of, 203, 400; granites of, 31, 324
Denner Hill sandstone quarries, 195, 316
Dentamaro limestone quarries, 414
Denwick sandstone quarries, 120, 271
Departmental Buildings, Ottawa, 141
Derbyshire, Carboniferous Limestone of, 117, 269; Keuper beds of, 296; Magnesium Limestone of, 144, 293; Millstone grit of, 127, 277, 278
Description of specimens, 2
Dettenhausen sandstone quarries, 385
Devil's Dyke, Cambridgeshire, 197
Devizes, chalk quarries near, 201, 318
Devon sandstone quarries, 136, 289
Devonian rocks of Belgium, 115, 373; Island of Bornholm, 115,

373; Canada, 116, 374; Denmark, 115, 373; Devonshire, 110, 264; the Eifel, 115, 374; Germany, 115, 374; New York State, 115, 374; the Rhine, 115, 374; United States, 115, 374
Devonport Docks, 20, 37
Devonshaw, dolerite quarries of, 259
Devonshire, Beer Stone, 200, 318; Cretaceous rocks of, 200, 318; Devonian rocks of, 110, 264; granite, colour of, 19; granites of, 16, 250; Metamorphic rocks of, 95, 260; Middle Chalk of, 200, 318; Permian rocks of, 142, 290; schist quarries of, 259; Tors of, 17
De Whepdale sandstone quarries, 292
Dhoon granite quarries, 22, 251
Dig Palace, 105
Dikili, lava quarries of, 84, 358
Dinorben limestone quarries, 272
Dionysus, theatre of, 234
Diorite of Egypt, 335; New Zealand, 60, 345; Queensland, 344; Russia, 331, 333; Singapore, 342; South Africa, 338; Transvaal, 96, 368
Djebal Djelond limestone quarries, 404
Djebal, Karrouba, limestone quarries, 404
Dock Offices of Leeds, 133
Docks of Avonmouth, 134; Barrow-in-Furness, 153; Bombay, 51; Brunswick (London), 129; Chatham, 30; Devonport, 20, 37; Hartlepool, 80; Keyham, 19, 20, 37; Liverpool, 28; Milford, 18; Milwall (London), 129; Pembroke, 19; Portsmouth, 23, 179; Swansea, 28
Doddington Hill sandstone quarries, 111, 265
Dodowa quartzite quarries, 96, 367
Doges, Palace of the, Venice, 219
Dolerites of Pennsylvania, 351; Scotland, 78, 259

Dolomites of Ceylon, 97, 368; Durham, 148, 294; Derbyshire, 144,293; Nottinghamshire,144, 292; Yorkshire, 147, 293

Donaghcumper limestone quarries, 125, 276

Doncaster, limestone quarries near, 294

Donegal,Carboniferous Limestone of, 126, 277; granites of, 30, 257

Donnybrook sandstone quarries, 191, 398

Donors of specimens, 2, 3

Doonagore sandstone quarries, 131, 282

Dorsetshire, Purbeck beds of, 183, 314; Portland Oolites of, 178, 313

Doulting Stone of Somersetshire, 167, 305

Dover, New Admiralty Harbour, 18; Pier, 183

Down Co., granites of, 29, 256; Keuper beds of, 157, 303

Downing College, Cambridge, 172

Downing Farm sandstone quarries, 149, 380, 381

Downton sandstone quarries, 100, 262

"Drag," used by masons, 10

Draycott sandstone quarries, 300

Dresden, buildings of, 205; granite used in, 34; Market Hall, 34; Technical Schools, 34

Drôme, Middle Jurassic rocks of, 187, 394

Drumaroon sandstone quarries, 127, 277

Drumkeelan sandstone quarries, 126, 277

Dublin, Christ Church Cathedral, 126; Museum, 29, 126; Parnell Monument, 30; Trinity College, 29; Wellington Monument, 29

Dublin Co., granites of, 28, 255

Ducal Palace of Venice, 203

"Ducka Stone" of Egypt, 414

Duke's sandstone quarries, 128, 278

Dumbartonshire, Old Red Sandstones of, 113, 266

Dumfries, sandstone quarries near, 302

Dumfriesshire, Triassic rocks of, 156, 301

Duna Almas limestone quarries, 228, 418

Dunbar, Parish Church, 113; sandstone quarries near, 112, 268

Dunblane Cathedral, 112

Dundee, College, 125; sandstone quarries near, 267

Dunedin, Bank of New Zealand, 225; First Church, 225; Government Life Insurance Buildings, 61, 163; Law Courts, 89, 225; New Railway Station, 88; Stuart Monument, 163; tuff quarries near, 363

Dunhouse sandstone quarries, 280

Dunkeld Cathedral, 112

Dunstable Priory, 200

"Duntrune Rock" of Forfarshire, 267

Durability of stone, 5

Durban, Bank of Africa, 35; Butchard's Buildings, 49; New Town Hall, 149; Standard Bank, 149, 161

Durham, Cathedral, 119, 185; County Council Offices, 130

Durham Co., Coal Measures of, 134, 288; Millstone grit of, 130, 280; Permian rocks of, 148, 294

Dutch occupation of Ceylon, 53

Dwyka series of Natal, 149

Dyce granite quarries, 24, 252

East Grinstead, sandstones of, 193, 315

East London, Cape of Good Hope, sandstone quarries near, 150, 382; Fire Station, 150

East Rand sandstone quarries, 150, 381

Eastern Quarries, basalt quarries, 356

Ecca beds of Natal, 149, 380; of Transvaal, 149, 381

Eccleshills sandstone quarries, 285

Eckstein's Buildings, Johannesburg, 104

Eddystone Lighthouse, 17, 182

Edinburgh, Calciferous Sandstone of, 122, 273; Fettes College, 124; Free Church Assembly Hall, 123; National Gallery, 144; Observatory, 112; Royal Infirmary, 123; St Giles' Cathedral, 124; St Mary's Cathedral, 124; sandstone quarries near, 122, 273; "Scotsman" Office, 120; Waverley Railway Station, 120; Wesleyan Memorial Hall, 112

Edithweston limestone quarries, 173, 309

Effingham sandstone quarries, 149, 381

Eggenburg, limestone quarries near, 217, 411

Egypt, "Batur Stone" of, 223, 414; "Chemari Stone" of, 414; Cretaceous rocks of, 207, 405; Diorite of, 335; "Ducka Stone" of, 414; Eocene rocks of, 220, 413; granites of, 42, 335; Nubian sandstone of, 207, 405; Nummulitic limestones of, 220

Egyptian granite in Rome, 45

Eifel, basalts of the, 79, 356; Devonian rocks of the, 115, 374; lavas of the, 79, 356; sandstones of the, 115, 374

Eighth Mile granite quarries, 342

Ekanäs granite quarries, 38, 331

Ekaterinoslav granite quarries, 40, 333

Eleanor Cross, Charing Cross, London, 147

Elginshire, Old Red Sandstone of, 114, 268; Triassic rocks of, 155, 300

Elkair limestone quarries, 404

Ellora, temples of, 50

El-Meaddasa Citadel, limestone quarries near, 414

Elterwater green slate quarries, 101, 262

Ely Cathedral, 143, 169, 170, 173; Lady Chapel of, 199

"Emerald Pearl" granite of Norway, 37

Emmanuel College, Cambridge, 172, 200

Emmanuel II Monument, 188

Emperor Nicholas of Russia, gift of, 39

"Emperor's Red" of Portugal, 188

Ennoggera granite quarries, 344

Entre Rios, sandstones of, 164, 390

Eocene rocks, of the Abruzzi, 219, 412; Antigua, 226, 417; Asia Minor, 224, 415; Austria, 217, 411; Burma, 224, 415; Egypt, 220, 413; Hungary, 217, 411; Istria, 218, 411; Italy, 218, 412; Jamaica, 226, 416; Lower Austria, 217, 411; Morocco, 223, 415; South Australia, 224, 415; Turkey, 220, 413; Tuscany, 219, 412; Venetia, 218, 412; Victoria, 224, 415; Western Australia, 225, 415

Ephesus, limestone quarries near, 235

Epidiorite of Sweden, 334

Equitable Life Assurance Offices, Sydney, 57

Erman, Adolph, 44

"Erratic Blocks" of Switzerland, 41, 334

Eskdale granite quarries, 21, 249

Estaillades limestone quarries, 401

Estancia Chapademalal quartzite quarries, 372

Estavayer sandstone quarries, 231, 422

Eton College, 177, 185, 196, 201; Ante-Chapel of, 178

"Euclid Blue Stone" of Ohio, 140

Euclid sandstone quarries, 140, 378

Euston Railway Station, London, 128, 129

Euville limestone quarries, 185, 392

Exchange Buildings, Glasgow, 124; Liverpool, 132; Manchester, 132; Newcastle-upon-Tyne, 135
Exeter Cathedral, 6, 167, 173, 200
Exhibition at Cork, 131
Eye sandstone quarries, 264

Faber, Mount, conglomerates of, 242; granites of, 54
Fairfield Co., granites of, 70
"False Bedded Old Red" of Hereford, 111
Fariolo, granite quarries near, 35
Farleigh Down limestone quarries, 176, 311
Fatehpur-Sikri, Palace of, 105
Fauldhouse, sandstone quarries near, 281
Favarage limestone quarries, 404
Faxe Chalk of Denmark, 203, 400
Federal Palace, Berne, 190
Felsite of Mysore, 86, 360
Feluy-Arquennes limestone quarries, 373
Ferric oxide in granite, 28
Fersenes sandstone quarries, 114, 268
Fettes College, Edinburgh, 124
Fiano tuff quarries, 83, 358
Fibyingyi limestone quarries, 371
Fifeshire, Calciferous Sandstone of, 124, 274
Fingal district of Tasmania, 151
Finhaut limestone quarries, 376
Finistère, granite of, 79, 356
Finland, granites of, 38, 331
Fire-resisting stone, of Madeira, 91, 364; Malmstone, 196; Merstham, 195, 316; Northumberland, 131, 135, 281, 288; St Christopher, 89, 363; St Helena, 92, 366; Windrush, 196, 317
Fire Station of East London, Cape of Good Hope, 150
Fire Stone of St Kitts, 89, 363
Firewood, sandstone quarries of, 315
First Church, Dunedin, 225
Fish Pond sandstone quarries, 286

Fitzwilliam Museum, Cambridge, 26, 182
Flagstone of Caithness, 114, 268
"Flecked Sandstone" of Cheshire, 154
Flint, knapped, 202, 318; pits near Greenhithe, 318
Florence, U.S.A., Post Office, 70
Focha lava quarries, 85, 359
Fokies lava quarries, 85, 359
Fontanelli tuff quarries, 83, 357
Fontveille limestone quarries, 401
Footprint Beds of Storeton Stone, 155
Fordell, sandstone quarries near, 125, 275
Forest of Dean, sandstones of, 133, 286
"Forest Sandstone" of Rhodesia, 161, 388
Forest Vale sandstone quarries, 388
Forêt de Brousses limestone quarries, 393
Forfarshire, Old Red Sandstone of, 113, 267
Forsyth Island, granites of, 76, 355
Fort of Galle, buildings of, 53; granite quarries near, 53
Forth Bridge, 25, 113
Fortifications of Vienna, 227
Fortuna Virilis, Temple of, 82
Forum of Augustus, 82
"Fossil Footprints" of Storeton, 155
Fougères granite quarries, 326
Fountain City, limestone quarries near, 108, 372
Foxdale granite quarries, 22, 251
Fox Island, granites of, 63
Foynes limestone quarries, 276
France, Cretaceous rocks of, 204, 400; granites of, 32, 324; Jurassic rocks of, 184, 391; Kirsantite of, 79, 356; Miocene rocks of, 419; Oligocene rocks of, 218, 411
Franko-Swabian rocks of Germany, 187
Frasnes-les-Couvin, limestone quarries of, 115, 373

Frederiks, sandstone quarries of, 369, 373

Frederiksstadt, granite quarries near, 329

Frederikshald, granite quarries near, 37, 328, 329

Free Church Assembly Hall, Edinburgh, 123

Free Library, Liverpool, 134

Freestones, 9

Freistadt, granite quarries near, 31, 323

Fremigère limestone quarries, 394

French Hoek Valley granites, 49

Freudenstadt sandstone quarries, 158, 383

Friborg Canton, Miocene rocks of, 422

Frogmore, Prince Consort Memorial, 27; Royal Mausoleum, 18

Frost, resistance of stone to, 6

Funchal, basalt quarries near, 91

Fundy Bay granite quarries, 76

Gabbros of the Isle of Man, 22, 251; Maine, U.S.A., 65, 349; Russia, 332; South Africa, 338; Sweden, 40, 334

Gabo Island, granites of, 56, 58, 343, 344

Gaj beds of Sind, 231

Galle, granite quarries near, 53, 341; Laterite quarries near, 429

Galloway, granites of, 28

Galveston, Court House of, 72

Galway, granites of, 30, 256

Gambetta, limestone quarries near, 235

Gamla Carlby granite quarries, 330

Gaol of Birmingham, 128; Leicester, 128

Gard, Upper Cretaceous rocks of, 402

Garnstone, quarries of, 264

Gas Company's Offices, Cambridge, 182

Gate of Honour, Gonville and Caius College, Cambridge, 172

Gatelaw Bridge sandstone quarries, 156, 301

Gateshead, sandstone quarries near, 288

Gawday à Samson limestone quarries, 373

Gawler Mountains, granites of, 59

Gaza, Crusaders' Mosque, 214; limestone quarries near, 239, 428

Gebel Auli sandstone quarries, 405

Gebel Silsella sandstone quarries, 405

Geikie, Sir Archibald, 112

Gelati granite quarries, 40

Gennesaret, Lake of, 85, 359

Genoa, cemeteries of, 35

Geological Survey, of Scotland, 156; of Wisconsin, 73

Georgia, granites of, 62, 346

Germany, basalts of, 79, 356; Bunter beds of, 158, 383; Cretaceous rocks of, 204, 402; Devonian rocks of, 115, 374; Franko-Swabian rocks of, 187; granites of, 33, 326; Keuper beds of, 158, 384; Kreuznach beds of, 148; lavas of, 79, 356; Permian rocks of, 148, 380; Pleistocene rocks of, 236, 426; Rhaetic beds of, 159, 386; "Schilfsandstein" of, 159; "Stubensandstein" of, 159; Trachyte tuffs of, 79, 356; Triassic rocks of, 157, 383

Gettysburg granite quarries, 70, 351

Ghizeh, limestone quarries near, 413; Pyramids of, 42, 217, 220, 221

Gib Mountain, syenite of, 57, 344

Gifnock Liver Rock quarries, 124, 274

Gilbert's limestone quarries, 417

Giuschi limestone quarries, 414

Gladstone Memorial Church, 154

Glais, sandstone quarries near, 287

Glamorgan, Rhaetic beds of, 155, 300

Glanmore marble quarries, 376
Glanton, sandstone quarries, 120, 271; Parish Hall, 120
Glasgow, Exchange Buildings, 124; Mitchell Library, 119; Municipal Buildings, 25; St Enoch's Railway Station, 156; University, 124; Western Infirmary, 136
Glaslacken flagstone quarries, 263
Glastonbury Abbey, 167
Glencullan granite quarries, 29, 255
Glendon Stone, quarries of, 171, 308
Glenville granite quarries, 29, 256
Gloucester Cathedral, 168
Gloucester Co., Canada, sandstones of, 140, 378
Gloucester, U.S.A., granite quarries, 67, 349
Gloucestershire, Coal measures of, 133, 286; Inferior Oolites of, 167, 306; Old Red Sandstones of, 111, 265; Pennant beds of, 133, 286; "Red Pennant Stone" of, 134
Gmünd granite quarries, 31, 323
Gneisses, of Ceylon, 52, 340; East Africa, 47, 337; Pennsylvania, 70, 351; Switzerland, 42, 335; West Africa, 46, 336; Wisconsin, 354
Gobanka limestone quarries, 217, 411
Gold Coast, granites of, 46, 336; quartzites of, 96, 367; Triassic rocks of, 160, 387
Gombio sandstone quarries, 382
Gonville and Caius College, Cambridge, 170, 171, 172, 174, 197
Goringler limestone quarries, 194, 315
Göteborg granite quarries, 40
Government Buildings, Buluwayo, 161
Government Departmental Buildings, Ottawa, 141
Government House, Pretoria, 104
Government Insurance Buildings, Dunedin, N.Z., 61, 163
Governor's Palace, Madeira, 91

Grafversfors granite quarries, 333
Grampian Hills, Victoria, Carboniferous rocks of, 139
Grand Rocque granite quarries, 257
Grand Trunk Railway, bridges of, 116
Granite Heights quarries, 73, 353
Granite Mountain quarries, 72, 352
"Granite, Stratified," 53
Granites of Aberdeenshire, 23, 254; Angola, 47, 336; Argentine Republic, 76, 355; Argyllshire, 27, 254; Austria, 30, 323; Bangalore, 51, 339; Black Forest, 34, 327; Calvados, 33; Canada, 75, 355; Cape of Good Hope, 49, 338; the Caucasus, 40, 333; Central Africa, 45, 336; Ceylon, 52, 340; China, 54, 342; Columbia, 70; Connecticut, 61, 345; Cornwall, 16, 247; Corsica, 33, 326; Cumberland, 21, 249; Denmark, 31, 324; Devonshire, 16, 248; Donegal, 30, 257; co. Down, 29, 256; Dublin, 28, 255; East Africa, 47, 336; Egypt, 42, 335; Finland, 38, 331; France, 32, 324; Gabo Island, 56, 343; Galway, 30, 256; Georgia, 62, 346; Germany, 33, 326; Gold Coast, 46, 336; Guernsey, 30, 257; Hong Kong, 55, 342; Hungary, 31, 323; India, 50, 339; Inverness-shire, 27, 254; Italy, 34, 327; Japan, 55, 343; Jersey, 30, 258; Kincardineshire, 27, 254; Kirkcudbrightshire, 28, 255; Leicestershire, 21, 250; Lower Austria, 31, 323; Lundy Island, 22, 252; Maine, 62, 346; Malabar Coast, 52, 340; Isle of Man, 22, 251; Maryland, 65, 349; Mashonaland, 50; Massachusetts, 65, 349; Malay Peninsula, 53; Moravia, 31, 323; Mysore, 51, 339; Natal, 49, 338; New Hampshire, 67, 350; New Jersey, 68, 350; New South Wales, 56, 343; New Zealand, 60,

345; North Carolina, 69, 351; North Wales, 21, 521; Norway, 36, 328; Pennsylvania, 70, 351; Portugal, 37, 330; Portuguese East Africa, 47, 336; Queensland, 57, 344; Rhode Island, 70, 351; Rhodesia, 50, 339; Russia, 38, 330; Saxony, 33, 326; Singapore, 53, 342; South Africa, 47, 337; South America, 76, 355; South Australia, 59, 345; South Carolina, 69, 351; Straits Settlements, 53; Sweden, 40, 333; Switzerland, 41, 334; Texas, 72, 352; the Transvaal, 47, 337; United States, 61, 345; Upper Austria, 31, 323; Vermont, 72, 352; Victoria, 58, 345; West Equatorial Africa, 46, 336; Western Australia, 59, 345; Westmorland, 20, 249; Wexford, 29, 256; Wicklow, 29, 256; Windsor Co., 72; Wisconsin, 73, 353; Würtemberg, 34, 327

Grant, General, sarcophagus of, 73

Grantham, limestone quarries near, 174, 310

Grasse, limestone quarries near, 187, 394; Bank of France Buildings, 187

Gravellona, granite quarries near, 35, 327

Great Canyon sandstone quarries, 378

Great Oolites of Wiltshire, 175, 311

"Great Sand Rock" of Yorkshire, 175

Great Southern and Western Railway of Ireland, 126

Great Wall of China, 54

Grecian Archipelago, Islands of, 84

Greece, conglomerates of, 234, 424; Miocene rocks of, 228, 419; Pliocene rocks of, 234, 424; Upper Cretaceous rocks of, 205, 403

Greek Church, of Kadekena, 224; of Mount of Olives, 211

Greenbrae sandstone quarries, 300

Greenhithe flint pits, 318

Greenmore sandstone quarries, 285

Green Point Savings Bank, New York, 61

Greensand, Lower, of Kent, 194, 316

Green Stone of Ning-po, 87, 362

Greg Malin sandstone quarries, 265

Grenada, basalt tuff of, 89, 363

"Grès de Fontainebleau" of Seine et Oise, 412

Grey granites of Aswan, 45, 335

"Grey Potstone" of India, 96, 368

"Grey Royal Granite" of Norway, 37

Greytown, sandstone quarries near, 149, 380

Grindelford, Millstone Grit of, 127, 277

Grindelwald, limestone quarries near, 237, 426

"Grindstone Post" of Durham and Northumberland, 135

Grinshill Stone of Shropshire, (Red), 154, 299; (White), 154, 299

Grinstead, limestone quarries near, 194, 315

Gröningen limestone quarries, 236, 426

Grotto of Jeremiah, near Jerusalem, 212

Grund granite quarries, 34, 327

Guadalix limestone quarries, 420

Guernsey, granites of, 30, 257; quartz diorites of, 257

Guildhall, London, 117

Guiting, limestone quarries near, 168, 306; Church, 168

Gunnislake granite quarries, 18, 249

Gussman limestone quarries, 414

Gwalior Palace, 105

Habergham sandstone quarries, 288

Haddingtonshire, Calciferous Sandstone of, 123, 274; Old Red Sandstone of, 112, 266

Hagar sandstone quarries, 409

"Hailes Sandstone" quarries, 123, 273, 274

Hainault, Devonian rocks of, 115, 373; Carboniferous Limestone of, 136, 375

Half Way Well sandstone quarries, 144, 292

Halifax, N.S., Legislature Buildings, 140

Halifax, Yorkshire, sandstone quarries near, 132, 283

Hall Dale sandstone quarries, 128, 278

Hall of Records, New York, 64

Hallowell granite quarries, 64, 348

Hamburg, Jews' Synagogue, 34; Town Hall, 34

"Ham Hill Stone" of Somersetshire, 166, 305

Hamilton (Bermuda), limestone quarries near, 430

Hamilton (Lanark) sandstone quarries, 113, 266

Hamilton series of Devonian rocks of U.S.A., 116

Hampshire, Malmstone of, 196; Upper Greensand of, 196, 317

Hampton Court, 195

Hands Hill limestone quarries, 231, 422

Handsworth sandstone quarries, 285

Hangingstone sandstone quarries, 129, 279

Hangö granite quarries, 39, 331

Hanina limestone quarries, 408

Harb, sandstone quarries near, 384

Harbour of Colombo, 53; Kingstown, 29; Laxoes, 38

Harbour Works, Lagos, 46; Simon's Bay, 104

Harcourt Hills, granite quarries of, 58, 345

Hard Blue York Stone, 133, 282

Hard Brown York Stone, 133, 282

Hardening of Stone, 11

"Hard Stone" of St Kitts, 89, 363

Hardwood Island, granite quarries of, 64, 347

Harmer Hill, sandstone quarries of, 154, 299

Harris, Court House of, 72

Harrow sandstone quarries, 194, 315

Harrow School, 111

Hartham Park limestone quarries, 311

Hartlepool, Docks of, 80

Hartshill Abbey, 99

Hartshill Freestone quarries, 98, 261

Hauraki Gulf, granites of, 60

Haslebury limestone quarry, 175

Haslingfield chalk quarries, 197

Hastings, sandstone quarries near, 194, 315; sea wall, 80

Hausen sandstone quarries, 158, 383

Haut Saône granite quarries, 33

Haute Savoie, granites of, 324; Miocene rocks of, 419; Upper Jurassic rocks of, 187, 395

Havana, limestone quarries near, 232, 423

Hawarden, Gladstone Memorial Church, 154

Hawes, sandstone quarries near, 270

Hawkesbury Bridge, Sydney, 57

Hawkesbury sandstone of N. S. Wales, 161, 389

Haydor limestone quarries, 174, 310

Haytor Rock granite quarries, 18

Headington limestone quarries, 177, 312

Heath Stone of Buckinghamshire, 316

Heaton Park sandstone quarries, 282

Heggadadevankote Potstone quarries, 96, 368

Heidelberg sandstone quarries, 382

Heights sandstone quarries, 144, 292

Heilbronn sandstone quarries, 159, 385

Heisterbach basalt quarries, 81, 356

Helidon sandstone quarries, 163, 389

Heliopolis, Obelisk of, 42

Helofaghton limestone quarries, 234, 424

Helsingfors, island near, 39, 332

Helvedsbakkerne granite quarries, 32, 324

Henley Hill Stone of Sussex, 315

Henry VI, 146, 198

Henry VII, 171; Chapel of, Westminster Abbey, 176, 196

Henry VIII, 179

Herald Offices, Chicago, 73

Hereford, Cathedral, 154; Old Red Sandstone of, 110, 264

Heréké conglomerate quarries, 224, 415

Herodes Atticus, Odeum of, 234

Herrenberg, sandstone quarries near, 384

Herrestad Manor granite quarries, 41, 334

Herzogwalden sandstone quarries, 403

Hesse, sandstones of, 148, 380

Heworth Burn sandstone quarries, 130, 280

Higgo granite quarries, 49, 338

High Island granite quarries, 63, 347

High School for Girls, Basle, 190

Higher Bebington sandstone quarries, 154, 300

Hiitis granite quarries, 331

Hill o' Fare granite quarries, 27, 254

"Hills and Holes of Barnack," 169

Hindu Temples of Mysore, 96

Hirshau sandstone quarries, 385

Hoar Edge Grit quarries, 99, 262

Hochdarf sandstone quarries, 384

Hoetjes sandstone quarries, 238, 427

Hofburg of Old Vienna, 217

Hofburg Theatre of Vienna, 203

Hofweller sandstone quarries, 374

Holborn Viaduct, 28

Hollington mottled sandstone, 153, 297; sandstone quarries, 153, 297

Holmesburg granite quarries, 70, 351

Homburg limestone quarries, 190, 397

Hong Kong, Bank of Shanghai, 55; granite quarries near, 55, 342; Law Courts, 87; porphyry quarries near, 362; Post Office, 55

Honister green slate quarries, 100, 262

Honister Pass, slate quarries near, 100

Hook, Dean, 185

Hope Bowdler, Parish Church, 99; sandstone quarries near, 99, 262

Hopeman, sandstone quarries near, 300

Hopton Wood limestone quarries, 117, 269

Horderley stone quarries, 100, 262

Horka granite quarries, 326

Hornton Church, 166; limestone quarries, 305

Horse Road sandstone quarries, 160, 387

"Horse's Teeth," 19

Horsforth sandstone quarries, 129, 280

Horsley Castle sandstone quarries, 297

Horsur porphyry quarries, 360

"Horton Flags" of Yorkshire, 101

Horton limestone quarries, 269

Hospital of Abuski, 46

Hôtel de Ville, Paris, 185

Houran basalt quarries, 85, 359

Houses of Parliament, London, 18, 27, 144, 181, 184, 200; Melbourne, 139; Ottawa, 107; Perth, W. Australia, 192; Quebec, 75; Toronto, 140; Vienna, 31

Hov granite quarries, 37, 329

Howley Park sandstone quarries, 132, 283

Hsipau sandstone quarries, 423

Huaris tuff quarries, 366

Huckford sandstone quarries, 286

Huddersfield, sandstone quarries near, 132, 284

Huddleston Hall, 146
Huddleston limestone quarries, 146, 293
Hughes, Professor T. McKenny, 101, 216
Huliyar, quartzite quarries near, 97, 368
Humbie sandstone quarries, 124, 274
Humbleton limestone quarries, 270
Hume, Dr W. F., 223
"Hummelstone Brown Stone" of Pennsylvania, 164; Hill sandstone quarries, 164
Hungary, Eocene rocks of, 217, 411; Miocene rocks of, 228, 418; granites of, 31, 323; Jurassic rocks of, 184, 391
Hunstanton, Red Chalk of, 196, 317; sandstone quarries near, 316
Hunter's Hill sandstone quarries, 162, 389
Hunwick, sandstone quarries near, 381
Hurricane Island, granites of, 63, 346
Hut dwellings of Cornwall, 17
Hyde Park, Transvaal, granite quarries, 48, 338
Hyderabad, granite quarries near, 339
Hydropathic of Ben Rhydding, 129
Hypabyssal rocks, 16

Ibn-son-Abed limestone quarries, 408
Iguesta de Andrés trachyte quarries, 365
Ilfracombe, St Peter's Church, 110
Ilkley, sandstone quarries near, 129, 279
Ille et Vilaine granites, 326
Immosthay limestone quarries, 313
Imperial Institute, London, 49, 117, 123
Imperial Palaces, St Petersburg, 38; Vienna, 227
India, basalts of, 86, 360; felsite

of, 86, 360; gneiss of, 340; granites of, 50, 339; Káimur group of rocks, 105; Laterite of, 240, 423; peridotite of, 340; Pleistocene rocks of, 239, 428; Porbander Stone of, 241, 428; porphyry of, 86, 360; Silurian rocks of, 105, 370; Vindhyan system of, 105
Indiana, Carboniferous rocks of, 139, 377; Oolites of, 192
"Indian Potstone," 96, 368
Indian Town, sandstone quarries near, 140, 378
Infantry Barracks, Madeira, 91
Infirmary, Newcastle-upon-Tyne, 133; Western, Glasgow, 136
Ingleborough, flanks of, 118
Injár, ecclesiastical buildings of, 164
Interlaken, buildings of, 190; Post Office, 190
Inverary Castle, 95
Inverness Castle, 114
Inverness-shire, granites of, 27, 254; Old Red Sandstone of, 114, 267
Iona Cathedral, 124
Iowa, U.S.A., Oolites of, 192
Ipplepen, limestone rocks near, 264
Ipswich, beds of, Queensland, 162
Irawadi River, limestone quarries near, 106
Irvine, sandstone quarries near, 136, 289
Ismid conglomerate quarries, 224, 415
Ispahan, buildings of, 97; schist quarries near, 97, 368
Istria, Eocene rocks of, 218, 411; Upper Cretaceous rocks of, 202, 399
Italy, "Boticino" of, 188; Cretaceous rocks of, 205, 403; Eocene rocks of, 218, 412; granites of, 34, 327; Jurassic rocks of, 188, 395; lavas of, 81, 357; "Peperino" of, 82, 356; "Piperno" of, 83, 357; Pleistocene rocks of, 236, 426; Pozzolano of, 82; Travertine

of, 83, 236, 426; volcanic tuff of, 81, 356

Jaimanito limestone quarries, 423

Jakkenhalli porphyry quarries, 361

Jamaica, Aqueduct, 226; Eocene rocks of, 226, 416; Miocene rocks of, 232, 423; Port Antonio Church, 233; Sugar Works, 226

James I of England, 179

Jamestown, buildings of, 92

Japan, Bank of, 56; granites of, 55, 343

Jardine Hall, Lockerbie, 156

Jasin limestone quarries, 408

Jaumont limestone quarries, 205

Jebel Mokhattam of Egypt, 220

Jebel Musa of Morocco, 223

Jeffrey Hale Hospital, Quebec, 141

Jeremiah's Grotto, near Jerusalem, 212

Jericho, sandstone quarries near, 213, 409

Jersey granites, 30, 258

Jerusalem, ancient buildings of, 211; Damascus Gate, 212; limestone quarries near, 210, 407; Russian Pilgrims' Society's Hospital, 211

Jesus College, Cambridge, 171

Jew's river limestone quarries, 223, 415

Jews' Synagogue, Hamburg, 34

Johannesburg, buildings of, 104; Carlton Hotel, 150; Eckstein's Buildings, 104; granite quarries near, 48, 338; St Mary's Church Parish Hall, 150; sandstone quarries near, 382

Jonesborough granite quarries, 65, 348

"Jonesboro' Red Granite," 65

Jones, Inigo, 179

Jonesport township granite quarries, 64, 347

Jönköping granite quarries, 40

Jurassic rocks of Austria, 184, 391; Dorsetshire, 178, 313; France, 184, 391; Germany, 187, 395; Gloucestershire, 167, 306; Hungary, 184, 391; Italy, 188, 395; Lincolnshire, 168, 174, 309; Morocco, 190, 397; Northamptonshire, 168, 307; Oxfordshire, 177, 312; Portugal, 188, 395; Rutland, 168, 309; Somersetshire, 166, 305; Switzerland, 166, 189, 391, 396; Victoria, 191, 398; Western Australia, 191, 398

"K Stone" quarries of New Zealand, 225, 416

Kadêhalli black trap quarries, 339

Kadekena Greek Church, 224; Post Office, 224

Kadur quartzite, 96, 368

Káimur group of rocks in India, 105

Kaiserstein limestone quarries, 418

Kajraha Temples, 105

Kakuli rocks of Palestine, 211, 409

Kalajoki, granite quarries near, 330

Kalamaki limestone quarries, 419

Kamenz, granite quarries near, 34, 326

Kandy, dolomite quarries near, 97, 368

Karabouran Peninsula lava quarries, 85, 359

Karáchi, limestone quarries near, 231, 422

Kärda granite quarries, 40, 334

Karguentah limestone quarries, 235, 425

Karnak, Obelisk of, 43

Karroo System of South Africa, 148, 380; Lower, of Natal, 149, 380; Lower, of Transvaal, 149, 381; Middle, of Cape of Good Hope, 150, 382; Upper, of Rhodesia, 160, 387

Káthiáwár, sandstone quarries near, 241, 428

Kato Liosia, conglomerate quarries near, 424

Keinton Mandeville, limestone quarries near, 304

Kellerberrin granite quarries, 60, 345

Kellett sandstone quarries, 226, 416

Kemnay granite quarries, 25, 252

Kempegandan Koppal quarries, 361

Kengal Koppal porphyry quarries, 361

Kennebec county, granites of, 64

Kent, Lower Greensand of, 194, 316; Upper Chalk of, 201, 318

Kentallen granite quarries, 254

"Kentish Rag" quarries, 194, 316

Kenton sandstone quarries, 130, 281

Kentucky, Jurassic rocks of, 192, 398

Kephisia, conglomerate quarries near, 424

Kersantite of France, 79, 356

Kersanton granite quarries, 79, 356

Kesalgiri porphyry quarries, 360

Keswick, Bank of Liverpool, 135; buildings of, 77, 100; granite quarries near, 250

Kettering, All Saints' Church, 172; limestone quarries, 171, 308; St Mary's Church, 172

Kettle River sandstone quarries, 107, 372

Ketton Stone of Rutland, 171, 172, 308

Keuper beds of Cheshire, 154, 299; Derbyshire, 296; Co. Down, 157, 303; Shropshire, 154, 299; Staffordshire, 153, 297; Worcestershire, 154, 298; Würtemberg, 159, 384, 385

Keyham Docks, 19, 20, 37

Khafra, tomb of, 42, 44

Khartoum, Gordon Memorial Cathedral, 208; sandstone quarries near, 208, 405

Khazistan sandstone quarries, 242, 429

Khufu, builder of the Great Pyramid, 42, 221

Kiang-su Prov., granites of, 55

Kildare, Carboniferous Limestone of, 125, 276

Kilkenny, Carboniferous Limestone of, 126, 276

Killingworth, sandstone quarries near, 288

Kilmarnock, Burns Monument, 157

Kincardineshire, granites of, 27, 254; Upper Carboniferous rocks of, 136, 289

Kincardine, sandstone quarries near, 136, 289

King's Cliffe limestone quarries, 172, 308

King's College, Cambridge, 125, 129, 143, 182; Chapel, 146, 147, 171, 173

King's Co., Carboniferous Limestone of, 125, 275

King's sandstone quarries, 120, 271

Kinross, dolerites of, 259

Kirkcudbrightshire, granites of, 28, 255

Kirkmabrede granite quarries, 255

Kirkstall Abbey, 129

Kirtle Bridge sandstone quarries, 157, 302

Kivelagher sandstone quarries, 150, 382

Kiveton Park limestone quarries, 293

Klerksdorp sandstone quarries, 150, 381

Klingenmunster sandstone quarries, 387

Klintellökke granite quarries, 324

Klippegaard granite quarries, 32, 324

Knapped flint, 202, 318

Kobe, granite quarries near, 55, 343

Kockemoed sandstone quarries, 150, 381

"Koblenkeuper" rocks of Germany, 158

"Kokanga Stone" of New Zealand, 88

Kokar granite quarries, 331

Königswinter, basalt quarries near, 81, 356

Kowloon granite quarries, 55, 342; porphyrite quarries, 87, 362

Kreise Wenew limestone quarries, 137, 376

Kreistadt, limestone quarries near, 376

Kreuznach beds of Germany, 148, 380

Kronoberg granite quarries, 40

Kroonstadt, sandstone quarries near, 161, 388

Kuh-i-Sufi mountain schists, 97

Kurla basalt quarries, 86, 360

"Kurla Stone" of Bombay, 86

Kursebi granite quarries, 40, 333

Kutais, buildings of, 206; granite quarries, 40, 333; limestone quarries near, 206, 404

Kwan-tung Prov., granites of, 55

Kyatanhalli porphyry quarries, 361

Labels on specimens, description of, 1

Laber granite quarries, 33, 326

Labradorites of Norway, 36

Ladies' College, Arequipa, 93

Ladoga Lake, granite quarries near, 39, 333

Ladybrand, sandstone quarries near, 160, 387

Lady Chapel, Ely Cathedral, 199

Lady Cromer Hospital, Cairo, 222

La Falda granite quarries, 76, 355

Lagos, granites of, 46, 336

Laguna Bay, Queensland, 162

Lake Cheviot area of Berwickshire, 112

Lake, Mr Philip, 240

Lake View Cemetery, Minneapolis, 75

Lakoma, Cathedral, 46; granite quarries, 46, 336; Island, 95

Lamberken sandstone quarries, 266

Lambeth Palace, London, 176

Lameiras limestone quarries, 396

Lamo granite quarries, 333

La Moie granite quarries, 30, 258

Lamorna granite quarries, 19, 247

Lanarkshire, Old Red Sandstone of, 113, 266; Upper Carboniferous rocks of, 136, 289

Lancashire, Bunter beds of, 152, 295; Coal Measures of, 134, 288; Millstone Grit of, 128, 279; Pebble beds of, 152, 295

Lancaster, New Town Hall, 128; sandstone quarries near, 279

Langton limestone quarries, 314

Langton, Sir John, 146

Lanna limestone quarries, 102, 369

"Lapis Albanus" of Rome, 82

"Lapis Tiburtinus" of Rome, 236

Larnaca, limestone quarries near, 234, 424

Larvik granite quarries, 36, 328

Larvikites of Norway, 36

Las Palmas, lava quarries of, 92, 365

Laterite of Ceylon, 241, 428; Malabar Coast, 239, 428

Laterite quarries near Cochin, 239, 428; Colombo, 241, 429; Galle, 241; of Malabar, 240

La Turbie limestone quarries, 187, 394

Launceston, Metamorphic rocks near, 94, 260

Lausanne, Cathedral, 230; sandstone quarries near, 422

Lavas of Asia Minor, 84, 358; Canary Islands, 92, 365; Germany, 79, 356; Italy, 83, 357; Madeira, 90, 364; Naples, 83, 357; Phlegræan Fields, 83; Siam, 87, 362

Laventille limestone quarries, 215, 410

Lavoux, limestone quarries near, 394

Law Courts of Brussels, 137; Dunedin, 89, 225; Perth, W. Australia, 192; Pretoria, 96, 150

Lawrence Co., limestones of, 139

Laxoes, Harbour -of, 38
Lazonby grit, 143
Lazonby Fell sandstone quarries, 143, 291
Leckhampton limestone quarries, 168, 307
Leebotwood, sandstone quarries near, 262
Leeds, Colosseum, 132; Dock Offices, 133; sandstone quarries near, 280, 284, 285; Town Hall, 133
Leek Island granite quarries, 76, 355
Leek, sandstone quarries near, 298
Lefke sandstone quarries, 138, 376
Legislative Council Buildings, Pietermaritzburg, 149
Legislative Buildings, Halifax, N.S., 140
Leibnitz, limestone quarries near, 419
Leicester Gaol, 128
Leicestershire, granites of, 21, 250
Leipzig, buildings of, 205; Market Hall, 34; Post Office, 34; University, 34
Leiria, limestone quarries near, 396
Leith Nautical College, 113
Leitha-Gebirge group, Miocene rocks of, 228, 418
Lengue granite quarries, 47, 336
Lens limestone quarries, 187, 394
Letterkenny Cathedral, 126
Leucite-tephrite of Naples, 357
Leucite lavas of Vesuvius, 83, 357
Liassic rocks of Oxfordshire, 166, 305; Somersetshire, 165, 304; Switzerland, 166, 391
Library, Central, Birkenhead, 155; Public, Sydney, 162
Lichen on building stone, 8
Lichfield Cathedral, 154
Liebig, Baron, 114
Life Insurance Buildings, Washington, 71
Lighthouse, Bass Rock, 124; Beachy Head, 19; Bell Rock,

25; Eddystone, 17, 19, 182; Margate, 183
Lignerolles limestone quarries, 393
Lilholt granite quarries, 37, 329
Lilla Bokult granite quarries, 334
Limasol, limestone quarries near, 235
Limbue soap stone quarries, 95, 367
Lime Street Railway Station, Liverpool, 155
Limerick, Carboniferous Limestone of, 126, 276
Lincoln, Cathedral, 174; limestone quarries near, 174, 310
Lincoln Co., Canada, limestone quarries, 116, 374
Lincolnshire, Inferior Oolites of, 168, 309; limestones of, 174, 309
Lincolnshire Limestone series, 168, 308
Lincolnville granite quarries, 348
Lindley limestone quarries, 310; sandstone quarries, 143, 290
Lindley Wood limestone quarries, 292
Line of bedding marked on stone, 10
Lineage and Hill Top sandstone quarries, 298
Linlithgowshire, Millstone Grit of, 131, 281
Lion's Head, Cape Colony, 103
Lion of Lucerne, 231
Liskeard, granite quarries near, 19, 248
Lithri lava quarries, 85, 359
Little St Mary's Church, Cambridge, 198
Liverpool, Cathedral, 111, 143, 152, 153, 154; Custom House, 155; Docks, 28; Exchange, 132; Free Libraries, 134; Lime Street Railway Station, 155; Museum, 134; St George's Church, 155; St George's Hall, 26, 128; Walker Gallery, 134; Wellington Monument, 155
Liverpool Street Railway Station, London, 157

Llanbedrog granite quarries, 21, 251

Llandaff Cathedral, 155, 167

Llwyndu sandstone quarries, 287

Loanda-Malanga Railway, 208

Loanda Municipal Buildings, 209

Lobito-Katanga Railway, 47

Lochabriggs sandstone quarries, 302

Loch Awe, granite quarries near, 255

Loch Fyne, St Catherine's Stone of, 95, 260

Lockerbie, Jardine Hall, 156

Lodden Valley, granites of, 58

Lohrville granite quarries, 354

Lombardy, Jurassic rocks of, 188, 395

London, Bank of England, 123; Banqueting Hall, Whitehall, 180; Bridge, 18, 195; British Museum, 123, 132, 182; Brunswick Dock, 129; Buckingham Palace, 123, 176, 185; Cleopatra's Needle, 42; Eleanor's Cross, Charing Cross, 147; Euston Railway Station, 128, 129; General Post Office, 182; Guildhall, 117; Houses of Parliament, 18, 27, 144, 181, 184, 200; Imperial Institute, 49, 117, 123; Lambeth Palace, 176; Liverpool Street Railway Station, 157; Millwall Docks, 129; Museum of Practical Geology, 146; New Law Courts, 143; New Sessions House, 133; New War Office, 132; Old Bailey, 133; Post Office Savings Bank, 132; Ritz Hotel, 41; St Pancras Railway Station, 143; St Paul's Cathedral, 12, 19, 122, 171, 177, 180, 181, 185, 194; Serpentine, Hyde Park, 179; Somerset House, 182; Thames Embankment, 128; Tower of, 201; Tower Bridge, 19; University College, 120; Vauxhall Bridge, 18; War Offices, 168; Waterloo Bridge, 18, 25; Waverley Hotel, 36

London and North Western Railway bridges, 143

Long Cove granite quarries, 65, 348

Long Lane sandstone quarries, 262

Longridge sandstone quarries, 128, 279

Lorraine, Lower Cretaceous rocks of, 204, 402

Lorrettorn limestone quarries, 419

Louvie limestone quarries, 400

Lovejoy granite quarries, 68, 350

Lower Austria, Eocene rocks of, 217, 411; granites of, 31, 323

Lower Carboniferous rocks of Canada, 140, 378; Scotland, 124, 274; United States, 139, 377

Lower Chalk of Bedfordshire, 199, 317; Cambridgeshire, 197, 317

Lower Chindwin sandstone quarries, 231, 422

Lower Greensand of Kent, 194, 316; Norfolk, 194, 316

Lucerne, Lion of, 231; sandstone quarries near, 422

Ludlow, buildings of, 100

Ludlow group of Upper Silurian rocks, 100; sandstone quarries near, 262

Lundy Island, granites of, 22, 252

Luston Stone, quarries of, 264

Lyde sandstone quarries, 264

Lydney, sandstone quarries near, 286

Lyons Railway Station, Paris, 186

MacDonald, Sir J. A., Monument of, 76

Maceira limestone quarries, 395

Machavie sandstone quarries, 381

Mackintosh, Messrs, business premises in Cambridge, 37, 40

Maddison, Wis., Science Hall, 74

Madeira, basalts of, 90, 364; Cathedral, 91; Governor's Palace, 91; Infantry Barracks, 91; lavas of, 90, 364; fire-resisting stone of, 91, 364

Madingley Hall, near Cambridge, 199

Madrid Prov., Miocene rocks of, 229, 420; buildings of, 229
Maggiore, Lake, granite quarries near, 35, 327
Magnesian Limestone of Derbyshire, 144, 293; Durham, 148, 294; Nottinghamshire, 144, 292; Yorkshire, 146, 293
Magnetic Island granite quarries, 344
Magstadt, sandstone quarries near, 159
Mahara granite quarries, 340
Maha Saeya Dāgoba, 52
Mahendragiri, Ganjam district, 51
Maine, U.S.A., black granite of, 65; gabbros of, 65, 349; granites of, 62, 346; number of granite quarries in, 62; quartz monzonite of, 64, 347
Maine et Loire, granite quarries of, 33, 325
Majestic Theatre, Chicago, 68
Malabar Coast, granites of, 52, 340; Laterite of, 239, 428
Malay Peninsula, granites of, 53
Malmesbury, Abbey, 175; Abbot of, 175
"Malmstone" of Hampshire, 196
Malta, Oligocene rocks of, 220, 413
Man, Isle of, Carboniferous Limestone of, 122, 272; Castletown limestone beds of, 122; Cathedral, 111; gabbros of, 22, 251; granite quarries of, 22, 251; Old Red Sandstone of, 111, 265
Manche granite quarries, 33, 326
Manchester Cathedral, 128, 153; Exchange, 132; Rylands Library, 153; Stock Exchange, 37; Town Hall, 132; Wholesale Co-operative Society, 128
Mandya porphyry quarries, 361
Manhalli potstone quarries, 96, 368
Mannerdorf limestone quarries, 418
Mansfield Stone (red), 143, 291; (white), 143, 290
Mansfield Woodhouse limestone quarries, 144, 145, 293

Manzinyama granites, 50
Marathon County Court House, Wausau, 74
Marbles, 15
Marble quarries of Bangkok, 106, 371
Marchalee Elm limestone quarries, 305
Mar del Plata, quartzite quarries near, 108, 374
Margate Lighthouse, 183
Maria Island, Tasmania, 151
Maria-Stiegen Church, Vienna, 227
Mariehamn granite quarries, 331, 332
Marino tuff quarries, 82, 357
Market Hall, Buluwayo, 161; Dresden, 34; Leipzig, 34
Marmora Island, limestone quarries, 220, 413
Marsden limestone quarries, 148, 294
Martigny and Chamonix Railway, 138
Maryland, U.S.A., granite quarries, 65, 349; Triassic rocks of, 163, 390
Marzana limestone quarries, 399
Mashonaland, granites of, 50
Masonic Temple, New York, 71; Philadelphia, 63
Massachusetts, granites of, 65, 349
Massangis limestone quarries, 186, 392
Matlock, Millstone Grit of, 127, 278
Matopo Hills, granites of, 50, 339
Matsáp beds of Cape of Good Hope, 104
Mauchline sandstone quarries, 157, 302
Maulbronn sandstone quarries, 158, 159, 384, 385
Mauritius, basalts of, 93, 366
Mauthausen granite quarries, 31, 323
May Memorial Chapel, Chicago, 62
McBride's basalt quarries, 363
Meckering granite quarries, 60, 345

Medarder sandstone quarries, 148, 380

Medical Schools, Cambridge, 165, 174

Meelin limestone quarries, 277

Méhaigne limestone quarries, 137, 375

Melbourne Cathedral, 224; Houses of Parliament, 139; Ormond College, 191; Town Hall, 151; University, 162; use of granite in, 25, 58

Melekeh rocks of Palestine, 211, 409

Melera limestone quarries, 203, 399

Melrose, trachyte rocks near, 78, 259

Memphis Junction limestone quarries, 398

Mendip Hills, Carboniferous Limestone of, 117

Menkaura, tomb of, 42

Mereuel limestone quarries, 393

Merligen limestone quarries, 190, 397

Merrimac River basin granite quarries, 68

Merrivale, granite quarries near, 18, 248

Mersey Dock and Harbour Board, 121

Merstham sandstone quarries, 195, 316

Mesques Mountain Railway, 40

Metamorphic rocks of British Isles, 94, 260; Central Africa, 95, 367; Ceylon, 97, 368; Cornwall, 94, 260; Mysore, 96, 368; Persia, 97, 368; South Africa, 96, 368; West Africa, 96, 367

Metz, limestone quarries near, 205, 402

Meuse limestone quarries, 115, 373; Middle Jurassic rocks of, 185, 392

"Meulière Blanc" of Seine et Oise, 218, 412

"Meulière Rouge" of Seine et Oise, 218, 411

Micaceous flakes in stone, 10

Mica talc rock of Mysore, 339

Middle Chalk of Devonshire, 200

Middlesex Co., Jamaica, sandstone quarries, 226

Middlesmoor sandstone quarries, 279

Midhurst sandstone quarries, 193, 315

Midland Railway Hotel, London, 21

Migne limestone quarries, 393

Milan Cathedral, 35; Cemeteries, 35

Mildenhall Railway, 197

Milford sandstone quarries, 196, 317

Milford Docks, 18

Milford, Mass., U.S.A., granite quarries, 67, 350

Milford, N. Ham, U.S.A., granite quarries, 68, 350

Milheiroz granite quarries, 330

Miller, Hugh, 114

Mill sandstone quarries, 152, 295

Mills quartz porphyry quarries, 358

Millstone Grit of Co. Clare, 131, 282; Derbyshire, 127, 277, 278; Co. Durham, 130, 280; Lancashire, 128, 279; Linlithgowshire, 131, 281; New Brunswick, 140, 378; Northumberland, 130, 281; Nova Scotia, 141, 379; Stirlingshire, 131, 281; Yorkshire, 128, 279

Millstone Hill granite quarries, 72

Millstone Meadow sandstone quarries, 275

Millwall Docks, London, 129

Minneapolis Lake View Cemetery, 75

Minnesota, Potsdam series of, 107, 372

Miocene rocks of Austria, 227, 418; Burma, 231, 422; Cuba, 232, 423; France, 419; Greece, 228, 419; Hungary, 228, 418; India, 231, 422; Jamaica, 232, 423; Northern Shan States, 232, 423; Sind, 231, 422; Spain, 229, 420; Switzerland, 230, 421

Miramichi sandstone quarries, 378

Mirzapore, Silurian rocks of, 370

Mission House of Onitsha, 238

Mitcham sandstone quarries, 371

Mitcheldean, Old Red Sandstones of, 111, 265

Mitchell Library, Glasgow, 119

"Mizzeh" rocks of Palestine, 210, 211, 407

Moat sandstone quarries, 143, 292

Mœander Valley limestone quarries, 235, 425

Mokhattam Hills limestone quarries, 220

"Molasse" of Switzerland, 230, 421

Monaco, Museum of Oceanography, 187; Prince of, 187

Monks of Canterbury, 185

Monks Park limestone quarries, 177, 312

Monmouthshire, Coal Measures of, 133, 286; Pennant beds of, 133

Monovar limestone quarries, 421

Mont Çuet granite quarries, 257

Monte Gerusalemme, limestone quarries of, 403

Monte Olibano lava quarries, 357

Monte Pedral granite quarries, 330

Montello granite quarries, 73, 353

Montgomery Co. sandstone quarries, 163

Monthey, "Erratic Blocks" of, 41, 334

Montjoie granite quarries, 326

Mont Mado granite quarries, 258

Montorfano granite quarries, 35, 327

Montreal, C. P. R. Telegraph Buildings, 140; limestone quarries near, 378; New Prison House, 75; Victoria Bridge, 116

Mont Salève limestone quarries, 187, 395

Monument, Emmanuel II, 188; Nelson, 19; Senator Sherman, 71

Moor granite quarries, 256

Moor Stone of Devonshire and Cornwall, 17

"Moor-Stone Rock" of Scotland, 131

Moose-a-bec red granite, 64

Morata de Tajuna limestone quarries, 420

Moravia, granites of, 31, 323

Moray Firth, Old Red Sandstone of, 114

Morley sandstone quarries, 283, 296

Morocco, Ape's Hill, 223; Eocene rocks of, 223, 415; Jebel Musa, 223; Jurassic rocks of, 190, 397

Morthoe, sandstone quarries near, 110, 264

Morton Corbet Castle, 154

Moreton Hampstead, granite quarries near, 19, 249

Morton sandstone quarries, 283

Moruya granite quarries, 57, 343

Moscow Government, Carboniferous rocks of, 137; limestone quarries, 148, 380; buildings of, 137

"Moscow White Stone," 137

Moses, Tomb of, 213

Mosselökke granite quarries, 32, 324

Mo-Tao granite quarries, 343

"Mottled Stone" of Staffordshire, 153

Mount Airy granite quarries, 69, 351

Mountcharles Sandstone, 277

Mount Crosby granite quarries, 344

Mount Faber conglomerate quarries, 242, 429; granites of, 54

Mount Lycabettus limestone quarries, 403

Mount of Olives, Greek Church on, 211

Mount Pagus lava quarries, 84, 358

Mount limestone quarries, 429

Mount Somers limestone quarries, 226, 416

Mountsorrel granite quarries, 21, 250

Mourne Mountains, 29

Mozambique granite, 47

Mrakotin granite quarries, 31, 323

Muir Central College, Allahábád, 105

Municipal Buildings, Glasgow, 25; Loanda, 209; Naples, 83; Pretoria, 48; Queenstown, Cape of Good Hope, 150; Windsor, 143; York, 130

Municipal Offices, Oran, 235

Municipal Water Works, Singapore, 54

Munster limestone quarries, 236, 426

Muraz, limestone quarries near, 39L

Murchison River, W. Australia, 59

Murchison, Sir R., 99, 114

Murom, limestone quarries near, 138, 376

Murphy's Creek sandstone quarries, 377

Murray Bridge, granite quarries, 59, 345; sandstone quarries, 225, 415

Murray Valley, sandstones of, 106

Musapet granite quarries, 339

Muschelkalk beds of Wurtemberg, 158, 383

Muscle Ridge Plantation granite quarries, 63, 347

Museum, British, London, 123, 132, 182; Dublin, 126; Geneva, 186; Liverpool, 134; Naples, 83; of Natural History, New York, 76; of Natural History, Paris, 186; of Oceanography, Monaco, 187; Ottawa, 140, 141; Perth, W. Australia, 192; of Practical Geology, London, 146; Sedgwick, Cambridge, 114, 173, 183; of South Africa, Cape Town, 238

Muttenz, limestone quarries near, 190, 397

Mutual Life Insurance Co.'s Offices, Philadelphia, 71

Myanaung, limestone quarries near, 224

Myaungmyau limestone quarries, 415

Mynyddyslyn sandstone quarries, 286

Mysore, bronzite quarries, 340; felsite of, 86, 360; granites of,

51, 339; Hindu temples, 96; mica-talc rocks of, 96, 368; New Palace, 51, 87, 96; Old Palace, 87; picrites of, 96, 368; porphyry of, 86, 360, 361

Mytilene Island lava quarries, 85, 359

Nabresina limestone quarries, 203, 400

Nailsworth Stone, quarries of, 168, 306

Namur, Carboniferous Limestone of, 137, 375

Naples, buildings of, 220; lava quarries near, 83, 357; leucite-tephrite of, 357; limestone quarries near, 403; Municipal Buildings, 83; National Museum, 83; "Piperno" of, 83, 357; Post Office, 83; trachyte-tuff of, 83, 357

Napoleon's Plateau limestone quarries, 187, 394

Napoleon's Tomb, Paris, 39

Narborough, granite quarries near, 250

Nari rocks of Palestine, 211, 409

Narva Cathedral, 39

Nashville Custom House, 192

Nassau limestone quarries, 243, 430

Natal, Dwyka series of, 149; Ecca beds of, 149, 380; granites of, 49, 338; Lower Karroo system of, 149, 380; Rosetta Stone of, 149, 380

National City Bank, New York, 67

National Gallery, Edinburgh, 144

National Museum, Naples, 83; Ottawa, 141; Washington, 69, 73, 163

Natural History Museum, New York, 76; Paris, 186

Nautical College, Leith, 113

Naval College, Dartmouth, 110

Nay, limestone quarries near, 204, 400

"Nebi Musa," rocks of Palestine, 213, 409

Nelson Monument, Trafalgar Square, 19

Nepheline-syenite of Norway, 329

Nerike, limestone quarries near, 102, 369

Nersac limestone quarries, 204, 401

Neschwitz, granite quarries near, 326

Nesscliffe sandstone quarries, 142, 290

Netherby, sandstone quarries near, 144, 292

Neuchâtel, Lake of, 207, 231; limestone quarries, 189, 396; Lower Cretaceous rocks of, 207, 404; Post Office, 189

New Admiralty Harbour, Dover, 18

Newark Cathedral, New Jersey, 64

Newbiggen sandstone quarries, 275

New Brunswick, Carboniferous rocks of, 140; granites of, 76, 355; Millstone Grit of, 140, 378

Newcastle, Co. Down, 153

Newcastle-upon-Tyne, buildings of, 120; Central Arcade, 130; Central Railway Station, 119; Exchange Buildings, 135; General Post Office, 119; Infirmary, 133; New Town Hall, 130

New Custom House, New York, 63

Newell granite quarries, 71, 351

New England granite quarries, 71, 352

New Government Offices, Westminster, 188

New Hampshire, granites of, 67, 350

New Jersey, granites of, 68, 350

Newlands granite quarries, 48, 338

New Law Courts, London, 143

New Locharbriggs sandstone quarries, 302

New Orleans, buildings of, 67

New Palace of Mysore, 51, 87, 96

New Post Office, Washington, 63

New Prison House, Montreal, 75

New Providence limestone quarries, 243, 430

Newry granite quarries, 29, 256

New Sea Wall, Singapore, 54

New Sessions House, London, 133

New South Wales, granites of, 56, 343; Hawkesbury sandstone of, 161, 389; Triassic rocks of, 161, 389

Newton sandstone quarries, 268

Newtownbarry, flagstones of, 101, 263; granite quarries, 29, 256

New Town Hall, Cape Town, 177; Lancaster, 128; Newcastle-upon-Tyne, 130

New York, American Bank, 71; Bank Note Buildings, 73; Chemical National Bank, 64; Columbia University, 62; Cotton Exchange, 192; Green Point Savings Bank, 61; Hall of Records, 64; Masonic Temple, 71; Museum of Natural History, 76; National City Bank, 67; New Custom House, 63; New Railway Station, 237; Obelisk, Central Park, 43; Pennsylvania Railway Station, 67; St John's Cathedral, 63; Standard Oil Buildings, 68

New York State, Potsdam series of, 107, 371

New Zealand, basalts of, 88, 363; diorite of, 60, 345; granites of, 60, 345; Oamaru Stone of, 225; Oligocene rocks of, 225, 416; tuffs of, 363; Triassic rocks of, 163, 390; volcanic agglomerate of, 363

Nexö, granite quarries near, 31, 324; sandstone quarries near, 369, 373

Nice, limestone quarries near, 187, 394

Nicosia, limestone quarries near, 234

Nidderdale, Millstone Grit of, 128, 279

Ning-po, granite quarries, 55, 343; green stone of, 87, 362; red stone of, 87, 362; tuff quarries near, 87, 362 ; white stone of, 87, 362

Nizhni Novgorod Government, Carboniferous rocks of, 137, 376

Nölbrod granite quarries, 329

Norcross granite quarries, 346

Norfolk, Cretaceous rocks of, 194, 316; Carstone of, 194, 316

Noric Alps, flanks of, 30

Normandy, Caen stone of, 184, 392

Normandy, granites of, 33

Northamptonshire, Inferior Oolites of, 168, 307; Lincolnshire limestones of, 168, 308

North Bridge sandstone quarries, 283

North Carolina, granites of, 69, 351

North Devonshire, Post Tertiary rocks of, 216, 319

Northern Shan States, sandstones of, 232, 423

North Heath sandstone quarries, 193, 315

North Queensland, granites of, 57, 344

Northumberland, Carboniferous Limestones of, 119, 270; Coal measures of, 134, 288; fire-resisting stone of, 131, 135, 281, 288; Millstone Grit of, 130, 281; Old Red Sandstone of, 111, 265

North Wales, Blue Slate rocks of, 101, 263; Cambrian rocks of, 101, 263; Coal measures of, 134, 287; granites of, 21, 251; Permian rocks of, 143, 291

Norton, limestone quarries near, 167, 305

Norway, granites of, 36, 328; Larvikites of, 36, 329; nepheline syenites of, 329; quartz syenites of, 329

Norwich Cathedral, 173

Nottinghamshire, Magnesian Limestone of, 144, 293; Permian rocks of, 142, 290

Nova Scotia, "Blue Stone" of, 141; Millstone Grit of, 141, 379

Novelda limestone quarries, 230, 421

Nubian sandstone of Egypt, 207, 405

Nummulitic limestones of Burma, 224; of Mokhattam Hills, 220

Nuneaton, Cambrian rocks near, 98

Nurmijarvi granite quarries, 332

Nürtingen sandstone quarries, 159, 386

Nyasa Cathedral, 95

Nyasa, Lake, granite quarries near, 46, 336; soap stone quarries near, 95, 367

Nystad, granite quarries near, 332

Oakham, limestone quarries near, 309

Oaks sandstone quarries, 278

Oamaru Stone of New Zealand, 225, 416

Obelisk of Central Park, New York, 42, 43; Cleopatra, London, 42; Heliopolis, 42; Karnak, 43 ; Place de la Concorde, Paris, 42, 43

Observatory, Edinburgh, 112

Oderheim sandstone quarries, 148, 380

Odeum of Herodes Atticus, 234

Ohio, Carboniferous rocks of, 140, 377

Öland Island limestone quarries, 102, 369

Old Bailey, London, 133

Old Brighton sandstone quarries, 131, 281

Old Cairo, limestone quarries near, 223, 414

Old Cummock sandstone quarries, 136, 289

Oldham, St Peter's Church, 154

Old London Bridge, 195

Old Palace of Mysore, 87

Old Red Sandstone, colour of, 114; of Aberdeenshire, 113, 267; Berwickshire, 112, 266; Caithness-shire, 114, 268; Dumbartonshire, 113, 266; Elginshire, 114, 268; Forfarshire, 113, 267; Gloucestershire, 111, 265; Had-

dingtonshire, 112, 266; Herefordshire, 110, 264; Inverness-
shire, 114, 267; Lanarkshire,
113, 266; Isle of Man, 111, 265;
Northumberland, 111, 265; Orkney Islands, 114, 268; Pembrokeshire, 111, 265; Perthshire, 113, 266

Old Town of Vienna, 217

Oligocene rocks of France, 218,
411; Isle of Wight, 216; Malta,
220, 413; New Zealand, 225,
416; Seine et Oise, 218, 411

Olivet, Mount, limestone quarries,
211

Olsbrucken sandstone quarries,
386

Onibury sandstone quarries,
262

Onitsha, Mission House, 238;
sandstone quarries near, 238,
427

Ontario, limestones of, 116, 374

Oolites, Bath, of Somersetshire,
175, 312; Wiltshire, 175, 311;
Great, of Oxfordshire, 177,
312; Inferior, of Gloucestershire, 167, 306; Lincolnshire,
168, 174, 309, 310; Northamptonshire, 168, 307; Rutland,
168, 309; Somersetshire, 166,
305; Yorkshire, 174, 311; Middle, of Oxfordshire, 177, 312;
Upper, Dorsetshire, 178, 313;
Wiltshire, 182, 314

Oporto, granite quarries near,
38, 330

Oran, limestone quarries near,
235, 424; Municipal Offices,
235; Police Station, 235;
Theatre, 235

Orange River Province, Triassic
rocks of, 160, 387

Orange River sandstone quarries,
382

Örebo, Silurian rocks of, 102

Oreja limestone quarries, 229,
420

Orkney Islands, Old Red Sandstone of, 114, 268

Ormond College, Melbourne, 191

Ornamental and Decorative
Stones, 15

Orsera limestone quarries, 411

Orta, Lake, granite quarries near,
35

Oscarhamn granite quarries, 334

Ostimundigen sandstone quarries,
421

Otago, Central, boulders of, 88

Otago Harbour, 89

Otero limestone quarries, 229

Otley, sandstone quarries near,
129, 280

Ottawa, Government Departmental Buildings, 141; Houses
of Parliament, 107; National
Museum, 141; Royal Victoria
Museum, 140

Overwood sandstone quarries,
289

Oxford, buildings of, 178; Christ
Church, 168; Clarendon Press
buildings, 178; St John's College, 168, 177

Oxfordshire, Great Oolites of,
177, 312; Liassic rocks of, 166,
305; Middle Oolites of, 177,
312; Oxford Oolites of, 177,
312; Upper Corallian beds of,
177

Paarl granite quarries, 49, 338

Paco d'Arcos limestone quarries,
396

Paerata, basalt quarries of, 88

"Paerata Bluestone" of New
Zealand, 88

Paignton breccia quarries, 290;
Parish Church, 142

Painswick Hill limestone quarries,
167, 306

Paisley Post Office, 156

Pakhanjee sandstone quarries,
422

Palace, Buckingham, London,
185

Palaces of Constantinople, 220

Palace of the Doges, Venice, 219

Palais de Justice, Brussels, 186;
Rome, 237

Palaiseau limestone quarries,
218, 411

Palatinate, Triassic rocks of, 158

Palestine, basalts of, 85, 359;
Cretaceous rocks of, 210;

Kakuli stone of, 211, 409;
Melekeh stone of, 211, 409;
Mizzeh stone of, 210, 407;
Nari stone of, 211, 409; Pleis-
tocene rocks of, 239, 428
Palmer granite quarries, 63, 346
Palomba limestone quarries, 414
Palotte limestone quarries, 393
Pamphili, Prince A. D., 219
Paradisbakkerne granite quarries,
324
Parcher granite quarries, 74, 354
Paris, Alexander III Bridge, 33,
186; Austerlitz Bridge, 185;
Hôtel de Ville, 185; Lyons
Railway Station, 186; Napo-
leon's Tomb, 39; Natural
History Museum, 186; Obelisk,
Place de la Concorde, 42; Quay
d'Orsay Railway Station, 186;
St Lazare Railway Station,
186
Parish Church of Ballymoney,
127; Clifton, 135
Parkmoor limestone quarries,
277
Park Rynie granite quarries, 49,
338
Park Spring sandstone quarries,
284
Parliament Houses, Cape Town,
49; London, 18, 27, 144, 181,
184, 200; Melbourne, 139;
Ottawa, 107; Perth, W. Aus-
tralia, 192; Quebec, 75; To-
ronto, 140; Vienna, 31
Parnell Monument, 30
Pasipas, sandstone quarries of,
388
Pass of Suram, 40
Pasture Hill sandstone quarries,
120, 271
Pau Cathedral Church, 187
Paul of Caen, 185
Paving and road-making mate-
rial, 15
Peasenhurst sandstone quarries,
287
Pebble beds of Lancashire, 152,
295
Pechaburi, lava quarries near, 87
Pedra Furada limestone quarries,
396

Peel, Castle of, Isle of Man, 111;
Cathedral, 111; sandstone
quarries near, 265
Pegu group of limestones, 224;
sandstones, 231
Pembroke College, Cambridge,
167, 173, 174
Pembroke Co., Tasmania, sand-
stones of, 151
Pembroke Docks, 19
Pembrokeshire, Carboniferous
Limestones of, 121, 272; Old
Red Sandstones of, 111, 265
Penkridge sandstone quarries,
154, 298
Penmon marble stone of Anglesey,
121, 272
Pennant beds of Gloucestershire,
133; Monmouthshire, 133;
South Wales, 133
Pennsylvania, dolerite of, 351;
gneiss of, 70, 351; granites of,
70, 351; Hummelston Brown
Stone of, 164, 390; Triassic
rocks of, 164, 390
Penobscot Bay, granites of, 63
Penrhyn, blue slate rocks of, 101,
263
"Penrith Sandstone" of Cum-
berland, 143, 291
Penrith War Memorial, 100
Penryn, granite quarries near,
19, 247
Penshaw sandstone quarries, 135,
288
Penzance, granite quarries near,
19, 247
Peperino of Italy, 82, 337
Peridotite of India, 340
Peristeri conglomerate quarries,
424
Perlonjour limestone quarries,
375
Permian rocks of Cumberland,
143, 291; Devonshire, 142,
290; Durham, 148, 294; Germany, 148, 380; North Wales,
143, 291; Nottinghamshire,
142, 290; Russia, 148, 380;
Shropshire, 142; 290; South
Africa, 148; 381; Tasmania,
150, 382; Westmorland, 143,
292

Permo-carboniferous rocks of Tasmania, 150, 382

Perrot and Chipiez, Art in Ancient Egypt, 45

Persia, schist quarries near Ispahan, 97, 368; Pleistocene rocks of, 242, 429

Persley granite quarries, 25, 253

Perth (W. Australia), Fine Art Gallery, 60; Law Courts, 192; Museum, 60, 192; Parliament Houses, 192

Perthshire, Old Red Sandstone of, 113, 266

Peru, "Sillar" of, 93; tuffs of, 93, 366

"Pest Basin" of Hungary, 217

Peterborough Cathedral, 168, 173

Peterhead, Blue granite quarries of, 26, 254; Red granite quarries of, 27, 254

Peterhouse College, Cambridge, 172, 197

Peter the Great, statue of, 39

Petit Granit of Belgium, 2, 136, 375

Petra sandstone of Syria, 212, 409; Theatre, 213

Petrel limestone quarries, 421

Petrological designation of rocks, 2

Pfalz, Triassic rocks of, 386

Pfrondarf sandstone quarries, 159, 386

Philadelphia, granite quarries near, 70, 351; Masonic Temple, 63; Mutual Life Insurance Company's Offices, 71; Western Savings Bank, 65

Phlegræan Fields, lavas of, 83

Physical character of Volcanic rocks, 77

Pianura, lava quarries of, 357

Pickwell Down stone quarries, 110, 264

Pictou, Bank Buildings of, 141; sandstone quarries, 141, 379

Piedmont, granites of, 35, 327

Pier of Dover, 183; Sunderland, 135

"Pierre Jaune" of Switzerland, 207

"Pietarsa" of Italy, 83

Pietermaritzburg, Colonial Office, 149; Legislative Council Buildings, 149; sandstone quarries near, 149, 381

Pike River granite quarries, 74, 354

Pilsen, granite quarries near, 31, 323

"Pink Granite" of Worcester, U.S.A., 67

Pink Hailes Stone quarries, 123, 274

Pipe and Lyde sandstone quarries, 264

Piperno of Naples, 83, 357

Piræus limestone quarries, 420

Pirna, limestone quarries near, 402

Pittsburg Colonial Trust Buildings, 71

Place de la Concorde, Paris, Obelisk of, 42, 43

Plane of the bedding of stone, 9

Platberg sandstone quarries, 160, 387

Plean sandstone quarries, 274

"Plean White Freestone," 124, 274

Pleasant River granite quarries, 65, 349

Pleistocene rocks of the Bahamas, 243, 430; Barbadoes, 242, 429; Bermuda, 243, 430; Cape of Good Hope, 238, 427; Ceylon, 241, 428; Germany, 236, 426; India, 239, 428; Italy, 236, 426; Malabar Coast, 239, 428; Palestine, 239, 428; Persia, 242, 429; Singapore, 242, 429; South Nigeria, 237, 427; Straits Settlements, 242, 429; Switzerland, 237, 426; Syria, 239, 428; Würtemberg, 236, 426; Zanzibar, 238, 427

Pliocene rocks of Algeria, 235, 424; Asia Minor, 235, 425; Cyprus, 234, 424; Greece, 234, 424

Plymouth Breakwater, 19

Podolosk, limestones of, 376

Poitiers, limestone quarries near, 186

Pola, limestone quarries near, 203, 399

Poland, Triassic rocks of, 159, 387

Police Barracks of Abuski, 46

Police Station, Oran, 235

Polmont Station, sandstone quarries near, 131, 281

Polyfant Stone quarries, 94, 260

Pompton Junction granite quarries, 68, 350

Pontefract, limestone quarries near, 147, 294; sandstone quarries near, 285

Pontypridd sandstone quarries, 287

Pool Bank sandstone quarries, 129, 280

Poolvash Black Marble, 122; limestone quarries, 122, 273

Poor Lots sandstone quarries, 279

Poortown granite quarries, 22, 251

Porebander Stone of India, 241, 428

"Poros Stone" of Greece, 228

Porosity of Stone, 5

Porphyry of China, 87, 362; Cape Colony, 84, 358; India, 86, 360

Port Antonio Church, Jamaica, 233

Port Chalmers, basalt quarries, 89, 363

Port Elizabeth Post Office, 209; sandstone quarries, 209, 406

"Porta Santa" of Chios, 210

"Portland Chalk Stone" of Jamaica, 233

Portland Island, 178, 179

Portland Oolites of Dorsetshire, 178, 313; Wiltshire, 182, 314

Portland Parish limestone quarries, 233, 423

Portland Stone, use of, 7, 17

Port Louis basalt quarries, 366; Post Office, 93

Porto de St Lourenço, tuff quarries near, 91

Porto Santo trachyte quarries, 91, 365

Portsmouth Docks, 23, 133, 179

Port of Spain, Cathedral, 215; limestone quarries near, 215, 410

Port Shepstone, granites of, 338

Portugal, "Emperor's Red" of, 188

Portugal, granites of, 37, 330; Lower Cretaceous rocks of, 206, 403; Upper Jurassic rocks of, 188, 395

Portuguese East Africa, granites of, 47, 336

Posilipo, Hill of, 84

Posta limestone quarries, 402

Postelwitz limestone quarries, 402

Post Office, Angoulême, 186; Auckland, New Zealand, 60; Baltimore, 67; Berlin, 34; Boston, 67; Cape Town, 49, 238; Florence, U.S.A., 70; Hong Kong, 55; Kadekena, 224; Leipzig, 34; General, London, 182; Savings Bank, London, 132; Naples, 83; Neuchâtel, 189; Newcastle-upon-Tyne, 119; Paisley, 156; Port Elizabeth, 209; Port Louis, 93; Saint Louis, 63; Sydney, 57, 162; Thun, 190

Post Tertiary rocks of Devonshire, 216, 319

Potchefstroom, sandstone quarries, 381

Potsdam Railway Station, Berlin, 34

Potsdam series of Minnesota, U.S.A., 107, 372; New York, 107, 371; Wisconsin, 108, 372

Pot Stone of Mysore, 96, 368; quarries of, 95

Pozzolana of Italy, 82

Pozzuoli, lava quarries near, 83, 357

Pre-Cambrian rocks of Siam, 106, 371; South America, 108, 372

Prestonpans, sandstone quarries near, 123

Preston, sandstone quarries near, 128, 279

Pretoria, diorite quarries near, 96, 368; Government House, 104;

granite quarries near, 48, 337;
Municipal Buildings, 48; New
Law Courts, 96, 150; Railway
Station, 48; sandstone quar-
ries near, 48, 382

Prince Christian Memorial,
Windsor, 173

Prince Consort, sarcophagus of,
Frogmore, 27

Prince of Monaco, 187

Princetown granite quarries, 18,
248

Priory of Barnwell, 199

Prospect sandstone quarries, 265

Prudham sandstone quarries, 119,
271

Pryor's, Messrs, business premises, Cambridge, 19; Ice
Manufactory, Cambridge, 27

Psychiko limestone quarries, 403

Public Buildings of St Christopher, 89

Pulo Obin Island granite quarries,
54, 342

Purbeck beds of Dorsetshire, 183,
314

Putney Bridge, 19

Pyramid granite quarries, 48,
337

Pyramids of Gizeh, 42, 217, 220,
221

Pyrmont, sandstone quarries of,
162, 389

Quarries, locality of, 2, 3

Quarry water in stone, 11

Quartz diorites of Austria, 323;
Guernsey, 257

Quartz Monzonite of Ireland,
256; Maine, U.S.A., 64, 347;
Singapore, 342; Vermont,
U.S.A., 352

Quartz-porphyry of Cape of
Good Hope, 84, 358; Ireland,
78, 259

Quartzite of Buenos Ayres, 108,
372; Gold Coast, 96, 367;
Mysore, 96, 368

Quartz-syenite of Norway, 329

Quebec, Jeffrey Hale Hospital,
141; Parliament Houses, 75

Quebec Province, granites of, 75,
355

Queensland, Blackwater Creek,
162

Queensland, Burrum beds of, 162;
Carboniferous rocks of, 138,
376; Darling Downs, 162;
diorite of, 344; granites of,
57, 344; Ipswich beds of, 162;
Laguna Bay, 162; sandstone
quarries of, 377; Triassic rocks
of, 162, 389

Queens' College Chapel, Cambridge, 199

Queen's College, Benares, 105

Queenstown, S. Africa, Municipal
Buildings, 150; sandstone
quarries of, 150, 382

Queen Victoria Memorial, 25

Querella Stone of South Wales,
155, 300

Quincy Common, boulders of, 65;
granite, 66, 349

Quinta Granda basalt quarries,
91, 364

Rabbaisbede granite quarries,
333

Radom Government sandstone
quarries, 159, 387

Railway Station, Antwerp, 137;
Basle, 190; Dunedin, 88; New-
castle-upon-Tyne, 119; New
York, 237; Lyons, Paris, 186;
Quay d'Orsay, Paris, 186; St
Lazare, Paris, 186; Pretoria,
48

Rainhill, sandstone quarries near,
152, 295

Ramsey Abbey, 169, 170

Rangoon, buildings of, 106, 224,
232

Raon l'Étape granite quarries,
33, 325

Ratio of absorption, 5

Ratschken granite quarries, 34,
327

Rattenburg sandstone quarries,
385

Rattlesnake Hill granite quarries, 68, 350

Ravenna in time of Honorius,
203

Raz-el-Ain limestone quarries,
235, 424

Reach, chalk quarries near, 197

Rebenacq, limestone quarries of, 400

Red Chalk of Hunstanton, 196, 317

Red Granite of Aswan, 45, 335

Red Hill, sandstone quarries near, 316

"Red Oriental" granite of Sweden, 40

"Red Pennant Stone" of Gloucestershire, 134, 286

"Red Stone" of Ning-po, 87, 362; St Helena, 92

Redstone granite quarries, 352

"Red Wilderness Stone," quarries of, 111, 265

Regulbium, Castrum of, 201

Reigate sandstone quarries, 196, 317

Rekawinkel limestone quarries, 217, 411

Remington sandstone quarries, 143, 291

Renfrewshire, Lower Carboniferous rocks of, 124, 274

Renningen sandstone quarries, 159, 385

Reppen limestone quarries, 399

Reso granite quarries, 333

Reval, limestone quarries near, 102, 369

Rhætic beds of Germany, 159, 386; Glamorganshire, 155, 300

Rhine, basalts of the, 79, 356; Devonian rocks of the, 115, 374

Rhode Island, granites of, 70, 351

Rhodes, Cecil, 49, 50

Rhodesia, "Forest Sandstone" of, 161, 388; granites of, 50, 339; Triassic, Upper Karroo system, 160, 388

Rhyolite quarries of Berlin, U.S.A., 74, 354

Rhyolitic tuffs of Asia Minor, 85, 359

Ribblesdale, Carboniferous Limestone of, 118, 269

Richmond Co., U.S.A., granites of, 70

Rifle Butts Kopje granite quarries, 50, 339

Ripon Cathedral, 147

Ritz Hotel, London, 41

Robert Burns statue, Barre, U.S.A., 72

Robin Hood Stone, quarries of, 284

Roche Abbey Stone, 147, 293

Rochester Castle, 194; Cathedral, 183

Rock Glen sandstone quarries, 374

Rockhampton, marble quarries near, 376

Rockport, granite quarries of, 66, 349; "Grey" granite, 66; "Green" granite, 66; sandstone quarries of, 379

Rokkosan, granite quarries of, 343

Roman Wall, remains of, 119

Rome, ancient buildings of, 82; Colosseum, 82, 83, 237; Courts of Justice, 237; Egyptian granite in, 45; "Lapis Albanus" of, 82; Marino quarries near, 82, 356; Palatine, 82; St Paul's Church, 36; St Peter's Church, 45, 237; Travertine quarries near, 236, 426; *Tufa Granulare* of, 81; *Tufa Litoide* of, 81, 356; Vatican, 45

Romsdal, granites of, 36

Romulus, Wall of, 82

Rönne granite quarries, 31, 324

Roofing slates, 15

Rora granite quarries, 26, 254

Rosbrien limestone quarries, 276

Rosebrae sandstone quarries, 268

Rosetta Stone of Natal, 148, 381

Ross, sandstone quarries of, 265

Ross of Mull, granites of, 27, 255

Rothwell Haigh sandstone quarries, 284

Rottenburg sandstone quarries, 159, 384

Rottnest Island sandstone quarries, 225, 415

Round Close Freestone, 135, 288

Rowali sandstone quarries, 370

Roxburghshire, trachytes of, 78, 259

"Royal Blue" granite of Norway, 37

Royal Infirmary, Edinburgh, 123; Newcastle-upon-Tyne, 133

Royal Mausoleum, Frogmore, 18

Royal Museum, Vienna, 227

Royal Oak limestone quarries, 126, 276

Royal Society of Denmark, 32

Royal Victoria Museum, Ottawa, 140

Ruabon Church, 134

Ruabon sandstone quarries, 134, 287

Ruapuke Island granite quarries, 60, 345

Rubislaw granite quarries, 25, 252

Rumbling Bridge, diorite quarries of, 259

Runcorn sandstone quarries, 154, 299

Rushen Abbey, 122

Ruskin, John, 29

Ruskeala granite quarries, 330

Russalka Monument, Reval, 39

Russia, amphibolite of, 331; analcime-dolerite of, 333; Carboniferous rocks of, 137, 376; diorites of, 331; gabbros of, 332; granites of, 38, 330; Permian rocks of, 148, 380; Silurian rocks of, 102, 369; Triassic rocks of, 159, 387

Russian Pilgrims' Society's Hospital, Jerusalem, 211

Rutland, Lincolnshire Limestones of, 172, 308; Inferior Oolites of, 168, 309; Kêtton Stone of, 172, 308

"Rutupiae" (Richborough), Kent, 201

Rylands Library, the, Manchester, 153

Sabino limestone quarries, 236, 426

Sackville, sandstone quarries near, 379

Safafa limestone quarries, 407

St Aidan's Church, Carlisle, 143

St Albans, Abbot of, 185; Cathedral, 185, 201

St Aldhelm, Abbot of Malmesbury, 175; Box Ground Stone, 175, 311

St Andrew's Church, Charmouth, 200

St Andrew's limestone quarries, 305

St Ann's Church, Saunton Sands, 216

St Austell, granite quarries near, 247

St Bees Head, sandstones of, 153, 296

St Benedict Church, Cambridge, 170

St Boniface limestone quarries, 195, 316

St Catherine's Island, Tenby, 23

St Catherine's Stone of Loch Fyne, 95, 260

St Cheron's sandstone quarries, 218, 412

St Christopher, basalts of, 89, 363

St Enoch's Railway Station, Glasgow, 156

St Gens granite quarries, 38, 330

St George, Grenada, basalt quarries, 363; buildings of, 90

St George, N. B., granite quarries, 76, 355

St George's Church, Liverpool, 155

St George's Hall, Liverpool, 128

St George's, Old Cairo, limestone quarries, 223

St George's River granite quarries, 65

St Giles' Cathedral, Edinburgh, 124

St Helena, fire-resisting stone of, 92, 366; tuffs of, 92, 365

St Helier, granite quarries near, 30, 258

St John's Church, Burham, 201; Dhoon, 22

St John's College, Cambridge, 199; Chapel, Cambridge, 24, 170, 174, 189

St John's College, Oxford, 168, 177

St John's, Is. of Man, granite quarries near, 251

St John, Antigua, Court House, 226; limestone quarries near, 226, 417

St John the Divine Cathedral, New York, 63

St Kitts, fire stone of, 89, 363; hard stone of, 89, 363

St Koloma limestone quarries, 148, 380

St Lawrence Co. Potsdam stone quarries, 107

St Lazare Railway Station, Paris, 186

St Louis Church, Bordeaux, 204

St Louis Post Office, 63

St Lucia, andesites of, 90, 364; dacites of, 364; tuffs of, 90, 364

St Margit limestone quarries, 228, 418

St Mark's Church, Venice, 219

St Martin, granite quarries near, 258

St Mary's Cathedral, Edinburgh, 21, 124

St Mary's Church, Johannesburg, Parish Hall of, 150; Kettering, 172; Shrewsbury, 154; Timaru, 89

St Maurice, granite quarries near, 41

St Oswald granite quarries, 323

St Pancras Railway Station, London, 21, 143

St Paul, Minnesota, buildings of, 108

St Paul's Cathedral, London, 12, 19, 122, 177, 180, 181, 185, 194

St Paul's Church, Bridgeton, N.J., 70; Cambridge, 202; Rome, 36

St Petersburg, buildings of, 125; use of granite in, 38

St Peter's Church, Ilfracombe, 110; Oldham, 154; Rome, 45, 237

St Sabinus' Church, Woolacombe, 110

St Stephan's Church, Stockholm, 102

St Thomas in the East sandstone quarries, 226, 416

Salcombe, diabase-schist of, 260

Saldanha Bay, quartz porphyry of, 84, 358; sand dunes of, 238

Salève Hills limestone quarries, 187

Salisbury, Barracks, 134; Cathedral, 183; Plain, Cretaceous rocks of, 200

Salisbury, U.S.A., granite quarries near, 69, 351

Salz limestone quarries, 190, 397

Samson Hill limestone quarries, 428

Sanchi, stupa of, 105

Sandberg granite quarries, 326

Sandringham, Norfolk, 143, 195

Sands granite quarries, 63, 346; sandstone quarries, 289

Sandstone "Potsdam Stone" quarries, 107, 372

Sandwich Islands, importing stone, 91

Sandwith sandstone quarries, 296

San Francisco, use of granite in, 55

San Lorenzo en Niletso lava quarries, 365

San Sebastiano lava quarries, 357

San Stefano limestone quarries, 203, 399

Sarakki granite quarries, 339

Sarakeni limestone quarries, 235, 425

Sarnath, stupa of, 105

Saturna Island sandstone quarries, 214, 410

Saunton Sands, sandstones of, 216, 319

Savigny sandstone quarries, 230, 422

Savoie, Upper Cretaceous rocks of, 204, 402

Saxony, granites of, 33, 326; Upper Cretaceous rocks of, 205, 402

Sayle & Co.'s, Messrs, business premises, Cambridge, 20
Scappling of stone, 10, 181
Scar sandstone quarries, 128, 279
Scarborough Town Hall, 147
Scares Liver Rock sandstone quarries, 136, 289
Scarlett limestone quarries, 122, 272
Schafat limestone quarries, 407
"Schilfsandstein" of Germany, 159, 385
Schillizze, Baron, Mausoleum of, 84
Schists of Persia, 97, 368
Schmöllen granite quarries near, 34, 327
Schneistheim limestone quarries, 395
Schöna, limestone quarries near, 205, 402
Schwab Hall sandstone quarries, 384
Science Hall, Maddison, Wis., 74
Scilly Islands, rocks of, 16
Sclattie granite quarries, 25, 253
Scotch sandstone quarries, 279
Scotland, Geological Survey of, 156
Scotsman Office, Edinburgh, 120
Scott, Sir Gilbert, 6, 189
Scrabo sandstone quarries, 303
Seacombe limestone quarries, 183, 314
Seasoning of Stone, 12
Seaton, limestone quarries near, 318
Seaton Mandeville limestone quarries, 165, 304
Sea Wall of Blackpool, 80; Clacton-on-Sea, 80; Hastings, 80; Singapore, 54, 242; Southend-on-Sea, 80
Sea Water, effect of on stone, 3
Sedbusk sandstone quarries, 118, 270
Sedden Monument, Wellington, N.Z., 60
Sedgwick Museum, Cambridge, 114, 173, 183

Sedgwick, Professor A., 17
Seeburg limestone quarries, 236, 426
Segovia Province, Miocene rocks of, 229, 420
Seine-et-Oise, "Grès de Fontainebleau" of, 218, 412; "Meulière Blanc" of, 218, 412; "Meulière Rouge" of, 218, 412; Oligocene rocks of, 218, 411
Selby Abbey Church, 147
Self-faced Flagstone of Yorkshire, 132
Selsfield sandstone quarries, 315
Senate House, Cambridge, 182
Seneca Creek sandstone quarries, 163, 390
Sens, William of, 185
Serge Island sandstone quarries, 226, 417
Seringapatam, Tippu's Mausoleum, 51; porphyry quarries, 360, 361
Serpentine, Hyde Park, London, 179
Serra de Lonsão, granites of, 37
Serra de Marão, granites of, 37
Servian Wall on the Aventine, 82
Settle, Blue flagstone of, 101, 263; limestone quarries near, 269
Setza, Prov. of, Japan, 56
Sevenacres sandstone quarries, 136, 289
Seven Oaks limestone quarries, 316
"Shamrock Stone" of Ireland, 131, 282
Shan States, sandstones of, 232
Shanghai, buildings of, 55
Shantalla granite quarries, 30, 256
Shap granite, colour of, 20; quarries, 20, 249; in Yorkshire, 20
Shápúr I, King of Persia, 242
Shawk sandstone quarries, 153, 296
Sheffield, Board Schools, 128; limestone quarries near, 293; sandstone quarries near, 285; Water Works, 127

Shepton Mallet Bastard Free-stone, 304; limestone quarries near, 165, 167, 304, 305

Sherman, Senator, monument of, 71

Shibden Head sandstone quarries, 283

Shipley sandstone quarries, 284

Shoeak Hill sandstone quarries, 371

Shrewsbury Abbey Church, 142, 154; St. Mary's Church, 154

Shropshire, Keuper beds of, 154, 299; Lower Silurian rocks of, 99, 261; Permian rocks of, 142, 290; Red Grinshill stone of, 154, 299; White Grinshill stone of, 154, 299

Shushter, sandstone quarries of, 242, 429

Siam, marbles of, 106, 371; Pre-Cambrian rocks of, 106, 371; tuffs of, 87, 362

Sidlingpur porphyry quarries, 361

Sidney Sussex College, Cambridge, 171, 198

Signa sandstone quarries, 412

Sileby, granite quarries near, 250

Silesia, Upper Cretaceous rocks of, 205, 403

"Sillar" of Peru, 93

Sills sandstone quarries, 143, 291

Silurian rocks of Burma, 106, 371; Cumberland, 100, 262; India, 105, 370; Russia, 102, 369; Shropshire, 99, 261; South Africa, 103, 369; Sweden, 102, 369; United States, 108, 372; Westmorland, 100, 262; Yorkshire, 101, 263

Simola granite quarries, 331

Simon's Bay, Harbour Works, 104; sandstone quarries, 104, 370

Simplon Railway, 166

Sind, Gaj beds of, 231; Miocene rocks of, 231, 422

Singapore, Breakwater, 54; conglomerate of, 242, 429; diorite of, 342; granites of, 53, 342; limestones of, 242, 429; Sea Wall, 54, 242; Water Works, 54

Sinkstone, 80

Sireuil limestone quarries, 401

Sites sandstone quarries, 214, 410

Sixth Mile Stone, porphyry quarries of, 361

Sjoehag Hill granite quarries, 41, 334

Slateford, sandstone quarries near, 273

Slates, roofing, 15

Slate Stone of Co. Wexford, 101, 263; of Honister Crag, 100, 262

Sleyib limestone quarries, 408

Smalund granite quarries, 334

Smaws limestone quarries, 147, 294

Smeaton's Eddystone Lighthouse, 17, 19

Smithsonian Institution, Washington, 163

Smoky towns, buildings in, 7

Smyrna, buildings of, 235; lava quarries, 84, 358

Snettisham sandstone quarries, 316

Soap Stone of Lake Nyasa, 95, 367

Sohl Prov. limestone quarries, 217

Soignies limestone quarries, 137, 375

Soldiers and Sailors Monument, Syracuse, N.Y., 61

Soleure, limestone quarries near, 189, 397

Solothurn, limestone quarries near, 189, 397

Somers Mount, limestone quarries near, 226

Somerset House, London, 7, 182

Somersetshire, Bath Oolites of, 175, 312; Carboniferous Limestone of, 117, 269; Doulting Stone of, 167, 305; Ham Hill Stone of, 166, 305; Inferior Oolites of, 166, 305; Liassic rocks of, 165, 304; Triassic rocks of, 300

Soochow, granite quarries near, 55, 343

Sordowala granite quarries, 330

Sóskút limestone quarries, 391

Soudley sandstone quarries, 99, 261

South Africa, Cape system of, 103, 369; "Cave Sandstone" of, 161; Cretaceous rocks of, 209, 406; diorites of, 338; "Forest Sandstone" of, 161, 388; gabbros of, 388; granite quarries of, 47, 337; Karroo system of, 148, 380; Permian rocks of, 148, 380; quartz-porphyry of, 84, 358; Silurian rocks of, 103, 369; Stormberg series of, 160, 388; syenites of, 337; Triassic rock of, 160, 387; Upper Karroo system of, 160, 387

South America, granites of, 76, 355; Pre-Cambrian rocks of, 108, 372; tuffs of, 93, 366

South Australia, Cambrian rocks of, 106, 371; Eocene rocks of, 224, 415; granites of, 59, 345

South Carolina, granites of, 69, 351

South Down sandstone quarries, 110, 264

South Elmsall limestone quarries, 294

South Milford granite quarries, 68, 350

South Nigeria, gneiss of, 336; Pleistocene rocks of, 237, 427

South Rand, sandstone quarries of, 382

South Shields, limestone quarries near, 148, 294

South Union Railway Station, Bolton, U.S.A., 62

South Wales, Carboniferous Limestone of, 121, 272; Coal measures of, 133, 287; Pennant beds of, 133; Querella Stone of, 155, 300

Southend-on-Sea, Sea Wall, 80

Southwell Church, 145

Spain, Miocene rocks of, 229, 420

Specific gravity of Stone, 4

Specimens, chemical composition of, 3; description of, 1

Sphinx in the Vatican at Rome, 45

Spiez, buildings of, 190

Spring Bay sandstone quarries, 151, 382

Spruce Head Island granite quarries, 63, 347

Spylaw sandstone quarries, 120, 271

Stabia limestone quarries, 206

Staffordshire, Keuper beds of, 153, 297

Stag's Fell sandstone quarries, 118, 270

Stamford, limestone quarries near, 308, 309

Stancliffe sandstone quarries, 127, 278

Standard Bank, Durban, 149, 161

Standard Oil Buildings, New York, 68

Stanstead granite quarries, 75, 355

Stanton sandstone quarries, 298

Stanwell sandstone quarries, 377

State Capitol Building, Austin, U.S.A., 72

State Life Insurance Offices, Copenhagen, 32

State Soldiers Monument, Chickamauga, Tenn., 75

Statue of Peter the Great, 39

Stawell sandstone quarries, 377

Staynerville granite quarries, 75, 355

Steinpan sandstone quarries, 388

Stephenson Memorial Monument, Washington, 63

"Stink Stone" of Shepton Mallet, 118, 269

Stirlingshire, Lower Carboniferous rocks of, 124, 274; Millstone Grit of, 131, 281

Stockern sandstone quarries, 421

Stock Exchange, Manchester, 37

Stockholm, buildings of, 103; St Stephan's Church, 102

Stocksfield sandstone quarries, 130, 281

Stoke Ground limestone quarries, 176, 312

Stoke-on-Trent, sandstone quarries near, 153, 297

Stone for paving and road making, 15

Stone House, sandstone quarries near, 289

Stone Mountain granite quarries, 62, 346

Stoneycombe limestone quarries, 264

Stoney Creek granite quarries, 62, 346

Stoney Stanton, granite quarries, 250

Stormberg series of South Africa, 160, 388

Storeton, "Fossil Footprints" of, 155; stone of Cheshire, 154, 300

Strain of Compression, 5

Straits Settlements, granites of, 53; Pleistocene rocks of, 242, 429

"Stratified Granite," 53

Street limestone quarries, 305

Strömstad granite quarries, 333

Stuart Co. sandstone quarries, 415

Stuart Monument, 163

"Stubensandstein" of Germany, 159, 386

Stuttgart, sandstone quarries near, 159

Sub-Carboniferous rocks of United States, 139, 377

Suffolk County Court House, Boston, U.S.A., 64

Sugar Works of Jamaica, 226

Sunderland, Piers, 135; Wear Commissioners' Offices, 130

Sun, effect of, on stone, 7

Suram, Pass of, 40

Surat sandstone quarries, 240, 428

Sur Baher limestone quarries, 409

Surra limestone quarries, 125, 275

Surrey Co., Jamaica, 226

Surrey, Upper Greensand of, 195,. 316

Sussex, Lower Cretaceous rocks of, 193, 315; Marble, 194, 315; Wealden Sandstone of, 193, 315

Suttö limestone quarries, 419

Svanike, granite quarries near, 324

Svartä granite quarries, 38, 331

Swabian Alps limestone quarries, 187

Swanage, limestone quarries near, 314

Swan Port, sandstone quarries near, 382

Swansea Docks, 28

Swarthmore blue flagstone quarries, 263

Sweden, amphibolite of, 334; "Black Granite" of, 40, 334; epidiorite of, 334; gabbros of, 40, 334; granites of, 40, 333; Lower Silurian rocks of, 102, 369

Swell-Tor granite quarry, 18

Swerthow sandstone quarries, 174, 311

Swinton sandstone quarries, 112, 266

Switzerland, Carboniferous rocks of, 138, 376; "erratic blocks" of, 41; gneiss of, 335; granites of, 41, 334; Jurassic rocks of, 166, 391; Liassic rocks of, 166, 391; Miocene rocks of, 230, 421; "Molasse," 230, 422; Pleistocene rocks of, 237, 426; Upper Jurassic rocks of, 189, 397

Sydney, buildings of, 57; Equitable Life Assurance Co.'s offices, 57; Hawkesbury Bridge, 57; Post Office, 162; Public Library, 162; sandstone quarries near, 162, 389; Town Hall, 162; Treasury Buildings, 56; University, 162

Syene granite quarries, 45, 335

Syenite quarries of Australia, 57, 344

Syenite of Gib Mountain, 57; South Africa, 48, 337

Syracuse, N.Y., Soldiers and Sailors Monument, 61

Syria, basalts of, 85, 359; Pleistocene rocks of, 239, 428; Upper Cretaceous rocks of, 210, 406

Taba-z-Induna sandstone quarries, 388

Table Mountain sandstone quarries, 103, 369

Tadagwade porphyry quarries, 361

Tadcaster, limestone quarries near, 146; Stone, 294

Tajao lava quarries, 92, 365

Ta-la-bi-eli limestone quarries, 407

Tally Ho limestone quarries, 306

Tangier, Bay of, 191, 397; limestone quarries near, 223, 415

Tanjore, Rajah's Palace, 51

Tanjong Pagar, Dock Board's Wharf, 54

Tansley, sandstone quarries near, 128, 278

Tapley Hill mudstone quarries, 371

Tarascon limestone quarries, 402

Tardree Mountain quartz-porphyry quarries, 78, 259

Tarqua granite quarries, 46, 336

Tarradale sandstone quarries, 114, 267

Tasmania, Permian rocks of, 150, 382; Permo-carboniferous rocks of, 150, 382

Tavistock, granite quarries near, 18, 249

Taynton limestone quarries, 177, 312

Technical Schools, Dresden, 34

Teheran, buildings of, 97

Tehognari, limestone quarries near, 206

Telegraph Offices, Adelaide, 225

Tell Zone of Algeria, 235

Temple of Fortuna Virilis, 82; of Vespasian, 83

Temples of Ellora, 50; of Kajraha, 105

Templeton sandstone quarries, 111, 265

Tenant Harbour, granite quarries near, 348

Tenby, buildings of, 111; limestone quarries near, 121, 272; St Catherine's Island near, 23

Teneriffe lava quarries, 365

Tenterfield, granite quarries near, 57, 343

Tercé Normandoux, limestone quarries of, 392

Tetchmakies, limestone quarries near, 416

Texas, granites of, 72, 352

Texture of Stone, 5

Thames Embankment, London, 18, 23, 128

Theatre of Dionysus, 234; of Oran, 235; of Petra, 213

Thebes, temples of, 208

Thesdale Stone, 146

Thorney Abbey, 170, 171

Thornhill, sandstone quarries near, 156, 301

Thornton sandstone quarries, 132, 282

Thorp sandstone quarries, 285

Thousand Islands, granites of, 76, 355

Three Castles limestone quarries, 126, 276

Three Saints limestone quarries, 223, 415

Threlkeld granite quarries, 250

Thun, buildings of, 190; limestone quarries near, 190, 397; Post Office, 190

Thursby, sandstone quarries near, 153, 296

Thurso, flagstone quarries near, 268

Ticino Canton granite quarries, 42, 335

Tiflis Cathedral, 206

Tilberthwaite green slate quarries, 263

Tillyfourie granite quarries, 25, 253

Timaru, basalt quarries near, 89, 363; St Mary's Church, 89

Tintern, Abbey, 134; sandstone quarries, 134, 287

Tippn's Mausoleum, Seringapatam, 51

Tisbury limestone quarries, 314

Tivoli travertine quarries, 236, 426

Tjylling, granite quarries near, 328

Tofsalo granite quarries, 332

Tomb of Moses, 213; of Napoleon I, 39

Torfou granite quarries, 325

Tor granite quarries, 18, 248

Toropto City Hall, 140; Parliament Buildings, 140

Torquay, buildings of, 110; limestone quarries, 110, 264

Torre del Greco lava quarries, 357

Tors of Devonshire, 17

Totternhoe Stone of Bedfordshire, 199, 317

Tower Bridge, London, 19

Tower of London, 201

Town Bush sandstone quarries, 149, 381

Town Hall, Birkenhead, 155; Bradford, 133; Copenhagen, 32; Durban, 149; Hamburg, 34; Leeds, 133; Manchester, 132; Melbourne, 151; Scarborough, 147; Sydney, 162; Walsall, 154

Townsville, granite quarries near, 58, 344

Trachytes of Canary Islands, 92, 365; of Madeira, 91, 365; near Melrose, 78, 259

Trachyte-tuff of Germany, 79, 356; of Naples, 83, 357

Trade name of a building stone, 11

Trafalgar Square, Monument of Nelson, 19

Tranent sandstone quarries, 124, 274

Transvaal, diorite of, 96, 368; Ecca beds of, 150, 381; granite quarries of, 47, 337; Lower Karroo system of, 149, 381; Silurian rocks of, 104, 370

Travertine of Italy, 83, 236, 426

Treasury Buildings of Sydney, 56

Trenton, gneisses near, 70

Triassic rocks, of the Argentine Republic, 164, 390; Ayrshire, 156, 302; Cheshire, 154, 299; Cumberland, 153, 295; Derbyshire, 296; Co. Down, 157, 303; Dumfriesshire, 156, 301; Elginshire, 155, 300; Germany, 157, 159, 383, 386; Glamorganshire, 155, 300; Gold Coast, 160, 387; Lancashire, 152, 295; New South Wales, 161, 389; New Zealand, 163, 390; Orange River Province, 160, 388; Pennsylvania, 164, 390; Poland, 159, 387; Queensland, 162, 389; Rhodesia, 161, 388; Russia, 159, 387; Shropshire, 154, 299; Somersetshire, 300; South Africa, 160, 387; Staffordshire, 153, 297; United States, 163, 390; West Africa, 160, 387; Worcestershire, 154, 298; Würtemberg, 158, 383, 384

Trient Valley, Carboniferous rocks of, 138, 376

Trier, sandstone quarries near, 374

Trieste, limestone quarries near, 203, 400

Trinidad, Cretaceous rocks of, 215, 410

Trinity Church, Burton-on-Trent, 154

Trinity College, Cambridge, 171, 173, 199; Dublin, 29

Trinity Hall, Cambridge, 197

Trofou limestone quarries, 189, 396

Tübingen, sandstone quarries near, 195, 385, 386, 426

Tuck Royd sandstone quarries, 132, 283

Tuedian group of rocks, 112, 119, 120

Tufa Granulare of Rome, 81

Tufa Litoide of Rome, 81, 356

Tuffs of, Asia Minor, 84, 85, 358, 359; Australia, 88, 362; China,

87, 362; Germany, 79, 356; Grenada, 89, 363; Madeira, 91, 364; Naples, 83, 357; New Zealand, 89, 363; Peru, 93, 366; Queensland, 88, 362; near Rome, 81, 356; St Helena, 92, 365; St Lucia, 90, 364; Siam, 87, 362

Tuffstein of the Bernese Oberland, 237, 426; Germany, 236, 426

Tula Government, Carboniferous rocks of, 137, 376

Tumkur Black Trap quarries, 51, 339; quartzite quarries, 97, 368

Tunbridge Wells Sandstones, 193

Tunis, Cretaceous rocks of, 207, 404

Tura Prisons limestone quarries, 220, 413

Turkey, Eocene rocks of, 220, 413

Turkey in Asia, lavas of, 84, 358

Turriff, sandstone quarries near, 113, 267

Turuvekere Stone of Mysore, 51, 339

Tuscany, Eocene rocks of, 219, 412

Tyrebeggar granite quarries, 25, 253

Uddevalla granite quarries, 333

Uitenhage series of Cape of Good Hope, 209, 406

Ulleberg granite quarries, 37, 329

United States, Devonian rocks of, 115, 374; granites of, 61, 345; Lower Silurian rocks of, 108, 372; National Museum, Washington, 67, 69, 73, 163; Naval Academy, Annapolis, 64; Potsdam series of, 106, 371; Sub-Carboniferous rocks of, 139, 377: Triassic rocks of, 163, 390; Upper Cambrian rocks of, 106, 371

University College, Bangor, 134; London, 120

University Library, Cambridge, 175, 182; Offices, 20

University of Glasgow, 124; Leipzig, 34; Melbourne, 162; Sydney, 162

Upper Austria, granites of, 31, 323

Upper Carboniferous rocks of Scotland, 131, 135, 281, 289

Upper Chalk of Kent, 201, 318

Upper Corallian beds of Oxfordshire, 177

Upper Cretaceous rocks of Denmark, 203, 400; Istria, 202, 399

Upper Greensand of Buckinghamshire, 195, 316; Isle of Wight, 195, 316; Surrey, 195, 316

Upper Jurassic rocks of Portugal, 188, 395; Switzerland, 189, 396

Upper Karroo system of S. Africa, 160

Upper Silurian rocks of Shropshire, 100, 262; Yorkshire, 101, 263

Urach, limestone quarries near, 426

Urban District Council of Ilkley, 129

Uttenhalli, granite quarries of, 389

Vagney granite quarries, 325

Valais, granites of, 334; limestones of, 138, 166, 376, 391

Valencia, buildings of, 229

Valetta, buildings of, 220

Valley of Ajagua trachyte quarries, 365

Vaucluse, Upper Cretaceous rocks of, 401

Vaud, Liassic rocks of, 166, 391; Miocene rocks of, 422; Upper Jurassic rocks of, 189, 396

Vauxhall Bridge, 18

Vehmo granite quarries, 332

Venetia, Eocene rocks of, 218, 412

Venice, Academy of Fine Arts, 219; Campanile, 219; Ducal Palace, 203; Palace of the

Doges, 219; St Mark's Church, 219

Ventnor, limestone quarries near, 316

Vermont, granites of, 72, 352; Quartz-Monzonite of, 352

Verona, limestone quarries near, 218, 412

Vespasian, temple of, 83

Vesuvius, leucite lavas of, 83, 357

Vezins granite quarries, 325

Victoria, B.C., Carnegie Library, 214

Victoria Bridge, Montreal, 116

Victoria, Carboniferous rocks of, 139, 377; Eocene rocks of, 224, 415; granites of, 58, 345; Jurassic rocks of, 191, 398

Victoria, Queen, Statue of, 57

Vienna, Aqueducts, 228; Central Railway Station, 31; College of Industry, 227; Fortifications of, 227; Hofburg Theatre, 203; Houses of Parliament, 31; Imperial Palace, 227; limestone quarries near, 217, 411, 418; Maria-Stiegen Church, 227; Old Hofburg of, 217; Old Town of, 217; Royal Museum, 227

Vienne, Lower Jurassic rocks of, 186, 392; Middle Jurassic rocks of, 186, 393

Vilhonneur Raillats limestone quarries, 186, 394

Villars limestone quarries, 186, 393

Villena limestone quarries, 421

Villeneuve limestone quarries, 166, 391

Villette limestone quarries, 391

Vinalhaven Island, granite quarries, 63, 346

Vincurial limestone quarries, 203, 399

Vindhyan system of India, 105

Vionnaz limestone quarries, 391

Vire, granite quarries near, 326

Volcanic agglomerate of N. Zealand, 363

Volcanic Ash of Borrowdale, 77, 259

Volcanic rocks, chemical composition of, 77

Volcanic tuffs of Italy, 81, 356

Voroklini limestone quarries, 234, 424

Vosges, granites of, 32, 325

Wády Músá, cliffs of, 212, 409

Waikola sandstone quarries, 163, 390

Waldoboro granite quarries, 64, 348

Wales, granites of, 21, 251

Walker Gallery, Liverpool, 134

Wallace, General, obelisk to, 71

Wallace sandstone quarries, 141, 379

Wall Dean sandstone quarries, 124, 274

Wall of Romulus, 82

Walsall Town Hall, 154

Walters Ash, sandstone quarries near, 316

Waltonville sandstone quarries, 390

Wankie district of Rhodesia, 50

Wansford limestone quarries, 171, 308

War Department buildings, Berwick, 121

War Memorial, Penrith, 100; York, 100

War Office, London, 132, 168

Waring and Gillow's sandstone quarries, 150, 381

Warkworth Castle, 131

Warkworth sandstone quarries, 131, 281

Warrigal Creek sandstone quarries, 377

"Warsaw Bluestone" of United States, 115, 374

Warsaw Cathedral, 39

Warwick (Queensland), sandstone quarries near, 162, 389

Warwickshire, Cambrian rocks of, 98, 261

Wasdale Fell granite quarries, 249

Washington Co., granites of, 72, 348

Washington Congressional Library, 68 ; Life Insurance Buildings, 71; National Museum, 67, 69, 73, 163; New Post Office, 63; Stephenson Memorial Monument, 63

Waterberg sandstone quarries, 104, 370

Waterfeet sandstone quarries, 380

Waterford, U.S.A., granite quarries, 61, 345

Water of imbibition, 12 ; saturation, 12

Waterloo Bridge, London, 18, 25

Waterval granite quarries, 48, 337

Water Works, Bristol, 18; Singapore, 54

Waupaca granite quarries, 75, 354

Waurn Ponds limestone quarries, 415

Wausau, Marathon County Court House, 74; granite quarries, 73, 353

Waushara Co., granites of, 75, 354

Waverley Hotel, London, 36 ; Railway Station, Edinburgh, 120

Wavertree Clock Tower, 153

Wealden sandstone of Sussex, 193, 315

Wear Commissioners' Offices, Sunderland, 130

Weight for cubic foot of stone, 4

Weilberg basalt quarries, 81, 356

Weldon limestone quarries, 147, 170, 308

Wellington's tomb, 19

Wellington Monument, Dublin, 29; Liverpool, 155

Wellington, N.Z., Bank of Australia, 60; Sedden Monument, 60

Wells Cathedral, 167

Wendelsheim sandstone quarries, 384

Wensleydale, Yoredale beds of, 118, 270

Weobley sandstone quarries, 264

Werth, Dr, Berlin, 239

Wesleyan Memorial Hall, Edinburgh, 112

West Africa, Cretacean rocks of, 208, 406; gneiss of, 46, 336; granites of, 46, 336; quartzite of, 96, 367; Triassic rocks of, 160, 387

West Equatorial Africa, granites of, 46, 336

Westerly, granites of, 71, 351; Blue Granite, 71; Red Granite, 71

Western Australia, Eocene rocks of, 225, 415; granites of, 59, 345; Jurassic rocks of, 191, 398; Museum of, Perth, 60

Western Province, Ceylon, granites of, 340

Western Savings Bank, Philadelphia, 65

"West Hall Rock" of Forfarshire, 267

West Hall sandstone quarries, 267

Westleigh Stone of Somersetshire, 269

Westminster Abbey, 146, 176, 183, 185, 196; Bridge, 19, 28, 182; New Government Offices, 188

Westmorland, granites of, 20, 249; Permian rocks of, 143, 292; Lower Silurian rocks of, 101, 262

Westmorland Co., Canada, sandstone of, 141

Weston limestone quarries, 313; sandstone quarries, 154, 299

Weston - super - Mare, sandstone quarries near, 300

Wexford Co., Cambrian rocks of, 101, 263 ; granites of, 28, 256

Whatstandwell, Millstone Grit of, 128

Whitby Abbey, 175

White granite of Connecticut, 61, 345

White Hailes Stone quarries, 123, 274

Whitehall Banqueting Hall, 180

Whitehaven sandstone quarries, 135, 288, 296

White Mountain limestone quarries, 414

White Stone of Kentucky, 192, 398; of Moscow, 137; of Ningpo, 87, 362

Whitsome Newton, sandstone quarries of, 112, 266

Wholesale Co-operative Society, Manchester, 128

Whyte Stone, 199

Wick, sandstone quarries near, 114, 268

Wicklow Co., granites of, 28, 256

Wight, Isle of, Oligocene rocks of, 216; Upper Greensand of, 195, 316

William the Conqueror, 169, 185

William of Sens, 185

Williamston, limestone quarries of, 121, 272

Willmanstvand granite quarries, 331

Wiltshire, Bath Oolites of, 175, 311; Chalk Stone of, 200; Great Oolites of, 175, 311; Lower Chalk of, 200, 318; Portland Oolites of, 182, 314

Winchester Cathedral, 167, 195

Windrush sandstone quarries, 196, 317

Windsor Castle, 153, 195, 201; Municipal Buildings, 143; Prince Christian Memorial, 173

Windsor Co., U.S.A., granites of, 72

Windy Nook sandstone quarries, 135, 288

Winnsborough, granite quarries near, 69, 351

Winsor's Hill limestone quarries, 269

Winterbourne, sandstone quarries near, 134, 286

Wisconsin, geological survey of, 73; gneiss of, 354; granites of, 73, 353; Lower Silurian rocks of, 108, 372; Potsdam series of, 108, 372; Upper Cambrian rocks of, 108, 372

Wise, Texas, Court House of, 72

Withington sandstone quarries, 265

Witley, sandstone quarries near, 196, 317

Wittkopje granite quarries, 48, 337

Woburn Abbey, 200

Woodburn sandstone quarries, 119, 270

Woodpoint sandstone quarries, 379

Woodstock granite quarries, 65, 349

Woolacombe, St Sabinus Church, 110

Wooler, sandstone quarries near, 265

Woolton sandstone quarries, 152, 295

Woolwich Arsenal Buildings, 133

Worcester Co., U.S.A., granites of, 67; Pink granite of, 67

Worcestershire, Bromsgrove Stone of, 154, 298; Triassic rocks of, 154, 298

Worth, Saxon Church of, 193

Worthsthorne sandstone quarries, 288

Wren, Sir Christopher, 12, 180, 194

Wrexham Parish Church, 134; sandstone quarries near, 134, 143, 291

Wright, Thomas, Freemason of London, 180

Würtemberg, granites of, 34, 327; Lower Bunter beds of, 158, 383; Lower Keuper beds of, 158, 384; Middle Bunter beds of, 158, 383; Middle Keuper beds of, 159, 385; Muschelkalk beds of, 158, 383; Pleistocene rocks of, 236, 426; Triassic rocks of, 157, 383; Upper Bunter beds of, 158, 383; Upper Jurassic rocks of, 187, 395

Wyborg, granite quarries near, 39, 332

Wychwood, limestone quarries near, 177, 312

Wycliffe sandstone quarries, 285

Wycombe sandstone quarries, 316

Yangan sandstone quarries, 162, 389
Yeovil, limestone quarries near, 305
Yonne, Middle Jurassic rocks of, 185, 392
Yoredale beds of Wensleydale, 118, 270; Yorkshire, 118, 270
York, Dean and Chapter of, 146; Minster, 7, 146, 173; Municipal Buildings, 130; York Stone, 132; War Memorial, 100
Yorkshire, Carboniferous Limestone of, 118, 269; Coal measures of, 132, 282; Horton Flags of, 101; Millstone Grit of, 128, 279; Inferior Oolites of, 174, 311; sandstones of, 118, 270; Upper Silurian Rocks of, 101, 263; Yoredale beds of, 118, 270

Ystehede, granite quarries of, 329

Zanzibar, Government Buildings of, 239; High Court, 239; limestone quarries of, 238, 427; Pleistocene rocks of, 238, 239, 427; Post Office, 239; sandstone quarries of, 239, 427
Zogelsdorf, limestone quarries of, 411

Printed in the United States
By Bookmasters